Hans Lengerer is an acknowledged authority on the Imperial Japanese Navy. A ~~~~~ entitled *Imperial Japanese Warships Illustrated* was published between 2016 an~~~~~ and II of *The Aircraft Carriers of the Imperial Japanese Navy and Army* in 2019 a~~~~~ respectively, and *Outline History of the Imperial Japanese Navy 1868–1945* in 2021. He is co-author, with Lars Ahlberg, of *Capital Ships of the Imperial Japanese Navy 1868–1945*, of which Vols I and III have been published.

Stephen McLaughlin retired in 2017 after working for 35 years as a librarian at the San Francisco Public Library. In addition to contributing regularly to *Warship*, he is the author of *Russian and Soviet Battleships* (US Naval Institute Press, 2003, reissued 2021) and is co-editor of an annotated version of the controversial *Naval Staff Appreciation of Jutland* (Seaforth Publishing, 2016). He is currently working on a book on Russian and Soviet cruisers.

Kathrin Milanovich has contributed a series of articles to *Warship* on the warships of the Imperial Japanese Navy during the period 1890–1945.

Jean Moulin has written widely on the French Navy, and is the author of several monographs on warships of the interwar and postwar periods. Jean is the co-author, with John Jordan, of *French Cruisers 1922–1956* (2013) and *French Destroyers 1922–1956* (2015). Books on the 'stealth' frigates of the *La Fayette* class and the Tripartite MCM vessels were published by Lela Presse in 2018 and 1921 respectively.

Dirk Nottelmann is a marine engineer by profession, and is currently working for the German shipping administration. He has contributed to *Warship*, *Warship International* and various German magazines, as well as being author of *Die Brandenburg-Klasse* (Mittler, 2002) and co-author of *Halbmond und Kaiseradler* (Mittler, 1999). His book *The Kaiser's Cruisers* (with Aidan Dodson) was published by Seaforth in late 2021.

John Roberts is a former editor of *Warship*. He is the author of many books and articles on the Royal Navy, including the seminal *British Battleships of World War Two* and *British Cruisers of World War Two* (Arms & Armour,1976/1980), both with Alan Raven. A noted draughtsman, John was responsible for *The Battlecruiser Hood* (1982) and *The Battleship Dreadnought* (1992) in the Conway Anatomy of the Ship series. Recent publications include *British Battlecruisers 1905–1920* (Seaforth, 2016), and *The Battlecruiser Repulse* (2019) and *The Destroyer Cossack* (2020) in Seaforth's 'original builders' plans' series.

Ian Sturton is a regular contributor of articles and illustrations to naval publications, including *Warship*, *Warship International* and *Jane's Fighting Ships*. He edited *Conway's Battleships: The Definitive Visual Reference to the World's All-Big-Gun Ships* (Conway Maritime Press 1987, updated 2008).

Conrad Waters is the author of numerous articles on modern naval matters and editor of *World Naval Review* (Seaforth Publishing). He also edited Seaforth's *Navies in the 21st Century*, shortlisted for the 2017 Mountbatten Award. His history of the Royal Navy's Second World War 'Town'-class cruisers was published in 2019 and a sequel on the *Fiji* class and their successors is due to be released in March 2024.

Jon Wise is a naval historian who worked in education for more than 30 years and is now retired. He has an MA in Maritime History from Greenwich University and a doctorate in Naval History from the University of Exeter. Published works include *Royal Fleet Auxiliary in Focus* (2002) and *Vickers, Barrow* (2008), both for Maritime Books, *The Role of the Royal Navy in South America, 1920–1970* (Bloomsbury, 2014) and most recently, *The Royal Navy and Fishery Protection: From the Fourteenth Century to the Present* (Seaforth Publishing, 2023). Jon is a regular contributor to *Warship*, *Ships Monthly* and *Warship World*.

WARSHIP 2024

WARSHIP 2024

Editor: **John Jordan**

Assistant Editor: **Stephen Dent**

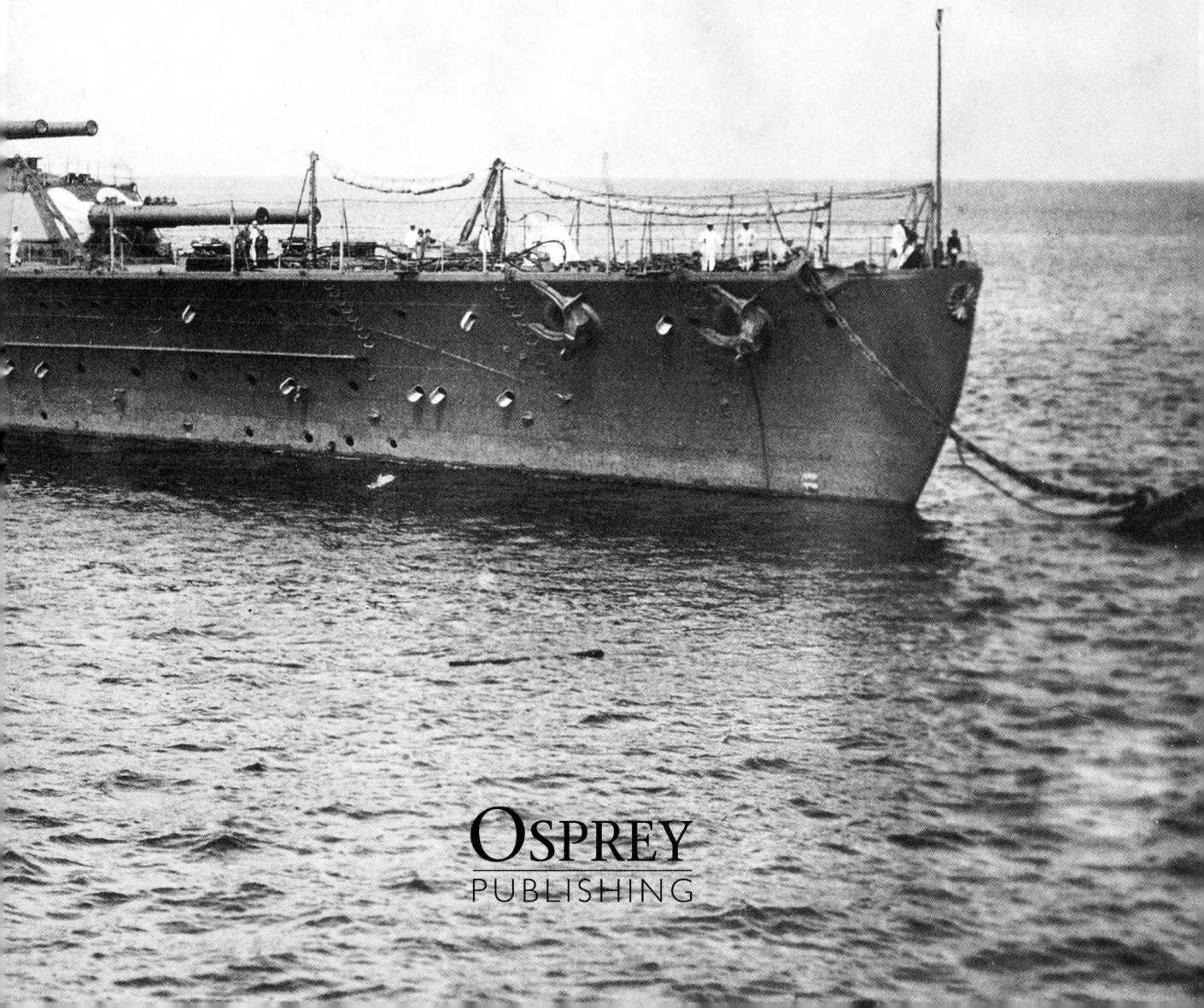

OSPREY
PUBLISHING

Title pages: The Japanese battleship *Nagato* in September 1924, following the addition of a tall clinker screen on the first funnel. *Nagato* and her sister *Mutsu* are the subject of our lead article this year, by regular contributor Hans Lengerer. (Lars Ahlberg collection)

OSPREY PUBLISHING
Bloomsbury Publishing Plc
Kemp House, Chawley Park, Cumnor Hill, Oxford OX2 9PH, UK
29 Earlsfort Terrace, Dublin 2, Ireland
1385 Broadway, 5th Floor, New York, NY 10018, USA
E-mail: info@ospreypublishing.com
www.ospreypublishing.com

OSPREY is a trademark of Osprey Publishing Ltd

First published in Great Britain in 2024

© Osprey Publishing Ltd, 2024

A catalogue record for this book is available from the British Library.

ISBN: HB 9781472863300; eBook 9781472863331; ePDF 9781472863324; XML 9781472863317

24 25 26 27 28   10 9 8 7 6 5 4 3 2 1

Cover design by Stewart Larking
Page layout by Stephen Dent
Printed and bound in India through Replika Press Private Ltd.

FSC MIX Paper from responsible sources FSC® C016779

Osprey Publishing supports the Woodland Trust, the UK's leading woodland conservation charity.

To find out more about our authors and books visit www.ospreypublishing.com. Here you will find extracts, author interviews, details of forthcoming events and the option to sign up for our newsletter.

CONTENTS

EDITORIAL

This year's annual leads with a major feature by Hans Lengerer on the battleships *Nagato* and *Mutsu*. These two vessels have received considerable coverage over the years; however, the focus has generally been on the ships as they were when Japan entered the war in December 1941, by which time they had been extensively rebuilt. During early discussions with the author, the Editor became increasingly aware of the scarcity of high-quality material on these ships as designed and first completed. In particular, the plans of the external configuration and general arrangements that have survived are in poor condition, being badly torn with faded and indistinct linework. This is a serious loss, as the design of these two ships was ground-breaking in a number of respects, notably their adoption of the 16in gun, a 'concentrated' protection system, and geared turbines. Their completion in 1920–21 caused major problems for the negotiations surrounding the Washington Naval Arms Limitation Conference of 1921–22; the retention of the second unit, *Mutsu*, which had been rushed to completion shortly before the conference was due to open, was used by the US Navy to justify the completion of two additional battleships armed with 16in guns and by the British to justify the building of the similarly-armed *Nelson* and *Rodney*.

In a second major feature, Przemysław Budzbon continues his series on the early Soviet Navy with a study of the fast flotilla leader *Tashkent*, purchased directly from the Odero-Terni-Orlando (OTO) shipyard at Livorno but armed with Soviet weaponry. This surprising collaboration between the Soviet Union and Fascist Italy benefited both parties: the USSR gained access to the latest European naval technology, while the Italians gained prestige for their naval shipyards and funding that helped to offset purchases of Russian grain. However, the collaboration proved to be extremely fraught due to the suspicions on both sides, which were exacerbated by the Spanish Civil War. While the Italians eventually delivered the desired fast ship, *Tashkent*'s entry into service was marred by delays in the development of the Soviet twin 130mm guns that were intended for her.

Our third feature is by a contributor who is new to *Warship*. Toby Ewin is a visiting fellow at King's College London (KCL) who has published a number of articles on the naval war in the Black Sea 1914–18. The subject of the present article is the brief and inconclusive action between the predreadnought battleships of the Russian Black Sea Fleet and the German 'battle cruiser' *Goeben* off the mouth of the Bosphorus on 15 May 1915. Using contemporary accounts Toby broadens the account to include a discussion of the relative capabilities of the Russian and the Turco-German naval forces in the theatre. It is hoped that this article will be the first of a series.

The following feature is a technical study of the 'missile frigates' *Suffren* and *Duquesne* coauthored by the Editor and Jean Moulin, whose authoritative book on these ships was published by Marines Edition in 1998. These were the first purpose-built missile ships to enter service with the *Marine Nationale*. The French took their inspiration for the design from the US Navy's DLGs of the *Farragut* and *Leahy* classes, but the weapon and sensor outfit was purely French in conception and manufacture. The *Marine Nationale* did not have the same benefits of series production as the US Navy, and *Suffren* and *Duquesne* were costly to build and to equip; only two units were completed, and the Navy had difficulty funding the necessary half-life upgrades to the Masurca area defence system and its associated electronics. Nevertheless, these were remarkable ships with a number of innovative features that influenced subsequent French ship design.

The remaining feature articles are the customary eclectic mix of navies and periods. Kathrin Milanovich's article on the destroyers of the *Matsu* and *Tachibana* classes highlights the struggles of the Imperial Japanese Navy (IJN) during the middle period of the Pacific War when confronted with the need to replace growing losses. The response adopted was to develop a simplified, robust design that could be built quickly and in numbers.

Dirk Nottelmann aims to correct the record with a detailed study of the 'prehistory' of the renowned German WWI commerce raider *Seeadler*, revealing in the process the considerations that influenced the IGN in its selection of a fully-rigged sailing vessel registered in the USA, *Pass of Balmaha*, and outlining the modifications that had to be implemented to equip her for the role. The introductory section of the article has an atmospheric account of *Seeadler*'s first sortie and her narrow escape from the British patrols.

Philippe Caresse completes his series on the French battleships of the *Flotte d'échantillons* with a study of *Bouvet*, generally considered the most successful of the five ships in service, but with all the defects and potential vulnerabilities of her half-sisters. During the Allied attempt to force the Dardanelles in March 1915, *Bouvet* would strike a mine and sink in less than a minute with heavy loss of life.

Since the early years of the 20th century the Italian *Regia Marina* had gained a reputation for innovative small craft which, when manned by courageous and skilled personnel, packed a punch out of all proportion to their size. Faced with the threat of a British Mediterranean Fleet superior in every category of major warship in the mid-1930s, Mussolini encouraged the development of small single-engine submarines that could be built cheaply, quickly and in numbers to provide an 'asymmetric' counter. The programme encountered serious technical difficulties, and with the tailing-off of the British threat in the late 1930s was not aggressively pursued, but it spawned some interesting concepts.

Durandal, the first in a series of French destroyers completed between the late 1890s and the Great War. *Durandal* and her 300-tonne sisters will be the subject of a major feature article by the Editor due to be published in *Warship* 2025. (Naval History and Heritage Command, NH 64459)

Regular contributor Jon Wise has been working on a book outlining the history of fishery protection in the Royal Navy. His article condenses much of the research for the book, and his account is illustrated with photographs and line drawings. The feature section of this year's annual then concludes with a short article by Stephen McLaughlin, who has detailed the modifications made to the Imperial Russian battleship *Orël*, captured by the Japanese in a damaged condition but largely intact after the Battle of Tsushima, to equip her for service with the Imperial Japanese Navy. The article is based on reports from US naval attachés and contemporary blueprints obtained from the Japanese that show the ship as rebuilt, and corrects a number of erroneous claims published in secondary sources.

There are some unusual contributions to Warship Notes. Enrico Cernuschi offers new insights into the fraught negotiations between Fascist Italy and Nazi Germany for access to German aircraft carrier technology in the form of lifts, catapults and landing gear; the high degree of dependency on the Germans even for suitable aero engines is something of a revelation. In the second of four Notes, Ian Sturton writes about French efforts to explore, occupy and ultimately colonise a large area of West Africa during the late 19th century using a combination of gunboats and barges on the upper reaches of the River Niger.

John Roberts relates an episode during the Spanish Civil War involving HMS *Cossack* and a British interpreter named Rapley, who broke his ankle when boarding the ship and was subsequently awarded compensation; in the process John highlights the extent to which references to the incident in secondary historical sources proved totally unreliable, even to the extent of the status and age of the individual concerned. Conrad Waters' account of a visit by the British constructor Ernest Guy Kennett to the recently-built design facilities of the French STCN (the so-called Balard complex) in the Paris suburb of Grenelle in the summer of 1939 has some valuable insights into the different practices of the two navies. Both of these last-named Notes are effectively 'spin-offs' from their respective authors' research for other projects: in the case of John Roberts his book *Destroyer Cossack* in Seaforth's 'original builders plans' series (2020, reviewed in *Warship 2021*), for Conrad his upcoming Seaforth book on the *Fiji* class (due March 2024). They reveal the value of following up oblique references to events that are not the main focus of the research undertaken.

The programme for *Warship 2025* is well advanced. Feature articles will include the first instalments of two new series on the German torpedo boats and the French destroyers designed and built prior to and during the Great War, by Dirk Nottelmann and the Editor respectively. There will also be two articles on battlecruiser designs: Stephen McLaughlin will focus on the plethora of designs submitted by domestic and foreign shipbuilders for the projected Imperial Russian battlecruisers of the *Izmail* class, while Michele Cosentino will look at battlecruiser studies for the Italian Navy during the interwar period. Przemysław Budzbon will continue his series on the early Soviet Navy with a study of the German-designed 'S'-class submarines, and his article will be complemented by a second submarine article by Tom Wismann on the Danish Navy's 'H' class, designed specifically for operations in the shallow waters of the Baltic. Coverage of the Imperial Japanese Navy will comprise an article by Hans Lengerer on French influence during the mid-/late 19th century, and a technical history by Kathrin Milanovich of the protected cruisers of the *Chikuma* class, the first major IJN warships to be propelled by turbines. *Warship 2025* will also have Part II of the article by Dirk Nottelmann and Aidan Dodson on the WWII German *Flak* ships, which has had to be held over from the current edition due to lack of space, and a major feature by Peter Cannon on the Royal Navy's 6in Mark XXI turret that armed the *Leander* and *Arethusa* classes.

John Jordan
1 April 2024

NAGATO AND *MUTSU*: THE 16in-GUN BATTLESHIPS THAT SURVIVED THE WASHINGTON TREATY

The IJN's *Nagato* and *Mutsu* were the first battleships in the world to have a main armament of 16in guns, and their early completion caused difficulties at the Washington Conference on Naval Arms Limitation in November 1921. **Hans Lengerer** looks at the principles behind their conception, and also explores the controversy surrounding authorship of their design.

The Imperial Japanese Navy (IJN) of the early 1900s was built up on the pattern of the Royal Navy, and with regard to warship construction the British style predominated. The IJN's first eight battleships were built in British shipyards, and from around 1904 the design, construction, and weapon technologies of Vickers at Barrow-in-Furness were adopted as standard for capital ships, culminating in the order for the 'battle cruiser' *Kongō* from Vickers in 1910.

In general terms, the construction of the hull and fittings as well as most of the equipment closely conformed to RN standards up to the *Ise* class. However, in the *Nagato* class the designers departed from the model in certain respects and implemented a number of striking innovations. *Nagato* and *Mutsu* were arguably the first battleships of purely Japanese conception, and spawned a series of unusual and idiosyncratic designs.

The key technical chacteristics of the new ships were as follows:

– They were the first battleships in the world to have a main armament of 16in guns, which were disposed in four twin turrets fore and aft with the second and third turrets superimposed.
– They were the first 'fast battleships' built for the IJN. Their designed speed surpassed that of the British *Queen Elizabeth* class and approached that of the new battle cruisers, while their overwhelming firepower was matched by heavy protection.
– High speed was realised by the adoption of geared turbines and a new type of small watertube boiler. The machinery generated almost twice the power of the preceding battleships, and the arrangement of the engine rooms was unique, the four main turbine sets being located in three compartments divided longitudinally with the turbines for the inboard shafts in the central compartment – an arrangement clearly influenced by the British *Queen Elizabeth*s. It was argued that in the event of one of the side compartments being flooded, three-shaft operation could be maintained and the list minimised.
– *Nagato* introduced a new system of underwater protection: a longitudinal torpedo bulkhead comprising three 1in (25.4mm) plates of HT steel was curved downwards from the lower edge of the waterline belt to the double bottom to form a protective barrier outboard of the ship's vitals (magazines and machinery spaces). According to *Kaigun Zōsen Gijutsu Gaiyō* ('Outline of Naval Shipbuilding Technique', Vol II, 204), this method of underwater protection had been the subject of experiments using a full-sized model, and the result confirmed that the bulkhead was capable of resisting the detonation of a 200kg (441lb) explosive warhead – the figure forecast at that time to be the bursting charge of future torpedoes.
– The *Nagato*s were the first Japanese battleships to adopt the 'all-or-nothing' system of the US Navy's *Nevada*. The length of the waterline belt was reduced to the length of the vitals, and casemate armour was abolished. The weight saved was used to augment the vertical and horizontal protection over the vitals. However, in contrast to *Nevada*, the main armoured deck remained at the at the level of the main deck, which was sloped downwards at the sides to meet the lower edge of the belt.
– The well-established tripod foremast was replaced by a new 'multi-post' type comprising a broad central post with a personnel lift inside supported by six angled struts to increase rigidity and eliminate vibration. The mast had platforms for fire control, command and communication equipment; it was adopted only after the construction of *Nagato* had begun (see below).
– The traditional clipper bow adopted for earlier battleships was discarded in favour of a new 'spoon-shaped'

A photo of *Nagato* during power trials in Sukumo Bay on 27 October 1920; the photo was released by the Navy Ministry. *Nagato* recorded an output of 85,500shp and attained the impressive speed of 26.7 knots. Lt-Cdr (Eng) Fujimoto Kikuo was on board for the trials, and proposed trunking back the first funnel to keep the exhaust gases clear of the bridge; the solution was rejected by Hiraga at the time but would be implemented during the mid-1920s following the failure of other measures. Note the torpedo nets (later removed) and the 'spoon'-type bow adopted to allow the ship to slide over the ropes of a moored mine. (Lars Ahlberg collection)

Nagato during firing trials in the Western Inland Sea on 24 October 1920. The photo was taken by Kure Navy Yard and shows that the 10-metre rangefinder on the forward part of the foremast has yet to be fitted. *Nagato*'s main gun turrets each had a 6-metre *Ha* (Bausch & Lomb) Type rangefinder. (Lars Ahlberg collection)

Mutsu: Profile & Plan

Note: The drawings are based on scans of the official 1919 plans supplied by the author. The plans do not show the position of the scuttles, which have been based on photographs, and the poor quality of the originals means that some of the other detail is unclear.

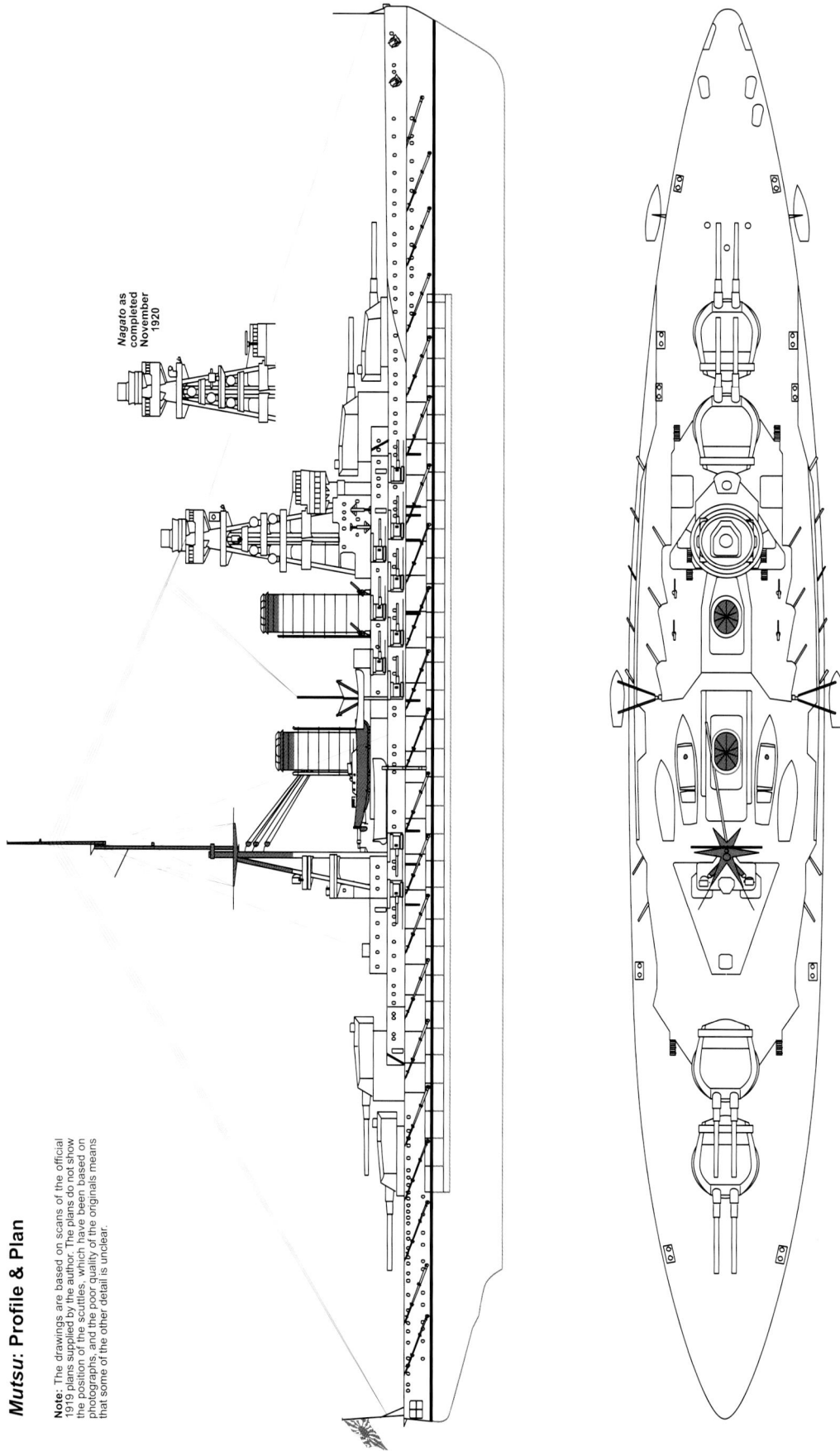

Nagato as completed November 1920

© John Jordan 2022

FEET
0   20   40   60   80   100

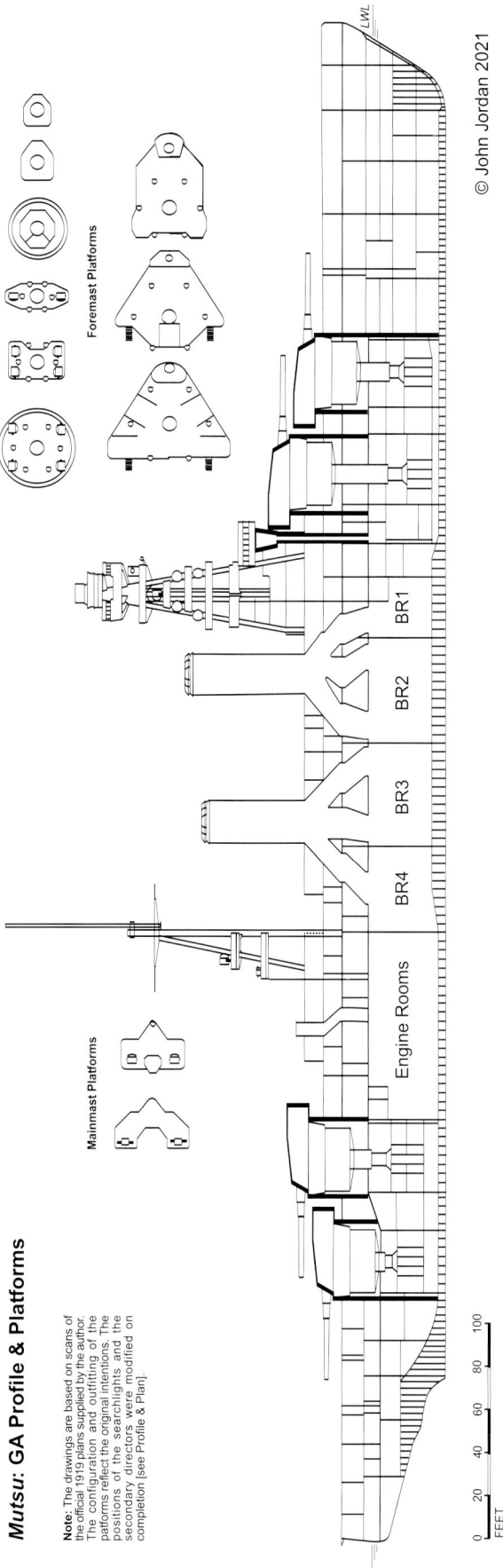

Mutsu: GA Profile & Platforms

Note: The drawings are based on scans of the official 1919 plans supplied by the author. The configuration and outfitting of the platforms reflect the original intentions. The positions of the searchlights and the secondary directors were modified on completion [see Profile & Plan].

Foremast Platforms

Mainmast Platforms

© John Jordan 2021

LWL

BR1

BR2

BR3

BR4

Engine Rooms

bow that prioritised defence against mines and had a broadly triangular vertical face that allowed for the mounting of the Imperial chrysanthemum (4ft in diameter). This proved to be more complex to construct, and also made for a wet forecastle – it would be modified in *Mutsu* during the 1930s – but the distinctive shape was adopted for all the ships of the Eight-Eight Fleet.

The Political Impact of the Ships

Nagato and *Mutsu* were the first two battleships of the IJN's 'dream' Eight-Eight Fleet and would be the only units completed. During the 'naval holiday', Japan built no battleships, so these two ships became the embodiment of her naval power from the conclusion of the Washington Treaty in February 1922 until the commissioning of *Yamato* in December 1941.

Nagato was completed in November 1920, and the construction of sister *Mutsu* was accelerated with a view to completion prior to the opening of the Washington Conference. Her ensign was hoisted in October 1921, but not all her weapons and other equipment had been embarked and trials were incomplete. This became a major issue at the conference. In the end the IJN was permitted to retain *Mutsu*, but in compensation was compelled to agree to the completion of a further two 16in-gun battleships for the US Navy (*Colorado* and *West Virginia*) and, more significantly, to the construction of two new British battleships with the maximum characteristics permitted by the treaty (*Nelson* and *Rodney*: 35,000 tons standard and 16in guns).

Had the IJN been compelled to scrap *Mutsu* and been permitted instead to retain *Settsu*, *Nagato* would have become a 'lone wolf' with her 16in guns and high speed of 26.5 knots, incapable of forming a tactical unit with the older battleships. Her military value would have been markedly diminished, and there would have been a further outcry from the Japanese naval 'hawks' for whom the Washington Treaty was a national humiliation.

The design of the *Nagato* class was complete and preparations for construction had already begun when the Imperial German Navy and Britain's Royal Navy fought the greatest sea battle of the First World War at Jutland. The lessons drawn from this battle were bound to influence the design of subsequent capital ships, and all the major navies were anxious to implement them as soon as possible. As an ally of Britain, Japan had observers aboard British warships and the IJN thereby obtained first-hand knowledge of the battle. As with the Royal Navy, modifications to ships whose design was already well advanced could not be realised in full, and in *Nagato* these modifications were limited to modest increases in protection and engine power.

For *Nagato*'s sister *Mutsu*, scheduled to be laid down some months later, a new design was drawn up to incorporate all the 'Jutland war lessons'. However, this A 125 design was rejected as construction would have been unduly delayed. Had it been approved, *Mutsu* would

FEET
0 20 40 60 80 100

Another official photo of *Nagato* during trials in Sukumo Bay on 27 October 1920. The 110cm searchlights have been fitted on the foremast and mainmast platforms, but not the 10-metre rangefinder. (Lars Ahlberg collection)

have still been on the building ways at the time of the Washington Conference, and it is doubtful that the IJN could have obtained permission for completion without even more serious concessions.

Origins of the Design

Taking into account the lessons of the Russo-Japanese War (RJW) of 1904–05, responsible officers of the Imperial Japanese Army (IJA) and IJN recognised the necessity for a national defence policy and planning for the necessary forces. After much heated discussion the emperor sanctioned a three-part policy document on 4 April 1907. The key naval element was to be a fleet composed of the eight most modern battleships and the eight most modern armoured cruisers ('battle' cruisers from 1912), which would serve in the front line for a maximum of eight years after commissioning before being replaced by new construction – the so-called Eight-Eight Fleet.[1]

Following the approval of this policy, however, the battleships that had fought in 1904–05 or had been laid down during or shortly after the war were rendered obsolescent by the commissioning of the 'all-big-gun' HMS *Dreadnought* in 1906, and the armoured cruisers suffered the same fate with the completion of HMS *Invincible* in 1908. To aggravate the situation further, Japan's naval budgets were constrained by the huge war expenses, the cost of repair of the Russian war prizes, and the building costs of the ships ordered during or shortly after the war.

On the other hand, the international situation, particularly the second revision of the Anglo-Japanese Alliance by which Britain, under US pressure, was freed of the

obligation to fight against the USA on the side of Japan in the event of war, required the urgent rearmament of the IJN. However, the domestic production of guns of greater calibre than 12in and the associated turrets, high-quality armour plates, and high-powered, reliable turbines was still not possible. The IJN therefore opted to order the battle cruiser *Kongō* directly from Vickers in 1910 with the goal of learning to design and build vessels of the 'super-dreadnought' type.[2] The contract included a substantial technology transfer element, with Japanese technicians and observers involved in the design and construction of the first ship, and *Kongō*'s three sister ships being built in domestic shipyards. This facilitated the realisation of a new generation of super-dreadnought battleships of the *Fusō* and *Ise* classes, all of which were designed and built in Japan.

In the meantime the Royal Navy had 'upped the ante' with the laying down of the fast battleships of the *Queen Elizabeth* class. The type was of particular interest to the IJN which, following the lessons of the RJW, placed a high value on tactical speed: a speed advantage of four knots was now seen as a key factor for defeating the US Navy in a decisive battle.[3]

The Early Design Studies

The preliminary design number of the battleships of the *Fusō* class was A 64; that of the modified type, the *Ise* class, was A 92 and was completed on 27 June 1914. The design of a more powerful class was then begun in response to the common trend in the world's major fleets towards an increase in gun calibre, which in turn brought with it the need to reinforce protection. The IJN was also looking to bring the speed of the new battleships more in line with the battle cruiser – a trend exemplified by the *Queen Elizabeth* class. A study of the battleships currently being built for Britain, the United States, Italy, France and Russia led to a decision to surpass these foreign counterparts both in gun power and speed and provide corresponding protection.

In pursuance of this policy, on 10 September 1915 Navy Minister Katō Tomosaburō submitted a new Warship Replenishment Plan[4] to the Cabinet according to which four battleships were to be built from fiscal year *Taishō* 5 (FY 1916) to *Taishō* 9 (FY 1920) with the aim of realising a fleet of eight modern battleships (including the *Fusō* and *Ise* classes). A National Defence Conference (*Bōmu Kaigi*) was convened to consider Katō's submission, and on 13 September agreed, in response to the political and economic situation, a partial rearmament plan that included one battleship of 32,000 tons to be laid down annually from 1916 to 1919.[5] Following the approval at the 37th session of the Imperial Diet of the warship building budget, with funding from FY 1916 to FY 1919, the Emperor gave his own approval, and the building programme was promulgated on 24 February 1916.

Following the decision of the National Defence Conference on 13 September 1915 that 'the laying down of the other ships should conform to the plan, with funding from FY 1917', Navy Minister Katō submitted to the Cabinet a new armament replenishment plan budget that included three battleships, to be laid down in 1917, 1918 and 1920 respectively with completion in 1920, 1921 and 1922, and this was agreed. This was the so-called 'Eight-Four Fleet Completion Plan', to be funded from FY 1917 to FY 1923. It was proposed by Katō to the 38th session of the Diet on 27 December 1916, but did not pass because of the dissolution of the Lower House on 25 January 1917. However, the situation changed dramatically in 1916 with the approval in the United States of the Naval Act of 1916, a three-year programme that included no fewer than sixteen capital ships, by both Congress and the Senate. If fully implemented it would have given the US Navy 100 per cent superiority over the IJN. In this situation Navy Minister Katō proposed an additional budget of Yen 261,521,160 as continuous expenses for building the ships in the original plan to the extraordinary 39th session of the Diet on 23 June 1917, and obtained agreement and the Emperor's approval on 19 July. Promulgation on the following day put in train the construction of the additional three battleships: *Mutsu* as sister to *Nagato*, and *Kaga* and *Tosa* as a new class based upon revised staff requirements.

It was initially envisaged that *Nagato* would be a modified *Ise* armed with twelve 14in 50-calibre guns of a new type to oppose the latest US battleships, which were likewise armed with 14in 50-calibre guns, or RN battleships armed with the 15in 42-calibre gun. However, Vickers was offering a battleship armed with eight 16in 40- or 45-calibre guns with a displacement of 30,000 tons and a speed of 25 knots, and the IJN also obtained plans of the *Queen Elizabeth* class. In consequence the 'modified *Ise* design' was revised, becoming a 32,500-ton ship with eight 16in guns in twin turrets superimposed fore and aft.

The 5.5in secondary guns were mounted on two levels, on the upper deck and the forecastle deck. The above-water torpedo tubes were raised to the upper deck and were trainable. Armour protection initially resembled that of the *Ise* class with a 12in belt at the waterline and an upper belt 4in thick. For horizontal protection there was 1in plating on the upper deck and 1½in on the main deck. The foremast and mainmast had the same tripod structure employed in *Kongō* and improved successively in the *Fusō* and *Ise* classes. With 60,000shp, speed was calculated as 24.5 knots.

After rejecting seventeen design studies (A 93 to A 109), design A 110 was adopted. The exact date when this design was approved is uncertain, but it must have been drawn up by the spring of 1916, because Kure Navy Yard received the order on 12 May 1916. A further uncertainty is the name of the chief designer. Hiraga Yuzuru was responsible for the modified design (A 112) but not for A 110, as claimed by some naval historians, as he was not in post until 15 May (see Appendix for time-line). It is therefore probable that the constructor who drew up the earlier design was Captain (Constructor) Yamamoto Kaizō.[6]

Revision of the Preliminary Design

In a lecture in the presence of the Crown Prince (the later Shōwa Tennō), on 18 December 1924, after his return from a tour of Europe and the USA, Rear Admiral Hiraga referred to himself as the 'chief designer of *Nagato* class'; however, he joined the Naval Technical Department (NTD) only on 15 May 1916, three days after the order for what would become *Nagato* was placed with Kure Navy Yard.

On 31 May and 1 June 1916 the Battle of Jutland took place. Three British battle cruisers, including the new *Queen Mary* – a contemporary of the Vickers-built *Kongō* – were sunk by magazine explosions that were initially attributed to insufficient horizontal protection.[7] After referring to the Washington Treaty and contemporary warship design in his lecture, Hiraga spoke about the capital ships currently under construction in Britain and the USA. He then focused on their protection, and summarised the influence of Jutland on the improvements incorporated in the design of the *Nagato* class.

High tactical speed was seen as having both offensive and defensive advantages, and the first modification made by Hiraga to the original design had been to increase maximum speed by 2 knots to 26.5 knots. This was to be achieved by the adoption of geared turbines – Hiraga claimed this was the first application of gearing in a major vessel[8] – and engine power was increased by 20,000shp to 80,000shp. Remarkably, this 33 per cent rise was achieved 'on the same weight and volume as a result of progress in boiler and engine design'. IJN studies of geared turbines had begun several months previously. The first iteration of the *Nagato* design had conventional direct drive, and Fukui Shizuo claims that the studies were undertaken with a view to adopting geared turbines for the following class of capital ships. The adoption of geared turbines for *Nagato* and *Mutsu* was made possible by the import of manufacturing expertise (and machine tooling) from the American Westinghouse Company.[9]

Hiraga than turned his attention to the protection system, and explained that the length of the waterline belt was reduced and the upper belt omitted in order to compensate for the increase in weight of the horizontal protection over the ship's vitals (magazines and the machinery spaces).[10] The basic structure of the hull and the waterline belt were left as they were;[11] only the horizontal protection was increased. Deck protection had proved to be the Achilles heel of the British battle cruisers; decks had been penetrated by projectiles fired from very long range and striking at high angles of obliquity. Reinforcement of the horizontal armour was therefore considered to be the most urgent improvement, together with flash protection measures inside the barbettes to prevent flames reaching the magazines.

Revisions to design A 110 had been requested by the Navy Ministry shortly after the reports of Jutland were received. These were to be executed quickly and with a maximum increase in displacement of 1,300 tons. The NTD had started work immediately and the main ship design section (now headed by Hiraga) submitted four alternative designs (A–D) in July, of which design C was preferred. Preliminary sketches and a table of the main characteristics were submitted to the Higher Technical Conference on 30 August and approved as design A 112 following intensive discussions in September. Kure NY received the order to lay down *Nagato* to the revised design on 28 October 1916.

The Final Design

In the event the keel of *Nagato* was not laid until ten months later on 28 August 1917. The delay was the result of further studies and revisions to the design drawings. The revision of A 110 to A 112 had been made in haste, and further studies resulted in some important changes, the most prominent of which was a new underwater protection system designed by Hiraga[12] based on a curved bulkhead composed of high-tensile steel plates. Revisions were so substantial that the building order was changed to design A 114 on 27 July 1917, one month before the laying down date.

The order for *Mutsu* was placed with Yokosuka Navy Yard four days later, on 31 July 1917. She was to be built as a sister to *Nagato* despite Hiraga's protests. His improved, post-Jutland A 125 design (flush deck, five 16in/41cm twin turrets,[13] inclined side armour, thicker horizontal protection, and a revised propulsion system featuring a reduction in the number of boilers with a single funnel) had been prepared, but was rejected by the NGS in favour of a tactical unit comprising two ships of uniform design.

Building Data and Principal Characteristics

The construction of *Nagato* in the building dock of Kure NY was without incident.[14] However, the launch (on 9 November 1919, weight 22,175 tonnes) was delayed for almost six months due to difficulties in obtaining materials and spiralling prices during the later years of the Great War. Once the hull was launched fitting-out proceeded at a surprising pace for such a large ship, and *Nagato* was completed just over a year later. Shipbuilding costs amounted to Yen 15,409,697 and the man-days (8.5 hours per day) were 1,960,319.

Table 1: Building Data

	Nagato	Mutsu
	(temporary No 7 Battleship)	(temporary No 8 Battleship)
Programme	1916 8-4 Flt Nomination Plan[1]	1917 8-4 Flt Completion Plan[2]
Budget	Armament Replenishment Expenses (*Gunbi Hojū Hi*)	[as *Nagato*]
Ordered	12 May 1916 (later revised)	31 July 1917
Builder	Kure Navy Yard	Yokosuka Navy Yard
Laid down	28 August 1917	01 June 1918
Launched	09 November 1919	31 May 1920
Completed	25 November 1920	24 October 1921[3]

Notes:
[1] The Navy Minister submitted the so-called Eight-Four Fleet Nomination Plan to the National Defence Conference (*Bōmu Kaigi*) on 10 September 1915 and requested:
(a) Confirmation of the 8-4 Flt Nomination Plan.
(b) Confirmation of the construction of more warships from FY 1917 onwards according to the plan drawn up by the Navy Ministry.
[2] The National Defence Conference confirmed the Nomination Plan on 13 September 1915, and added Yen 260,000,000 to the budgets for FY 1917–1923 for the construction of a further eight battleships and two battlecruisers.
[3] The actual completion date of *Mutsu* was 23 November 1921; the 'official' date was announced with the upcoming Washington Naval Arms Limitation Conference in view.

New Foremast Structure

The initial design of the *Nagato* class featured the same tripod masts as in the *Ise* class. When *Nagato* was laid down in August 1917, *Ise* and *Hyūga* were being fitted out and were almost complete (December 1917 and April 1918 respectively). The late Fukui Shizuo explains that the distinctive multi-post foremast of *Nagato*, which would have also been a feature of the later battleships and battlecruisers of the Eight-Eight Fleet, was adopted during her construction. He attributes the concept to then-Captain (later Vice Admiral) Kaneda Hidetarō (1872–1925), who argued forcefully for its adoption. Fukui further states that he learned this from Hiraga directly, who was by then in charge of the project. Kaneda himself was not a technical officer (*gijutsu shōkō*) but a line officer (*heika shōkō* = graduate of the Naval Academy).[15]

The purpose of this mast structure was to eliminate the vibrations that impacted on the accuracy of target designation, rangefinding, spotting and other highly sensitive instruments, and to permit the continuation of these operations even when the heavy guns were fired. The 16in guns of the *Nagato* class had a greater range than the 14in guns of the *Kongō*, *Fusō* and *Ise* classes. Moreover, engagement ranges during the First World War were much greater than had been expected, and at these longer ranges the effectiveness of the main guns depended on the quality of the fire control in order to hit first and keep hitting.

The most important presupposition to attain this goal was to provide a rigid platform for mounting these instruments high up. Increased range required higher positions for optical instruments, otherwise the enemy would remain below the horizon due to the curvature of the earth. An enemy vessel that had its rangefinders and other optical instruments mounted high would have the advantage and would outrange its opponents – a situation that had to be avoided at all costs.

The modifications to the tripod mast realised in the preceding classes had not brought about the expected results, and the only solution was seen to be a completely new structure. Kaneda's proposed multi-post mast comprised a central post of large diameter supported by no fewer than eight equally-spaced inclined support struts that met the central post high above the waterline.

On 21 November 1917, three months after *Nagato* was laid down, a Technical Conference (*Gijutsu Kaigi*) of the NTD was held at which the main item for discussion was Kaneda's proposal. A fierce debate took place, with major concerns being raised by Hiraga regarding the weight of the mast and its impact on stability. In the end Hiraga was prepared to accept the structure provided the number of support struts was reduced from eight to six, at the expense of a degree of rigidity.

In its final form the mast was composed of a main post 6ft 3in in diameter fitted internally with a personnel lift connecting the middle deck with the spotting and control top, supported by six inclined support struts 3ft in diam-

Table 2: Characteristics as Completed

Displacement:		
normal	(N) 33,870 tons	
	(M) 34,116 tons	
standard	(N/M) 32,300 tons	
full load	(N) 40,204 metric tons	
	(M) 39,082 metric tons	
Length:		
pp	660ft 7in (201.35m)	
waterline	700ft 9in (213.60m)	
overall	708ft (215.80m)	
Beam:		
waterline	95ft (28.96m)	
maximum	95ft (28.96m)	
Depth:		
to upper deck	42ft (12.81m)	
to f'c'sle deck (mean)	48ft 4in (14.73m)	
Draught:		
forward	29ft 6in (9.00m)	
aft	30ft 1in (9.17m)	
mean	29ft 9in (9.08m)	
Freeboard:		
forward	26ft (7.93m)	
amidships	21ft 6in (6.55m)	
aft	16ft 6in (5.03m)	
Coefficients:		
block (Cb)	0.583 (0.595)	
midship (Cf)	0.982 (0.984)	
prismatic (Cp)	0.604 (0.602)	
waterline (Cwl)	0.692 (0.693)	
Ratios:		
length/beam (L/B)	7.373 (7.375)	
draught/length (d/L)	0.042 (0.043)	
beam/draught (B/d)	3.189 (3.167)	
length/draught (L/d)	13.6	
depth/draught (D/d)	1.73 (1.72)	
tons per cm immersion	44	

Sources: Fukuda Keiji, *Outline of the Fundamental Design of Warships* (*Gunkan Kihon Keikaku Shiryō*), 1, 32; Fukui Shizuo, *Japanese Warships* (*Nihon no Gunkan*), *History of Shipbuilding in Shōwa Era*, Vol 1, 435–36,776, 779; *Weight of Centre of Gravity Data for Miscellaneous Warships by the Preliminary Design Group* (Lt [naval architect] Tōyama, Eng Imai, Asst Engs Takahashi and Ogino), second revision, October 1941; Makino Shigeru, Fukui Shizuo (Ed) *Outline of Naval Shipbuilding Technique*, Vols I, II and V.

Notes:
1 For Coefficients and Ratios, data in parentheses are planning values for *Mutsu* from Fukuda.
2 The bow was the so-called 'spoon' type (this refers to the double curvature with the large cut-up of the forefoot). This particular bow shape was adopted to permit the ships to slip over the ropes of tethered mine No 1 (= *dai ichi gō renkei kirai*). This mine was classified as a military secret (*gunki*) and many of them were to be laid in front of the approaching enemy force. The bow shape of the *Nagato* class was the template for the ships of the Eight-Eight Fleet.
3 The appearance of the stern was simplified by abolishing the stern walk.

eter with 60-degree spacing and combining at the upper end to form a conical structure. The lower platforms were wrapped around the mast. Searchlights were mounted on each of the open platforms above, and the more substantial upper platforms housed the fire control and rangefinding equipment; the hexagonal structure was therefore only visible in the centre part of the mast. The height from the upper deck to the upper platform was 14 floors.

The mast was described by the IJN as *yagura* ('pagoda') style and formed the basis for the 'pagoda masts' that were a characteristic feature of Japanese interwar battleships prior to the adoption of a tower-type foremast in the *Yamato* class in the mid/late 1930s. Even before the reconstruction of *Nagato* and *Mutsu* in the late 1930s there were constant additions and modifications with the mounting of new weapons and FC equipment.

Machinery[16]

The machinery adopted for *Nagato* is of particular interest for three reasons:

– Geared turbines were adopted for the first time in a major IJN warship.
– The turbines were designed independently by the Design Group of No 5 Division of the *Kaigun Gijutsu Honbu*[17] using the revolutionary Triple Flow Turbine concept developed by the American Westinghouse Company (see below).
– The reduction gearing for *Nagato* was purchased directly from Westinghouse, but the gearing for her sister *Mutsu* was manufactured under licence by Yokosuka NY using a hobbing machine purchased from the US company.

Steam for the operation of the main turbines and the greater part of the auxiliary machinery was supplied by fifteen *ro gō* Kampon-type oil burning boilers[18] with superheaters generating saturated steam, and six boilers with mixed firing.[19] There were animated discussions about the continued use of boilers with mixed firing alongside boilers fired exclusively by oil, but the debate was concluded by the head of No 5 Division, who argued that coal firing was still necessary due to the high cost of imported heavy oil.

There were four sets of Gihon impulse-type turbines each comprising a single high-pressure (HP) and a single low-pressure (LP) turbine in a cross-compound arrangement, the shafts from the turbines being combined via reduction gearing to drive a single propeller shaft using the principle of the American Westinghouse triple flow system. The steam passed through the first impulse stage in the HP turbine and was then divided into three parts: one part (approximately one third) continued its way through the group of rotor blades in the HP turbine casing, and the remainder entered the casing of the LP

Nagato at Kure in December 1920. *Nagato* was completed on 25 November and a delivery ceremony was held. On 1 December she became flagship of the 1st Fleet (1st Squadron), and she is here flying the flag of Admiral Tochinai Sōjirō. The long-base 10-metre rangefinder, which was mounted on rails, is now fitted (3rd platform from the top), as are the 5th Year light rangefinders for the secondary 14cm guns (forward edge of the middle searchlight platform). Beyond the aftermost gun turrets can be seen the battleship *Ise*. (Lars Ahlberg collection)

Nagato 1921: Machinery Layout

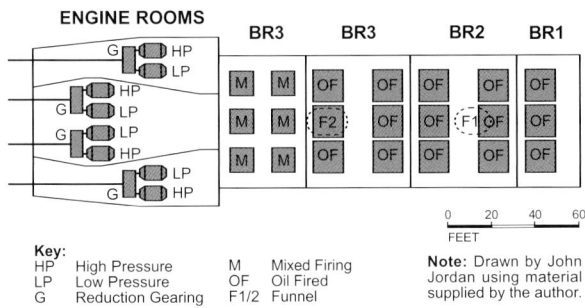

ENGINE ROOMS

| | BR3 | BR3 | BR2 | BR1 |

Key:
HP High Pressure
LP Low Pressure
G   Reduction Gearing

M   Mixed Firing
OF Oil Fired
F1/2 Funnel

0 20 40 60
FEET

Note: Drawn by John Jordan using material supplied by the author.

Table 3: **Machinery Spaces**

	Boiler Rooms	Engine Rooms
Machinery	15 oil-fired, 6 mixed Kampon *ro gō*	4 sets of Gihon geared turbines
Length	163ft 0in (49.68m)	88ft 0in (26.82m)
Width	54ft 6in 16.61m)	63ft 6in (19.34m)
Height	28ft 0in (8.53m)	31ft 0in (9.45m)
Floor area	8,884ft^2 (825.30m^2)	5,588ft^2 (516.20m^2)

Source: Fukuda, *op cit*, 147.

turbine at the half-way point, where it was divided equally, passing through the two groups of rotor blades in opposite directions (see drawing). When the ship was running at cruise speed the flow of steam to the LP turbine was cut off and only the HP turbine operated. There was no cruise stage in the main turbines nor was there a separate cruise turbine. The turbine blades were of the Gihon type developed by the IJN from the Brown-Curtis system, and spawned the later Kampon-type blades. The primary aim of the triple flow system was to reduce steam (and hence fuel) consumption, but results proved disappointing. A peculiarity of these ships was the mounting of an auxiliary turbine to drive the centre shafts at synchronous speed when cruising on the wing shafts in order to eliminate 'drag' when the shafts were feathered. The diameter of each of the four three-bladed propellers was 13ft 9in, pitch 14ft 5½in. Twin rudders of the balanced type were fitted.

Mutsu during 10/10 (full power) trials off Tateyama on 19 October 1921, during which she recorded a speed of 26.73 knots with 87,494shp. The high speed of these ships was a well-kept secret; the published figure was 23 knots. (Lars Ahlberg collection)

Nagato & Mutsu: Kampon-Type Impulse Turbine

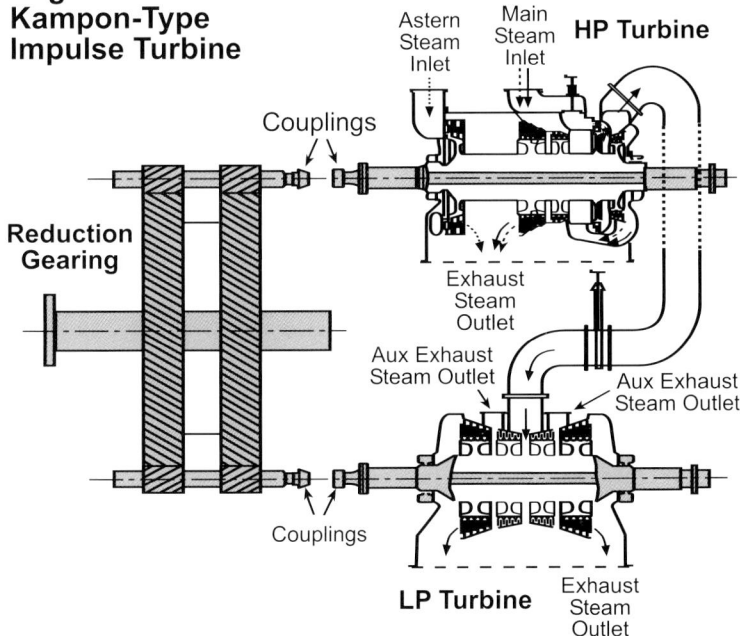

Astern Steam Inlet
Main Steam Inlet
HP Turbine
Couplings
Reduction Gearing
Exhaust Steam Outlet
Aux Exhaust Steam Outlet
Aux Exhaust Steam Outlet
Couplings
LP Turbine
Exhaust Steam Outlet

RPM
HP Turbine	2730
LP Turbine	2730
Propeller shaft	230

Gearing
Main Wheel
No of teeth	392
Inclination	30°
Pinions
No of teeth	33

KEY
← Full Power
←- - HP Full Power [overload]
←---- Astern

Note: Drawn by John Jordan using material supplied by the author.

Another photo of *Mutsu* on 19 October 1921. In order to save her from being scrapped under the Washington Treaty she was rushed to completion. The official completion date was 24 October 1921; however, much work on the ship remained to be done, and neither the searchlights nor the rangefinders are fitted in this early view. (Lars Ahlberg collection)

Reduction gearing allowed the adoption of light-weight, fast-running turbines. The rotational speed of both the HP and the LP turbines was 2,730rpm, while the propeller shafts turned at a maximum of 230rpm. The gearing for *Nagato* was manufactured by the American Westinghouse Company,[20] and was of the 'floating frame' type; *Mutsu* had a more conventional I-beam type.[21] The reduction gearing of the latter ship was fabricated at Yokosuka Navy Yard using a hobber purchased from Westinghouse. The wheel had 392 teeth cut with 30-degree inclination and each of the pinions had 33 teeth; there was no difference between the HP and LP turbine gearing. The reduction gearing mounted in US Navy destroyers was particularly noisy, and following the decision to instal Westinghouse gearing in *Nagato* particular attention was paid to this aspect of performance, but the gearing proved to be relatively quiet in operation and good results were obtained.

The weight of one set of Gihon turbines was 50.8 tons, that of the reduction gearing 54 tons, making a total of 419.2 tons for all turbines. The remarkable figure of 190.8shp per ton of engine weight for this class compared to 53.3shp/ton for *Fusō* with and 56.1shp/ton for *Hyūga*, both of which had conventional 'heavy', slow-running direct drive turbines. With 3,400 tons of heavy oil and 1,600 tons of coal, endurance was estimated at 5,500nm at 16 knots.

Electricity was supplied by four 250kW turbo-generators (1,110A), two 150kW reciprocating generators (667A), and one 26kW semi-diesel generator (115A), giving a total generating capacity of 1,326 kW. Voltage was 225V with a uniform DC current.

Main Guns

The studies which ended with the adoption of the 16in (41cm) gun for the battleships to follow the *Fusō* class[22] were begun in 1913. At that time the mounting of 15in (38cm) guns in the newly-designed British fast battleships of the *Queen Elizabeth* class had already been announced, and the mounting of 16in guns on US battleships was considered to be only a matter of time. However, the characteristics of the larger-calibre gun were yet to be defined, and the increase from 14in to 16in was considered a major challenge for the IJN from a technical point of view. A comparative study of the British 15in gun and the proposed 16in gun was undertaken in order to determine the next step.

The key questions that the study was intended to address concerned the length in calibres (40- or 45-calibre), the weight (and penetration) of the projectile, and the calibre of the gun (15in/38cm or 16in/41cm). The ballistic requirements were that the remaining velocity of the projectile at a range of 8,000 yards should be 1,900ft/sec with a maximum barrel pressure of 18.6 tons/in².

The advantages of the 40-calibre gun were: a shorter barrel length (and therefore less weight), a reduction in vibration, and slightly smaller dimensions of the turret and the diameter of the roller path. The advantages of the 45-calibre gun were: a smaller powder chamber, less erosion (hence longer barrel life), and increased range. The latter characteristics were considered decisive, and a length of 45 calibres was proposed and accepted.

As for the projectile, three types were investigated for each of the gun calibres considered:

Table 4: **Main Guns**

Designation	41cm 45-cal 3rd Year gun Model II
Calibre	41cm › 40cm nominal (406.4mm actual)[1]
Construction	Wire-wound: four layers at muzzle and breech
Barrel length	18.84m
Bore length	18.29m
Length of rifling	15.84m
Length of wire-wound part	15.63m
Centre of gravity (from breech face)	5.88m
Breech mechanism (BM)	Interrupted screw[2]
Weight (incl BM)	102 tons
No of grooves / depth x width	84 / 4.1mm x 8.75mm
Twist	uniform 1/28
Chamber dimensions	2.43m x 440–513.8 mm
Chamber volume	480 litres
Pressure at breech	3,000–3,070kg/cm[2]
Projectile weight[3]	1,000kg AP, 936kg Common
Powder charge[4]	219kg cordite[5] (4/4 bagged charges)
Muzzle velocity[6]	790m/s
Maximum range	30,200m at 30°

Notes:

[1] As these guns were built after 5 October 1917, the designation and official data were in metric, not imperial. The first official metric designation was 41cm, but this was revised downwards to 40cm, probably for 'political' reasons as the official metric equivalent of 16in in the wording of the Washington Treaty was 406mm.

[2] In contrast to the interrupted screw breech of the Vickers type, the fore and after parts were inclined at 5 degrees to facilitate opening and closing. In profile the middle section appeared to be bulged. Devised by the later Vice Admiral (weapons production) Dr Arisaka Kinzō, the breech was officially adopted on 21 August 1908 and used from the 3in (8cm) gun upwards for nearly all Japanese naval guns as the Type 41 (1908).

[3] Figures for the Type 88 projectile. The Type 91 'boat tail' projectile that entered service in 1931 had a weight of 1,020kg.

[4] Three types of charges were used: full (4/4), reduced (3/4), and weak (2/4). There was also a reinforced charge (full + up to 20%), but this was used only for trial purposes.

[5] Figure from *Shōwa Zōsen-shi*, Vol 1, 701; other sources give 222kg or 224kg.

[5] Muzzle velocity depended on the weight of shell and the extent of barrel wear.

	15in	16in
Lightweight	1,720lbs	2,090lbs
Intermediate	1,820lbs	2,210lbs
Heavyweight	2,090lbs	2,560lbs

The lighter the projectile, the smaller the powder chamber and the less the erosion. However, weight also impacted on penetration, and if the lightweight projectile were to be adopted there was little space for improvement should a higher muzzle speed be required in the future. On the other hand, a heavyweight projectile would result in significantly greater barrel wear; it was therefore proposed to adopt a projectile of intermediate weight.

When the committee came to study the calibre to be adopted, the key advantages of superior penetration, a larger burster (hence greater destructive power), and increased firing range carried the argument in favour of the 16in gun. A further attraction was that of putting the IJN ahead of the game, as with the adoption of the 14in gun for the *Kongō* class.

According to Horikawa Kazuo and Ondo Shinsaku (*Notes on the History of Early Military Steel Technology*), the prototype of a 'lightweight' 16in gun

Table 5: **Ballistic Data**

Elevation/Range	Angle of fall	Striking speed	Penetration (AP)
15,000m			406mm
20,000m	21°40'		272mm
20° = 24,100m	30°26'	420m/sec	211mm
25° = 27,400m	37°59'	420m/sec	
30° = 30,300m	44°38'	434m/sec	

Source: Fukuda, *op cit*, 159: penetration of vertical Vickers Cemented (VC) armour. *Umi to Sora* 5/1958, 51, gives higher penetration figures that may relate to later models of shell; the latter source gives a flight time for the projectile of 59 secs at 30,000m range with 780m/sec mv.

was completed in 1914 and then entered series production, with a number of improvements, as the 41cm 3rd Year Model II gun (for key technical data see Tables 4 and 5).

According to the late Rear Admiral Takasu Kōichi a total of 38 guns were produced by the Weapons Division of Kure Navy Yard. The Technical Vice Admiral (weapons production) Noda Yoshikawa noted that the

Steel Production Division of Kure NY made the ingots for 58 1A, 50 2A and 51 4A tubes for 40cm guns up to 1921. Horikawa Kazuo and Ondo Shinsaku (*op cit*) refer to tests carried out on 77 2A tubes (from the first to the final stages) between 1919 and 1922, which ended with the acceptance of only 23 tubes (32 per cent!). This illustrates the considerable difficulties experienced in the production of these tubes, which were in excess of 17 metres (55ft) long, including test edges.

Main Gun Mountings

The mounts were basically copies of the British-built turrets designed for the *Kongō* class. The principal improvements evolved by the Japanese designers were greater elevation, loading at all angles, improved flashtightness in the gun house and working chamber,[23] and safer loading cages for the upper hoists.

Power for training and elevation of the guns was hydraulic on the British model. The training gear used a worm drive that suffered from wear; elevation was by cylinder and piston attached to the rear of the slide. The turrets could be trained at 1°35' per second and the guns elevated at 8°5'/sec. The split hoists were likewise on the British model: dredger cage hoists each raised a single shell and four quarter charges from the handing room to the working chamber. The complete round was then raised from the working chamber to a position behind the breech by a hydraulically-powered upper cage hoist. Ramming of the shell and the charge was by power-operated chain, the rammers being mounted on the continuation of the slide. Although in theory the guns could be loaded at any angle up to 20 degrees; the breech could not be opened above 25 degrees of elevation.[24] The firing cycle was approximately 1.5 rounds per minute.

Two types of turret were built. The first was for the *Nagato* class and allowed a maximum angle of elevation of 30 degrees; the second, intended for the *Kaga* class, permitted elevation up to 35 degrees. The weight of the *Nagato* turret was 892 tons.

Eighteen turrets (of which six for the *Nagato* class) were produced by the Weapons Division of Kure NY and four (two for the *Nagato* class) at Yokosuka NY.[25] Of the

Mutsu fires the guns of No 4 turret during an exercise in Tōkyō Bay on 11 March 1922. The front plate of the turret was angled 40 degrees and had a thickness of 12in (305mm). Each of the main gun turrets was fitted with an 8-metre *Bu* (Barr & Stroud) Type rangefinder. Note the 3.5-metre 5th Year rangefinder for the secondary guns in its protective housing on the lower platform of the mainmast. (Lars Ahlberg collection)

fourteen turrets produced for the *Kaga* class, six were transferred to the IJA to be employed in coast defence batteries. The remaining eight turrets were modified to give an elevation angle of 43 degrees, and replaced the original turrets of *Nagato* and *Mutsu* during the reconstruction of the mid-1930s.

Other Weapons

As secondary armament *Nagato* and *Mutsu* were fitted with twenty 14cm (5.5in) 50-calibre 3rd Year guns in lightly-protected casemates arranged on two levels (forecastle and upper decks), sixteen of which faced forward and four aft. Four 8cm (3in) 40-calibre 3rd Year HA guns were mounted abreast the first funnel, and there were three 6.5mm 3rd Year-type machine guns and eight Type 41 short 8cm (3in) guns for saluting.

The torpedo armament comprised four Armstrong-type 21in (53cm) underwater broadside fixed torpedo tubes and four above-water trainable torpedo tubes with 14–16 6th Year torpedoes.

Fire Control Systems

The fire control systems for the main and secondary batteries installed in *Nagato* and *Mutsu* featured the latest equipment, as these ships were essentially prototypes for the new generation of battleships and battlecruisers of the Eight-Eight Fleet. The main fire control station was located on the upper platform of the new heptapod foremast, and was equipped with a Type 13 director. There were auxiliary control stations atop the after superstructure and in No 2 turret.

At the forward end of the platform below the spotting and control top, a rangefinder of a new long-base 10-metre type was fitted shortly after completion,[26] and there were rangefinders purchased from the US company Bausch & Lomb (*Nagato*, 6-metre *Ha* type) and Barr & Stroud (*Mutsu*, 8-metre *Bu* type) atop each of the four main turrets. Gun range was supplied from 1921 by a Vickers Type 10 range clock (*kyōri dokei*) located in the main transmitting station (*sokuteki*),[27] with transmission to the main guns using follow-the-pointer systems, and there was a range rate panel (*henkyoritsuban*) on the foremast.

For the 14cm secondary guns there were *Bu*-type 3.5-metre 5th Year rangefinders, with gun range being supplied by two Type 10 range clocks. Two 2/2.5-metre 5th Year *Bu*-type rangefinders were provided for the 8cm HA guns, which later had Type 89 FC computers. Much of the fire control equipment for the secondary and HA guns was developed during the 1920s, after the completion of the ships.

For night action there were no fewer than ten 110cm searchlight projectors: eight on the foremast platforms and two on the upper mainmast platform. Remote searchlight control panels were mounted on the foremast.

Nagato: **Protection Scheme**

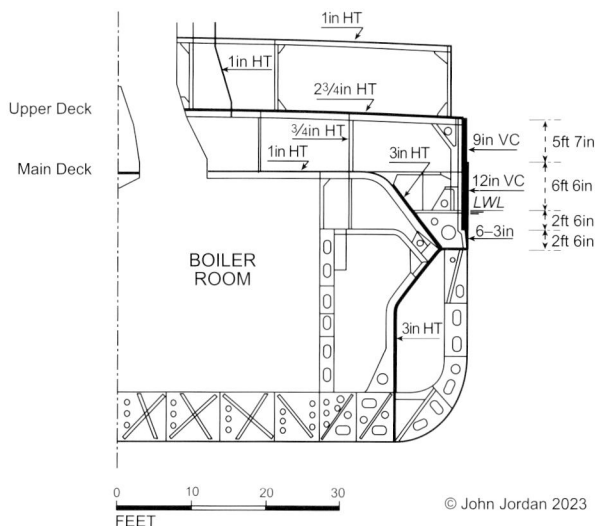

© John Jordan 2023

Protection

The protection system represented a compromise between conventional systems and the 'all-or-nothing' system employed for US Navy battleships from *Nevada* onwards. There was a heavy, deep waterline belt that extended from the magazine for No 1 turret to the magazine for No 4 turret, but no protection for the ends; the weight thereby saved was invested in horizontal protection over the vitals. In contrast to the US Navy's scheme, however, there were two armoured decks: the upper deck and the main deck, which had a conventional flat section on either side of the middle line with added HT plates on the slopes.[28]

Hull

The main belt comprised 12in (305mm) Vickers Cemented (VC) armour and extended from 6ft 6in above the waterline to 2ft 6in below; its upper edge was just above the flat of the main deck. Above it was a 5ft 7in strake of 9in (229mm) VC armour which extended to the upper deck. Below it was a wedged-shaped strake of armour which was 6in (152mm) thick where it joined the main belt, tapering to 3in (76in) at the bottom edge. All these plates were seated on a wooden backing, with armour bolts securing them to the shell.

The armoured bulkheads that enclosed the citadel were as follows: 9in (229mm) VC fore and aft above the main/middle decks; 13in (330mm) VC forward at the level of the lower deck, and 9in (229mm) VC aft, tapering to 3in (76mm) Vickers Non-Cemented (VNC) at its lower edge.

The armoured citadel was completed by two protective decks. The upper deck rested on the upper edge of the belt, and comprised three plates of HT steel totalling 2¾in (70mm). The main deck was 1in (25.4mm) on the flat reinforced by 2in (51mm) plates of HT steel on the slopes.

The torpedo bulkhead, the upper section of which was

curved outboard to meet the lower edge of the belt and extended down to inner floor of the double bottom, was composed of three plates of 1in (25.4mm) HT steel, for a total thickness of 3in (76mm).

The steering compartment had only ¾in (19mm) HT protection at the sides. This came to be seen as a potential weakness, and it was considerably enhanced when the ships were reconstructed during the 1930s. There was also 1in (25.4mm) HT plating on the sides of the funnel uptakes, with ¾in (19mm) HT on the transverse bulkheads.

Turrets

The main gun houses had 12in (305mm) VC armour on their faces, which were inclined by 40 degrees, and 9in–7in (229–178mm) VC on their sides, which were inclined 30 degrees; the rear wall was of 7½in (190mm) VC. The turret roofs were of 6in–5in (152–127mm) HT steel and the floors of 4in (102mm). The barbettes were protected by VC plates 12in (305mm) thick, declining to 9in (229mm) below the upper deck.

The 14cm (5.5in) secondary guns had only light splinter protection, as in contemporary US Navy ships with the 'all-or-nothing' protection system. They had 1½ (38mm) HT gunshields, and there were bulkheads of 19mm HT steel between the casemates. The roofs and sides of the casemates had only light (1in/25.4mm?) plating.

Conning Tower

The conning tower had 13in–10in (330–254mm) VC plates on the face and sides, a 7in (178mm) VC roof and a 3in (76mm) floor. The communications tube had 5in (127mm) plating above the armoured decks, reducing to 3in (76mm) below.

During trials it became apparent that, due to the proximity of the first funnel to the heptapod foremast, the upper platforms on which the optical fire control instruments were mounted were likely to be blinded by funnel smoke. As designed the first funnel of these ships was raised compared to the second but this measure proved to be insufficient, and shortly after the ships entered service the funnel was fitted with a tall clinker screen. The paired rangefinders atop the bridge in this photo of *Nagato* are a 'heavy' 3.5-metre 5th Year model. The smaller rangefinder on the platform above is the 2.5-metre 5th Year model. (NHHC, NH 111587)

Nagato in July 1927, with a floatplane atop No 2 turret. She still has the tall clinker screen on the first funnel. Note the chrysanthemum emblem at the bow. (NHHC, NH 111604)

Appendix: Hiraga Yuzuru

Lieutenant Hiraga joined the NTD on 3 February 1909 and worked from the following day in the Third Division (Shipbuilding), the head of which was Fukuda Umanosuke, the father of Fukuda Keiji, famous as chief designer of the *Yamato* class. Hiraga was involved in the preliminary design for the *Fusō* class.

On 5 August 1912 he joined the Shipbuilding Department of Yokosuka Navy Yard and had to discontinue his lectures for the Technical Course at Tōkyō Imperial University which he had begun on 25 September 1909. On 26 August 1912 he was appointed head of the drawing section and supervised the construction of the battleship *Yamashiro*, the battle cruiser *Hiei*, and the destroyer *Kaba*. On 10 June 1913 he also became head of the shipbuilding section of Yokosuka NY and combined both functions. On 14 November 1913 he delivered a lecture about the use of special steel in recent warships to the Shipbuilding Association, for which he received a fee on 23 October 1915.

In the meantime he acted as member of the trial committees of *Hiei* (February 1914) and *Kaba* (February 1915), and became a member of the committee for the 50th anniversary of the founding of Yokosuka NY on 22 September 1915. On 3 November 1915 the battleship *Yamashiro* was launched, and on 26 January 1916 the opening ceremony for the great dock of Yokosuka NY took place.

The clinker screen proved relatively ineffectual in shielding the upper foremast platforms from smoke, and from 1926–27 the fore-funnel was trunked back in both ships in a distinctive 'S' configuration. This is *Nagato* off Sarushima in 1931. (Lars Ahlberg collection)

On 7 April 1916 Hiraga ended his dual duties at the naval dockyard, and was transferred to the NTD (then *Kaigun Gijutsu Honbu*) as Superintendent of Shipbuilding on 15 May 1916. He was appointed to the Fourth Division (Shipbuilding), with Fukuda Umanosuke as overall head and Asaoka Mitsutoshi as head of the Preliminary Design Section. From that time he was in charge of the preliminary design of the capital ships of the Eight-Eight Fleet.

Immediately prior to this, on 12 May 1916, the order for the battleship *Nagato* had been placed with Kure NY. On 31 May/1 June 1916 the Battle of Jutland was fought, and on 11 August[29] Hiraga submitted revised design A 112 on the basis of the lessons learned. The modifications focused on protection and speed. The Navy Minister ordered the new plans to be implemented on 28 October 1916. After that Hiraga worked on successor designs for high-speed battleships and battlecruisers, all of which featured an inclined armour belt and 16in guns, the number of guns being increased to ten from preliminary design A 119 onwards.

In January 1917, following the loss of the battle cruiser *Tsukuba* to a magazine explosion,[30] he was appointed to the Investigation Committee (chairman: Rear Admiral Katō Hiroharu – known as Kanji; the report was submitted on 23 July), was promoted to Captain on 1 April 1917 and proposed the so-called 'Modified *Mutsu* design' (A 125) on 12 June 1917.

On 14 July 1917 the budget for the Eight-Four Fleet was approved by the Diet. On 26 July the former chief of the Preliminary Design Section, Asaoka Mitsutoshi,[31] became head of the Fourth Division, and Yamamoto Kaizō took over his former post and function.

On 1 December 1920 Yamamoto was promoted head of the Fourth Division, and Captain Hiraga was appointed head of the Preliminary Design Section. He was promoted to Rear Admiral on 1 June 1922 but was dismissed from his post on 1 October 1923 and was despatched on a tour of Europe and the USA in order to investigate the shipbuilding situation following the conclusion of the Washington Treaty. His dismissal was triggered by disagreements with the Naval General Staff regarding the fitting of torpedo tubes in the new 'Treaty' cruisers of the *Myōkō* class, but this was only one example of his resistance to staff requirements he deemed inappropriate.

Prior to this, on 13 August 1923, Vice Admiral Yamamoto had been replaced by Suzuki Keiji as chief of the Fourth Division of the NTD. Hiraga was therefore the only head of Preliminary Design not to be promoted head of the Third/Fourth Division. It is reported that he regretted this all his life.

Endnotes:
1 See the author's article in *Warship 2020*, 28–47, and his more detailed account in *Outline of the History of the Imperial Japanese Navy 1868–1945*, Despot Infinitus (Zagreb, 2021), 142–161.

[2] See the author's article in *Warship 2012*, 142–151.

[3] The Battle of Tsushima, in which the Second and Third Russian fleets were more or less annihilated, became the template for war against the USA. The IJN adhered to this strategy until 1944, irrespective of the fact that the preconditions had changed so decisively that a repeat of Tsushima was no longer on the cards.

[4] This and subsequent plans are described in greater detail in Chapters 32–34 of *Kaigun Gunbi Enkaku* ('Outline of Naval Armament').

[5] Other ships in the programme were two 3,500-ton small cruisers, one large 1,222-ton destroyer, three 700-ton submarines and one special service ship.

[6] The correct rank of Yamamoto was *Zōsen Daikan*, but in September 1919 this was changed to *Zōsen Daisa* (Captain Constructor). By that time Yamamoto had been promoted to Rear Admiral and he retired as a Vice Admiral.

[7] As the IJN's capital ships currently under construction had RN-type armour protection, the NGS required that this aspect of the design be revisited, and also argued that the increased speed of the *Queen Elizabeth* class had been shown to be insufficient.

[8] In reality the first application in a major vessel was in the British 'Large Light Cruisers' *Glorious* and *Glorious*, which were laid down in the spring of 1915 and which had a doubled-up light cruiser plant featuring small-tube boilers and geared fast-running turbines designed to develop 90,000shp. Reduction gearing was also planned for the Italian battleships of the *Caracciolo* class, the first of which was laid down in October 1914; however, construction was subsequently suspended.

[9] Westinghouse would supply the plans for geared turbines to the Swedish Navy in 1917 for their 7,000-ton coast defence ships *Gustaf V* and *Drottning Victoria*. However, their proposals for geared turbines for US Navy battleships had been rejected in favour of turbo-electric propulsion, which secured the desired operational efficiency and economical range, albeit at the expense of weight and volume.

[10] The Japanese term for this arrangement of the armour was 'concentrated protection'. The system was perfected in the *Kaga* class by the adoption of inclined belt armour topped by a heavy armoured deck and was a feature of Hiraga's capital ship designs. It was a feature of his *Kongō* replacement design of the later 1920s, which created a considerable stir, and was also employed in the *Yamato* class, for which Hiraga acted as adviser to Fukuda Keiji.

[11] This could not be permitted as it would have delayed completion by about a year, thereby rendering most of the preparatory work useless.

[12] Torpedo hits during the early war period had demonstrated either insufficient or non-existing underwater protection in existing battleships.

[13] Hiraga's preference was for a triple turret fore and aft with twin turrets superimposed, and this was supported by the gunnery division. However, this solution was rejected because development and production of the triple turret might have delayed the completion of the ships.

[14] *Nagato* was the second capital ship after the battleship *Fusō* to be built in this dock, and was followed by the battle cruiser (later aircraft carrier) *Akagi* following the extension of the dock and the strengthening of the cranes. In November 1937 the battleship *Yamato* would be laid down in the same dock.

[15] Kaneda is referred to in several postwar Japanese books as a 'weapon production officer' (*zōhei shōkō*), but according to Hiraga this is an error. Fukui Shizuo claims that Kaneda was an 'authority on gunnery techniques', and when he died in June 1925 he was a member of the First Division of the NTD as well as an instructor at the naval academy. A further claim that the multi-post mast was adopted prior to the start of the construction of *Nagato* has likewise been dispelled by Hiraga. The magazine *Senpaku* had published the correct information before the Pacific War, but this appears to have been forgotten after 1945.

[16] Sources: *Kaigun Kikan-shi*, Vol II, 520–28; *Shōwa Zōsen-shi*, Vol 1, 664–72, *Kaigun Zōsen Gijutsu Gaiyō*, Vol VII, 1671–72, 1758.

[17] The Naval Technical Department (NTD – *Kaigun Kansei Honbu*) was separated into a technical (*Kaigun Gijutsu Honbu* = Gihon) and administrative Warship Equipment Division (*Kaigun Kanseibu*) from 1 October 1915 until 30 September 1920 due to the so-called Siemens Affair (a bribery case). The two branches were subseqently again combined.

[18] Each of the oil-fired boilers generated sufficient steam for 4,450shp, making a total 66,750shp.

[19] Each of the boilers with mixed firing generated steam for 2,200shp, making a total 13,250 SHP.

[20] Gear cutting would remain a problem in the IJN for a considerable time. Hobbing (teeth-cutting) machines were imported first from Britain (David Brown & Sons) and then Germany (Reinecker,) but it took until the end of the 1930s before the quality was satisfactory.

[21] According to *Kaigun Kikan-shi*, 526, *Nagato*'s LP turbines were not initially equipped with this 'free gear' type at first, while *Mutsu*'s LP turbines had it. However, comparative experiments demonstrated a saving of 10 per cent of fuel in *Mutsu*, and this led to the retro-fitting of the gearing in *Nagato*.

[22] At that time four ships of *Fusō* class were to be built.

[23] Two longitudinal flash-tight bulkheads were fitted between the guns with about 0.5m distance between them. Passage from one side of the gun house to the other was by means of flash-tight doors through these bulkheads. The working chamber was divided longitudinally in a similar fashion. The post for the turret trainer in the working chamber was fully enclosed and comparatively soundproof – important for the transmission of verbal orders.

[24] *Umi to Sora*, 6/1958, 76.

[25] Report O-47(N)-1 states that both guns and turrets were produced in Kure and Muroran but the latter claim could not be confirmed by the Japanese sources used by the author.

[26] It was mounted on rails to enable it to be trained in any direction.

[27] Despite its close links with the Royal Navy, the IJN was not given access to the Dreyer Table nor any equivalent. After the First World War the IJN adopted an analogue computer developed by Barr and Stroud (see Friedman, *Naval Firepower*, Seaforth Publishing, 268).

[28] In *Nevada* the main armoured deck, which had a thickness of 3in, sat on the upper edges of the belt to form an armoured *caisson*. There was a light 'splinter' deck beneath at the level of the middle deck.

[29] Naitō Hatsuho, *Life of Hiraga Yuzuru*, 581. Hiraga (*On the Post-War Development of the Ships of the Imperial Navy*) states that the new design was not completed until September, but this may refer to the date it was approved by the Higher Technical Conference.

[30] See Kathrin Milanovich, *Warship 2019*, 122–126.

[31] Replaced by Yamada Sukehira on 1 December 1917.

THE BEGINNINGS OF SOVIET NAVAL POWER

THE FLOTILLA LEADER *TASHKENT* AND HER WOULD-BE SUCCESSORS

The collaboration of Fascist Italy with the Soviet Union on a number of naval projects is one of the more surprising aspects of prewar history. In the latest of his series of articles on the early Soviet Navy, **Przemysław Budzbon** looks at the purchase of the flotilla leader *Tashkent* and the impact she had on subsequent Russian design and construction.

The creation of a warship, like most sophisticated engineering products, is a process requiring balance, cooperation, and an exchange of information between all the participants. If we narrow our focus to the building of a single 'product' (ship), it allows us to treat the process as linear. The role of the naval advisory board is to specify the desired characteristics, that of the ship design team is to draw up plans that enable the shipwright to assemble the 'product' from available components, and that of the navy to operate it.

However, when we look at the development and improvement of ships from one series to the next the process is not linear but cyclic. To close the cycle, we need to add in the navy's experience with operating the ship ('feedback') and use it to refine the future requirements of the navy, to improve the product by the design team and to improve the manufacture and components to be used by the shipwright. Thus, from one cycle to the next the experience of all the participants in the process grows, and the final 'product' is improved and updated. The incremental increase in experience is key to the process. Any interruption to this process causes experience to evaporate and risks to materialise. The stability issues of the *Farragut* class of the mid-1930s resulted directly from the twelve-year hiatus in destroyer design in the US Navy, and similar challenges were experienced with the design process of the British nuclear submarines of the *Astute* class.

Balance is as much the key to success as the development cycle described above; any imbalance in the process could impact adversely on the quality of the final product. For example, the *Tomozuru* Incident, in which a Japanese torpedo boat capsized during trials in March 1934 because of her excessive topweight, was the result of an imbalance in the process between the Japanese Naval Staff, which insisted on an increase in the number of guns and torpedo tubes, and the design team – which was relatively inexperienced – who bent the calculation rules to satisfy the Naval Staff.

Nearly every fleet has experienced similar mishaps, but the Soviet Navy of the 1920s and 1930s experienced multiple instances simultaneously. It should be acknowledged that the starting base was not of the highest order, as Russian naval shipbuilding had generally lagged behind that of other naval powers since the times of Peter the Great. Any progress made was largely due to imports of skilled workmen and technology, or the purchase of samples which were then copied. This was largely due to the fact that the Russian system of decision-making and internal economic relations was over-centralised. After the fall of the Tsar, the system imposed by the Soviets served to further rigidify the environment, and the only stimulus to development was the imposition of terror.

The first result of attempts to resume naval shipbuilding in Russia after the Civil War – the escort ships of the *Uragan* class – was a complete failure. A paper prepared by the Soviet naval authorities in 1938 stated that: 'The escort ships do not represent a step forward in terms of their armament or the technical solutions applied, as compared with the torpedo boats of the

The escort ship *Smerch* (Yard Number *S-322*) of the *Uragan* class during trials in 1933. The first surface warships built in the Soviet Union did not gain much respect, even in Soviet naval circles. (Author's collection)

The flotilla leader *Leningrad*, the lead ship of her class, in 1938 – still in the refining phase, six years after she had been laid down. The failure of this project, which was apparent practically from the beginning, was a direct reason to seek for foreign assistance. (Boris Lemachko collection)

Sibirskij strelok class of 1906 – having the same speed and carrying in essence an identical armament – the escort ships have only one advantage: a lighter displacement, albeit at the expense of a reduction in fuel stowage.'

The design and construction of the *Leningrad*-class flotilla leaders (see *Warship 2022*) also turned out to be a failure, as was becoming apparent during the early 1930s. However, by that time the Soviets were increasingly able to obtain access to Western technology.

Soviet-Italian Relations

As Soviet Russia was initially considered a pariah, no import of technology, especially military technology, was legally possible. In particular, large and conspicuous objects such as ships were out of the question for a long time. Empty Soviet coffers did not help: the Tsar's gold reserves were quickly exhausted, as were the jewels plundered from the aristocracy, which were used to fund illegal imports.

The decision of the Allied Supreme War Council on 16 January 1920 to lift the economic blockade of Soviet Russia marked a turning point. The first Western country to make a serious attempt to renew commerce was Italy. In 1914 grain from Russia had accounted for 40 per cent of deliveries, and this had to be replaced with more expensive imports from the USA and Argentina. Moreover, Italian heavy industry, which had grown significantly during the Great War, needed markets. In 1922 a Russian-Italian joint venture company named Russitatorg was established (ARCOS in Britain was similar) to facilitate trade, with a hidden Soviet agenda of espionage.

The building of mutual ties accelerated after Benito Mussolini became Prime Minister of Italy in October 1922. The Soviet Union was formed in December of the same year, and recognition by the British government on 1 February 1924 and the desire to compete with France

Table 1: Mutual Courtesy Visits of Major Warships 1925–1933

Name	Colours	Type	Port	Date
Vorovskij	Soviet	Border Guard patrol ship	Naples	August 1925
Pantera Tigre Leone	Italian	Destroyers	Leningrad	June 1925
Nezamozhnyj Petrovskij	Soviet	Destroyers	Naples	October 1925
Palestro Calatafimi	Italian	Destroyers	Odessa	March 1926
Frunze Nezamozhnik	Soviet	Destroyers	Naples	September 1929
Parizhskaia kommuna Profintern	Soviet	Battleship Cruiser	Cagliari Naples	January 1930
Chervona Ukraina Nezamozhnik Shaumian	Soviet	Cruiser Destroyers	Messina	October 1930
Tricheco Delfino	Italian	Submarines	Batum	May 1933
Krasnyj Kavkaz Petrovskij Shaumian	Soviet	Cruiser Destroyers	Naples	October 1933

spurred Mussolini into an acceleration of formal links: Italy and the USSR signed a treaty week later. The warming of military relations followed step-by-step those of economic relations, as can be seen from the number of naval visits paid to the ports of both countries (see Table 1).

This went much further when Mussolini, guided by the naive belief that Soviet military intelligence was working against Britain, decided not to obstruct such activity in Italy during the late 1920s. This enabled the Soviets to obtain a good grasp of the potential for links with the Italian military industry.

Frustrated by the failure of the early Soviet naval

The cruiser *Krasnyj Kavkaz* during her 1933 Mediterranean deployment, which included a visit to Naples. (US Naval History & Heritage Command, NH 86807)

Table 2: **Warships Visited by the Soviet Delegation in 1930**

Name	Type	Place	Remarks
Bolzano	Washington cruiser	Genoa	On the stocks
Gorizia	Washington cruiser	Livorno	On the stocks
Trento	Washington cruiser	Naples	Manoeuvres
Zara	Washington cruiser	Muggiano	Fitting out
25 de Mayo	Cruiser	Livorno	Argentinian, fitting out
Alberto di Giussano	Light cruiser	Genoa	Fitting out
Bartolomeo Colleoni	Light cruiser	Genoa	On the stocks
Giuseppe Miraglia	Seaplane tender	La Spezia	In service, seaplane operations
Antonio Da Noli	Large destroyer	La Spezia	In service
Antoniotto Usodimare	Destroyer	Genoa	Nearing completion
Freccia	Destroyer	Genoa	Fitting out
Balilla	Submarine	La Spezia	In service
Luciano Manara	Submarine	La Spezia	In service
MAS 423– 426	MTBs	La Spezia	Manoeuvres

Table 3: **Shipyards, Factories and Naval Facilities Visited by the Soviet Delegation in 1930**

Place	Date	Establishment	Remarks
Livorno	29 Sept–2 Oct	Naval training facility	
		Naval academy	
		Naval radio establishment	
		Shipyard	Orlando
La Spezia	2–8 Oct	Naval base	
		Torpedo workshop	San Bartolomeo
		Artillery test fields	Castagna, Viareggio
		Naval arsenal	
		MTB Flotilla	
		AA battery	
		Naval training facility	
		Comms and torpedo school	
		Torpedo test facilities	
		Ordnance factory	Odero-Terni
		Gunpowder factory	Val de Locci
Genoa	9–17 Oct	Shipyard	Ansaldo
			Riva Trigoso
		Optical and mechanical plant	San Giorgio
		Radio equipment works	
		Naval hydrographic command	Ansaldo
Florence	17 Oct–5 Nov	Optical and mechanical plant	Officine Galileo
		Mine factory	Penione
Fiume	17 Oct–5 Nov	Torpedo plant	Whitehead
		Shipyard	Quarnaro
Trieste	17 Oct–5 Nov	Shipyard	San Marco
		Machine factory	Sant'Andrea
Monfalcone	17 Oct–5 Nov	Shipyard	Adriatico
Milan	17 Oct–5 Nov	Precision machinery plant	Olap
		Radio-electronical plant	Morelli
		Motor and machinery plant	Isotta-Fraschini
Venice	17 Oct–5 Nov	MTBs shipyard	SVAN
		Motormen school	
Naples	5–14 Nov	Torpedo factory	Silurificio Italiano
		Shipyard	Castellamare
		Ordnance factory	Ansaldo (Puzzoli)
Taranto	17–25 Nov	Shipyard	Franco Tossi
		Naval base	
		Naval radio school	
		Destroyer COs school	

The Italian *esploratore leggere Antonio Da Noli*, which was
visited by the Soviet delegation at La Spezia on 3 October 1930.
She apparently inspired the order of a ship in Italy. (US Naval
History & Heritage Command, NH 86331)

construction programmes, Stalin dismissed Viacheslav
Zof[1] from the position of Chief of the Naval Forces
(*Nachal'nik morskikh sil*, acronym *Namorsi*) in 1926,
and replaced him in this role by Romuald Muklevich.[2] It
appears that his background as a mechanician made him
appear more professional in the eyes of Stalin and thus
predestined him to become the CO of the Navy. This
change in personnel was also related to the opportunities
to renew foreign contacts, which made the industrialisa-
tion process possible. In order to obtain funds for the
purchase of the latest machinery and equipment from
abroad, raw materials, including timber from the
Siberian forests and grain from Ukraine, were exported
on a massive scale. The primacy of Western technology
was not questioned and was even apparently praised by
Stalin himself, and access to it gave rise to the hope of
copying the purchased designs and launching their large-
scale production. Kliment Voroshilov in November 1927
praised the British submarine *L-55* (sunk in 1919 and
raised by the Soviets) which, he stated, 'represents the
latest technology of the leading naval power and
embodies the war experience of the victorious Royal
Navy' (see *Warship 2020*). A similar attitude was echoed
by Muklevich, who in his letter to Stalin of March 1931
wrote: '... the Italians have a very high level of marine
technology, both in shipbuilding and in naval weapons ...
whereas our own industry is unsuccessfully trying to
master problems that have long been resolved abroad.'

Such a decisive statement by the *Namorsi* resulted from
the knowledge acquired by Soviet naval and industrial
experts during an official visit to Italy which took place
from 23 September to 25 November 1930. The Soviet
delegation, headed by Aleksandr Sivkov,[3] visited 37
leading industrial plants and several warships. The delegates observed naval exercises on 4 October (4 MTBs),
and on 12 November 1930 (3 cruisers, 6 destroyers,
2 submarines) on board the new Washington Treaty
cruiser *Trieste*. The friendly atmosphere was not even
spoiled when the ship's band on board *Trieste*, in honour
of the Soviet delegation, played the anthem of Imperial
Russia ('God save the Tsar').

The openness and hospitality of Mussolini had two
drivers: one was a desire to secure lucrative military

contracts and bind the Soviet naval authorities for years
to come; the second was rivalry with France. Stalin had a
very different agenda: to acquire as much knowledge as
possible and to instigate the cheap production of copies
of Italian ship designs and equipment in the Soviet
Union; and to play Italy and France off against the other
in order to acquire as much influence as possible.

Sivkov had one particular item on his own agenda: to
probe the Italians on the possibility of purchase of a
warship, either one nearing completion or one already
completed. During his meeting on 8 November 1930
with Italian Naval Minister Ammiraglio Giuseppe
Sirianni his request was refused, with the Italians citing
Chapter I Paragraph XVIII of the Washington Treaty.
However, the construction of new warships was offered
instead, and it was made clear that Soviet orders were
more than welcome.

The Flotilla Leaders

The Leader I
It took almost half a year for the Soviet bureaucracy to
allocate the budget and dispatch to Italy a negotiating
team headed by Paul Oras,[4] who had gained experience
during negotiations with the Germans on assistance with
construction of submarines five years previously.
However, the negotiating style of the Soviet *apparatchiks*
proved to be quite different that of Italian *uomo d'affari*
(businessmen).

The first to experience this were the companies
Officine Galileo and San-Giorgio, the manufacturers of
precision optical and fire control computers. The latter
were of special interest to the Soviet naval experts, who
were completely unaware of recent progress and were
instructed to make acquisition of this equipment a
priority, first to fit on *Proekt 1* flotilla leaders (then in the
design phase), with a view to studying and copying the
technology and starting production in the Soviet Union.
The aggressive negotiating tactics of the Russians, who
attempted to play off these two contractors against one
another in a bid to lower prices, was alien to their Italian
counterparts, who were unused to such dealings in the
case of governmental contracts. On the other hand, the
Soviet negotiators were under severe pressure from Stalin
to show tenacity in the face of the 'capitalist blood-
suckers', so the talks took place in a very tense atmosphere. Moreover, the current trade imbalance with

Table 4: Preliminary Characteristics of *Leader I*

Full load displacement	3,000 tonnes
Length oa	137.0m
Breadth	13.0m
Mean draught	3.8m
Weights:	
hull	1,055t
machinery & fuel	1,605t
Machinery	115,000–120,000shp = 44.5kts max

Table 5: **Soviet Naval Orders from Italy – Status April 1934**

Plant	Specification
Ansaldo	2 patrol vessels of the *Dzerzhinskij* class
	cruiser-size geared turbines building plant
	technical assistance in design and construction of the *Kirov* class cruisers
OTO	12 twin 100mm AA mountings
San Giorgio (Sestri)	46 4-metre rangefinders
Officine Galileo	4 complete fire control systems for destroyers
	15 attack periscopes for submarines
	14 10-metre & 8-metre rangefinders
	100 searchlight mirrors
Silurificio Italiano	95 21in torpedoes, two 21in triple torpedo mountings
Silurificio Whitehead	80 21in, 15 450mm torpedoes
Isotta Fraschini	51 marine motors Asso 1000 MAD

The French *contre-torpilleur Le Terrible* photographed on 10 May 1935, shortly after her commissioning. (US Naval History & Heritage Command, NH 86331)

Russia prompted the Italian government to attempt to reduce imports, resulting in a ban on any purchases by the Soviet Trade Ministry, which effectively killed the possibility of a deal. Ironically Oras, who was charged with continuing the negotiations, was not informed about the suspension of purchases.

Following the intervention of Stalin, negotiations resumed a year later but the lost time could not be made up, while the Italians became increasingly reluctant to cooperate. A break in the calendar of ship visits during 1931–32 is indicative of the cooling of Italian-Soviet relations during this period. In the meantime, on the eve of the 15th anniversary of the Bolshevik Revolution, on 5 November 1932, Politburo member Sergei Kirov drove in the first rivet of the keel of the *Projekt 1* flotilla leader, the name-ship of the *Leningrad* class. However, it was now becoming clear to the Soviet leaders that industry was unable to deliver the machinery, fittings and weaponry required for these and other new ships. An alternative foreign design which might be copied was needed, and negotiations for this had already been opened two years earlier by the Italian Admiral Sirianni. At the request of Sivkov, the characteristics of a ship of Italian origin (hence the code *Leader I*) were issued by Vladimir Nikitin,[5] who was in charge of the *Proekt 1* design team (see Table 4).

Armed with these characteristics, Sivkov attempted to initiate discussions with the Italian shipyards Ansaldo and Adriatico early in 1933, inviting their representatives to Berlin where he was on a business trip. However, his invitation was refused, forcing him to travel to Italy, where he was received quite coldly. In the end the Italians were ready to provide a preliminary design, nothing more, which was insufficient for the Soviets as they wished to purchase a detailed technical design, a complete set of machinery, and technical assistance

during construction. A similar attempt was made by Muklevich, head of the Shipbuilding Directorate, in France: in July 1934 the Soviets started negotiations with the shipyard Ateliers et Chantiers de France at Dunkirk, with a view to purchasing the design of a flotilla leader derived from the *Le Fantasque* class. In February 1935, following the extremely positive speed trials results of *Le Terrible* (42.92 knots), the French raised the initial price by 15 per cent, but subsequently withdrew, apparently because of a lack of enthusiasm on the Soviet side. Lenin's dictum 'The capitalists will sell us the rope with which to hang them' failed to work in this case.

Willingly or unwillingly, the Soviets made a further approach to the Italians, the move being approved by Voroshilov (and therefore almost certainly backed by Stalin) at the request of *Namorsi* Vladimir Orlov.[6] This was a more favourable time than 1933 because of the significant volume of Soviet naval orders already placed in Italy.

The characteristics of the *Leader I* were modified, and the shipyards of Ansaldo, Odero-Terni-Orlando and Adriatico submitted tenders with with delivery terms similar to those offered by the French; the bid from OTO, which at 19m lire was little more than half that of Ansaldo, was selected. The Severnaia verf' (the 'Northern Shipyard', which would be renamed in 1935 after Andrei Zhdanov and numbered in 1936 as Works No 190) at Leningrad was selected to build the ship. The flotilla leader *Leningrad* was still there, undergoing static trials during which serious problems emerged. Despite the pitiful state of the shipyards the Soviets were reluctant to have their ships built entirely in foreign yards, in part because of their hidden 'copy and paste' agenda but also because of the desire to upgrade their own shipbuilding capabilities, which would be necessary for the giant armaments programmes envisaged for the future.

Moscow's negotiations with the Italians for the supply of materials and expertise, despite the support of the Italian government, were prolonged and difficult. The Italian shipyards, for obvious reasons, were primarily interested in the sale of complete warships. The most aggressive sales pitch was adopted by OTO, which

Table 6: Characteristics of *Leader I*, May 1935

Standard displacement	2,600 tonnes
Speed	42 knots
Range	6,000nm at 20 knots
Armament	6–130mm (single/double mounts)
	2/3–3in HA, 3/4 0.5in MG
	6–21in (2 x III)
Seaplane	one
Complement	200

proposed a price of 54 million lire for a complete ship – a figure only 25 per cent higher than the French price for documentation, technical assistance and machinery. The Italian shipyards were asked again to tender, this time for a complete ship (without armament), a second set of machinery and technical assistance in the construction of this second unit in Russia. Following an agreed reduction in range from 6,000nm to 4,500nm requested by the shipyards, tenders were again submitted, with the one from OTO being the lowest (54.6 million lire). The contract was duly signed in Leningrad on 9 September 1935, with delivery of the ship scheduled within 24 months at a cost of 44 million lire – the additional set of machinery was finally rejected.

Proekt 20I

The technical design, which was coded *Esploratore Veloce U.R.S.S.* ('USSR Fast Scout') by the Italians and designated *Proekt 20I* by the Soviet Navy, was delivered by the shipyard in February 1936, two and a half weeks ahead of schedule. This was in marked contrast to current Soviet experience, with projects involving local design and construction being delivered some two and a half years late. However, this was not the end of the surprises. The Soviet military bureaucracy, which had been holding up projects for approval for months, dealt with the documentation within three weeks, appending remarks and requests for changes. In fairness it must be admitted that the Soviet authorities and their officials had an interest in implementing the project as soon as possible, driven by pure professional curiosity. And the Italians, for their part, were interested in the smooth implementation of the contract and, more importantly, keeping it alive in the face of the rapid deterioration of relations with the international community following their invasion of Abyssinia. In addition, they skilfully used the advantage of their cultural capital, which eased potential tensions.

There were only three important issues to be addressed:

– maximum sustainable speed
– range
– the location of torpedo tubes atop the deckhouse.

Both sides reached a compromise through concessions made by the Russians, which was unusual in the Soviet era. The Soviets accepted that the maximum sustained

The torpedo tubes of *Tashkent*, the location of which was a concern for *Namorsi* Orlov. (Author's collection)

speed at a normal load of 3,116 tonnes should be 42.5 knots with a machinery output of 100,000shp. In calculating a maximum range of 5,000nm at 20kts, which required 1,200 tonnes of oil fuel, the Russians also accepted a deep load displacement of 4,175 tonnes; this was normal in Soviet practice – the overload condition was termed *peregruz*. The location of the torpedo tubes atop the midship deckhouse was raised by Orlov himself, who questioned such a high position above the waterline. Orlov was concerned that this would not guarantee the successful launch of a torpedo into the waves at a relative bearing of 35–40 degrees to the ship's axis with a launch velocity of 12m/sec, which was the maximum for the planned 39-Iu type mounting. He insisted that trials be conducted on the cruiser *Krasnyj Kavkaz* and the flotilla Leader *Leningrad*, and requested a delay in any decision until 1 October 1936.

In the event the trials would not take place: *Leningrad* had not even left the shipyard by this time, while *Krasnyj Kavkaz* could only handle 450mm torpedoes. The torpedo tubes for *Proekt I* therefore remained in their original positions. The technical documentation was finally approved by Orlov and Muklevich on 11 March 1936 – the Soviet position on the location of the torpedo tubes was reserved. In contrast to Soviet practice, the

Italians would not consider starting construction before all the documentation was signed off, so the keel would not be laid until February 1937.

During the detailed design work, the Italians proposed main guns in twin mountings, which they considered better suited to the ship's architecture than the B-13 single mountings in service with the Soviet Navy. This was accepted, and the preliminary design of the twin 130mm/50 B-31 mounting was prepared by ANIMI (Artillery Naval Scientific-Research Institute), which estimated weights and the centre of gravity for the purpose of stability calculations and also for the stiffening requirements and the magazines and ammunition hoists. To reduce the weight of the mountings both guns were to be mounted in a single cradle; in consequence the mounting received the designation 'Italian style'. It was also planned to embark these lightweight mountings on the *Proekt 45* experimental destroyer *Opytnyj*. Progress was slow, and in October 1938 the design was abandoned in favour of the B2-LM mounting destined for the *Proekt 30* destroyers and the *Proekt 48* flotilla leaders. This change required a substantial reworking of structures due to the increase in weight and recoil forces and the different ammunition supply, as the construction of the *Proekt 20I* ship was well advanced at that time.

Tashkent

Following laying down, construction of the hull proceeded smoothly and the ship was named *Tashkent* on Voroshilov's order of 29 April 1937. However, when a month later the turbine rotors were rejected, the completion date was moved to mid-February 1938, which was considered over-optimistic by the experienced Russians. They were proved right, as in March 1938 *Tashkent* was still at the fitting-out quay and her delivery had become questionable. Soviet-Italian relations were already strained by the Spanish Civil War, and incidents such as the sinking by an Italian destroyer of the Soviet freighter *Timiriazev* on 31 August 1937 did not help the situation. A full-scale bombing offensive targeting maritime traffic begun by the *Regia Aeronautica* on 15 March 1938 led to a grave deterioration in relations and the suspension of fitting-out work on *Tashkent*. These tensions eased in July, so construction was resumed with the aim of prompt delivery. On 15 November 1938 the ship ran her first speed trials, easily achieving 36.08 knots with moderate loading of the power plant. During a preliminary full speed trials run on 14 December 1938 a speed of 42.72 knots was achieved with 105,460shp, which was sufficient to meet the contractual requirements.

In the meantime, matters on the Soviet side became complicated. In February 1938 Sivkov and Muklevich were executed, joined by Orlov in July and his successor Flagman 1st Rank Mikhail Viktorov[7] in August 1938, while Army Commissar Pyotr Smirnov[8] (Victorov's

Tashkent in Autumn 1938, before her shipyard trials. Note the Russian custom of displaying the names of lesser warships at the bow, in addition to the stern as in other nations. (Author's collection by courtesy of Aldo Fraccaroli)

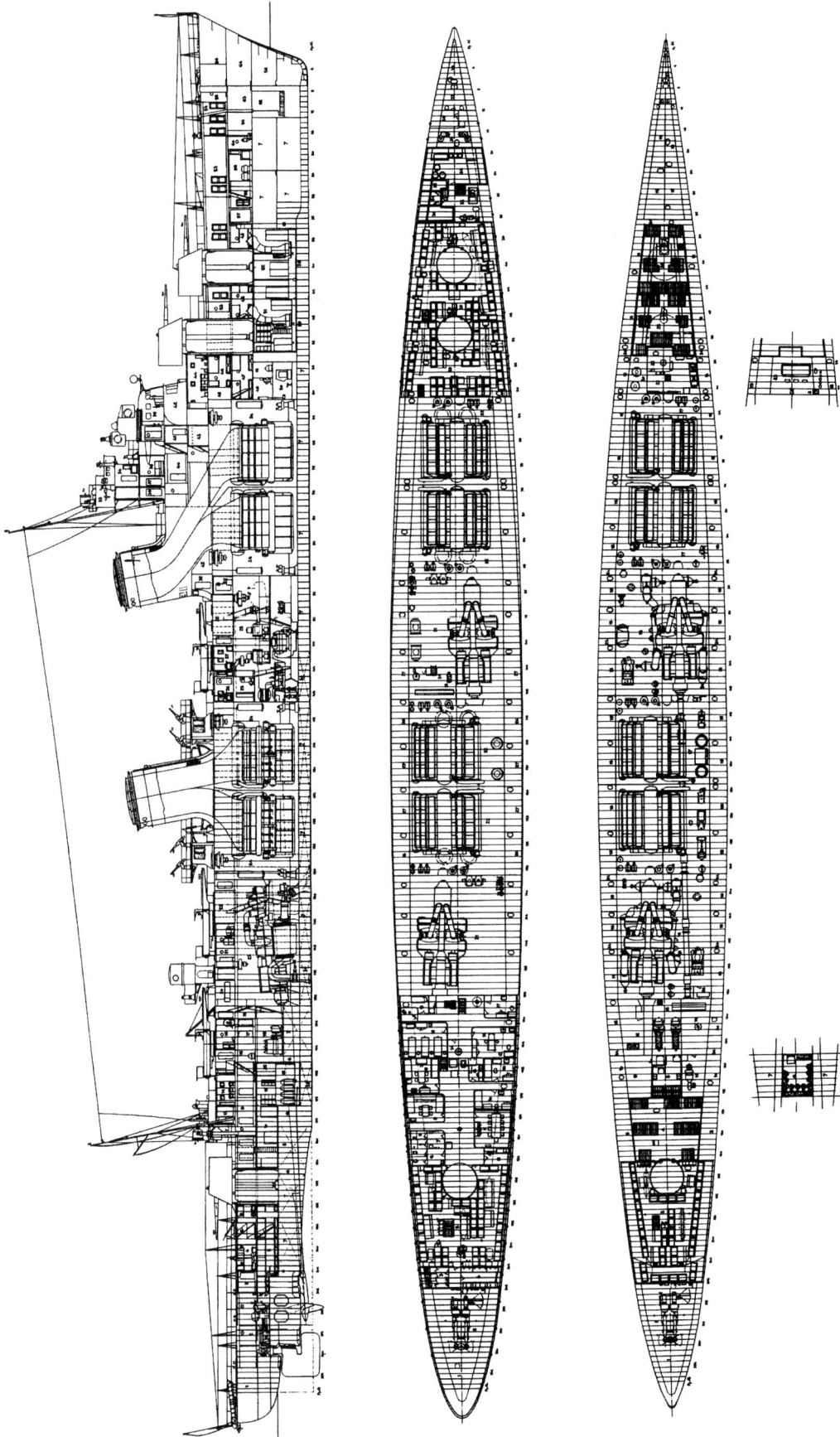

General arrangement of *Tashkent*, from the contractual documentation. (Author's collection)

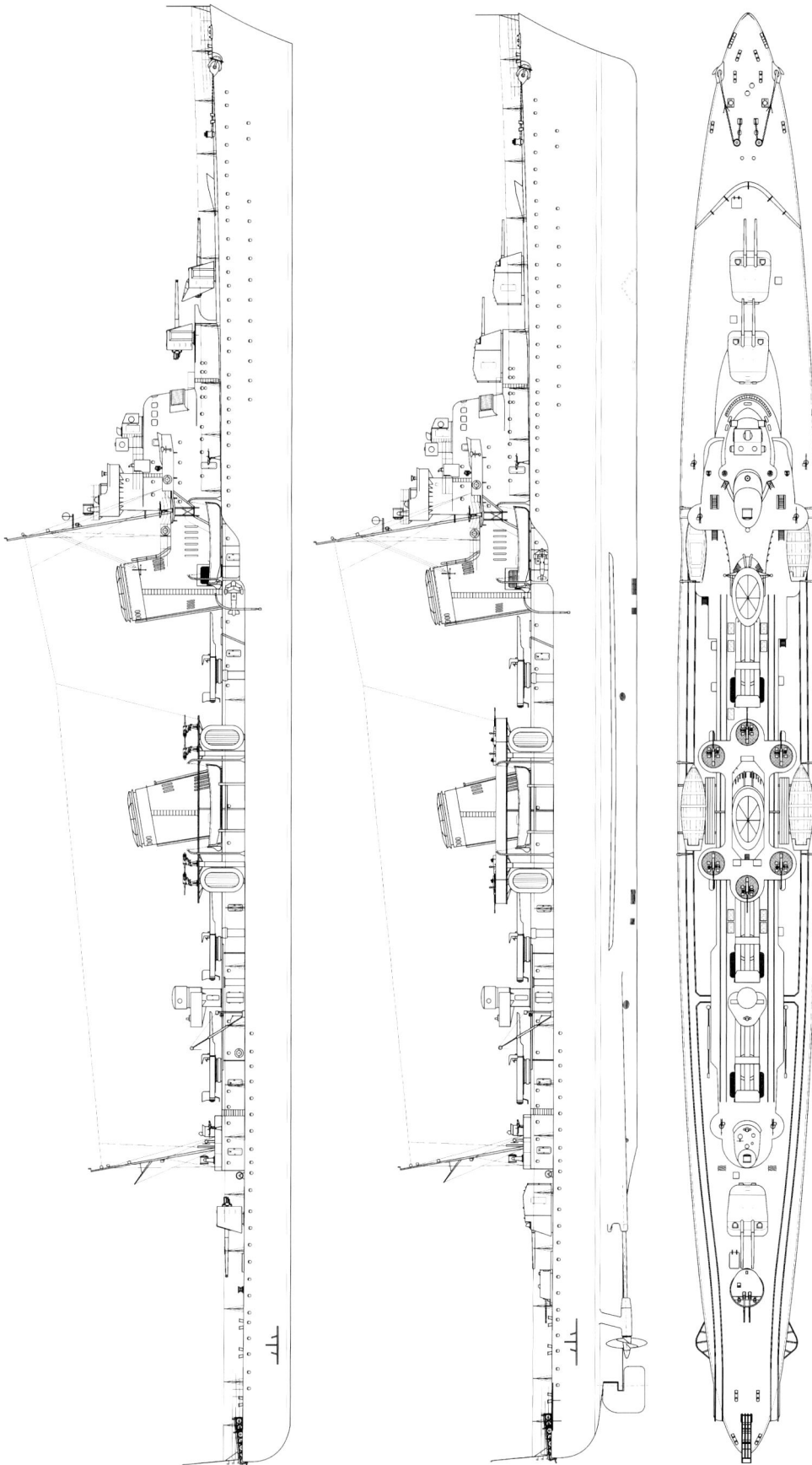

Tashkent as commissioned with provisional armament, October 1940 (top), and her final configuration in November 1941 (bottom). Note the exceptionally fine lines of the stern section, which was designed to secure high speed. (Drawn by Tomasz Grotnik)

The freshly painted *Tashkent* on the eve of her handing over to a Soviet crew. (Author's collection by courtesy of Aldo Fraccaroli)

successor) was jailed from June 1938, awaiting the firing squad. He was replaced by Komkor 1st Rank (promoted two notches to the rank of Komandarm 1st Rank – still an Army, not a naval rank) Mikhail Frinovskij,[9] whose lengthy service with the NKVD possibly explains the exaggerated security that accompanied *Tashkent's* handover.

The first difficulty experienced was the issue of bringing the ship to Leningrad by an Italian crew, as per the contract. As this was now likely to be in the winter, icing and the related difficulty of conducting speed trials provided a good excuse for delay. At that time the Great Seagoing and Ocean Navy construction programme was running at full speed, and this needed to be kept away from prying eyes. When it turned out that the transfer of the ship would take place later, the danger of passing through the Gibraltar Strait and the Sound was raised as, due to the Spanish Civil War, these 'choke' points could be a good place for a potential sneak attack on the Soviet ship. Odessa in the Black Sea was therefore proposed, as this was far enough from the Black Sea shipyards of Sevastopol and Nikolayev.

Following the arrival in Italy of the Soviet acceptance commission, the 6-hour full speed trials were run on 17 March 1939: an average of 43.56 knots were achieved at 3,422 tonnes displacement with a machinery output of 125,000shp at 378rpm. Following re-calculation for the

Tashkent in a bow quarter view, provisionally armed with single 130mm B-13 mountings. Note the box-shaped extensions for the trainer on the right side of the gunshields. (Boris Lemachko collection)

contractual trial displacement of 3,216 tonnes, the result was 44.6 knots, which meant a 5 per cent bonus for the shipyard. Some members of the commission attempted to question the calculations but desisted after being reprimanded by Moscow, which was interested in taking over the ship as soon as possible. Following signature of the acceptance document *Tashkent* was taken over by a Soviet crew on 6 May and, assisted by an Italian skeleton crew, left for Odessa. Despite the conclusion of the Spanish Civil War a month earlier, exaggerated security measures were undertaken during this passage. To fool foreign observers, canvas awnings with painted portholes were stretched between the superstructures to give the impression of a merchant ship. (The late naval historian Vitalii Kostrichenko confirmed this in the 1980s during an interview with a member of the first crew.)

By 9 May three sorties from Odessa had been completed with one run at 41 knots, and the following day the Italian crew members, who were not welcome on Soviet soil, were released. The new acquisition caused a lot of interest among residents, attracting attention because of its size, modern silhouette and blue-grey paint scheme, which prompted the nickname 'Blue Cruiser'. The problem was the lack of the designed armament, because the B-2-LM twin mountings had yet to start proving trials. *Tashkent* was therefore provisionally armed with three B-13 type single mountings of the first production series. These were distinguished by their rectangular shields with a prominent box-shaped extension for the trainer on the right-hand side (see drawings). The full designed armament was installed during a February–July 1941 refit at Works No 198 (the Andre Marti Yard) at Nikolaev, and the proving trials of the B-2-LM mountings were concluded in May 1941.

The forward pseudo-turrets were modified to conform to the Italian practice of separating the projectile and powder magazines and re-designated B-2-LM-I, while the after turret had the common propellant and projectile magazine standard in the Red Fleet. The outdated 45mm/46 21-K guns were replaced by the brand-new model 37mm/67.5 70-K in single mountings. Three pairs of single 0.5mm DShK MG were installed: on the forecastle deck abaft the DCT, on platforms extending from the bridge wings, and on the after deckhouse close to the mainmast. The original electronic equipment was changed to the current Soviet models.

Following her first serious battle damage by a German bomber on 30 August 1941 *Tashkent* spent six weeks in the Sevastopol Naval Yard (*Morskoj zavod*), where in addition to repairs she received a twin 3in/55 39-K mounting on the quarterdeck. The mounting was countersunk into the general quarters below so as not to obstruct the arcs of fire of the after 130mm mounting.

Tashkent never had the opportunity to fire her guns against comparable opponents but proved particularly successful in the fire support and fast transport roles. She finally succumbed to the *Luftwaffe* on 2 June 1942 at the port of Novorossiysk. There is insufficient space here for her wartime story, which merits a separate account.

An aerial view of *Tashkent* with her provisional armament. Note the blast shield protection for 'A' mounting. (Author's collection)

Close-up of the forward deckhouse. Note the details of 'B' mounting with the trainer extension. (Vitaliy Kostrichenko collection)

The bridge of *Tashkent* photographed from one the ship's boats. (Author's collection)

Proekt 20

The frequently-changing Soviet naval construction programmes during the 1930s and 1940s were consistent in one respect: they provided for the construction of about a dozen leaders of a new type. There was provision in the *Tashkent* contract for Italian technical assistance in the construction of a further unit in a Russian shipyard and exclusively for the Red Navy. Moreover, during the construction of *Tashkent*, the Soviets were permitted to maintain up to 37 people at any one time at the OTO yard with access to all the workshops and factories of the subcontractors. Thus, in addition to the formal technical support mentioned earlier, the Soviets had the theoretical opportunity to spend more than 13,000 man-hours acquiring information; this constituted real technology transfer with invaluable consequences for future naval design and construction.

The original plan envisaged that *Tashkent's* sister would be bult at Leningrad in Works No 190 (Zhdanov Yard, formerly Severnaia verf'), which had painful recent experience of construction of the *Proekt 1* prototype *Leningrad*. The second ship of the class, to which the name *Baku* was for a time assigned, was to be constructed there, with a third unit to be built at Works

No 198 (Marti Yard), Nikolaev (see Table 14). However, this would turn out to be wishful thinking. Differences between the technology used in the former Russian shipyards and the fittings of Soviet construction, which were generally much heavier, meant that significant documentation had to be supplied by the Italians. In addition, all these changes required coordination with OTO engineers, and in those days documentation had to be submitted on paper and physically sent to and from Russia by secret post. Also, the status of these documents was different in both countries – secret or top secret in Russia, open-source in Italy – meaning that this exchange of documents took time.

Despite the above, the design was sufficiently advanced to lay down the hulls. However, that would imply access by Italian engineers to the heart of Soviet naval secrets: Leningrad and the nearby Kronshtadt naval base. This meant that professionals, who would spend at least three years there – a construction time that was probably over-optimistic – would be able to disclose the size and number of capital ships building at the yard, which exceeded by a large margin the limits agreed by the Treaty of London signed on 17 July 1937 with Great Britain. In addition, the paranoid state of the Soviet social system meant that any contact with foreigners was by definition suspicious – a characteristic that has defined the Russian state from Ivan the Terrible until today. Taking all this into consideration, Frinovskij ordered cancellation of this part of the contact. This decision also ended speculation about the possibility of adding a 130mm gun turret aft at the cost of one of the three triple torpedo mountings. Thus, the construction of ships of *Proekt 20* lost all meaning.

Proekt 48

After the arrest of Muklevich (May 1937) and Orlov

(June 1937), the construction of *Tashkent* began to be perceived as an experiment of doubtful value, perceptions being influenced by the propaganda which promoted the achievements of the (failed) Second Five-Year Plan. The Defence Committee of *Sovnarkom* (Council of People's Commissars – the Soviet government headed by Molotov at that time) on 15 August 1937 issued a decree concerning naval shipbuilding, which stated:

- The *Leningrad* class flotilla leaders possess good tactical and technical characteristics.
- Construction of this class should be continued until a better design becomes available.
- The following modifications to the existing *Proekt 38* design should be implemented:
  - main armament of three twin 130mm type B-31 turrets
  - long-range anti-aircraft armament of one twin 3in HA enclosed mounting with separate fire control
  - a range of 4,000nm
  - protection for torpedo tube sights, DCT and main command post
  - ice strengthening of the hull.
- The *Proekt 20I* flotilla leader has no particular advantages over the modified *Leningrad* class, so construction of the follow-on ships is to be suspended until results of trials with *Tashkent* can be obtained.
- The *Proekt 20* hulls (Yards Nos 511 and 512) are to be replaced with two hulls of the modified *Leningrad* class.
- The main guns and machinery plant should be uniform for all flotilla leaders and destroyers.

Characteristics based on this decree were prepared by the Naval Staff of the RKKA (Workers' and Peasants' Red

Table 7: **Characteristics of *Tashkent*, 1942**

Displacement	2,840 tons standard, 3,216 tonnes normal, 4,163 tonnes full load
Dimensions	length oa 139.7m, beam (max) 13.7m, draught at full load 4.2m
	depth to the main deck 7.5m; to forecastle 11.2m
Boilers	four Yarrow three-drum watertube, 28kg/cm^2, 340°C
	80t feed water, one evaporator 60t/h
Fuel	Marine Oil Fuel: 1,072t full load, 1,200t max
Main machinery:	two sets of Parsons geared turbines, 124,500shp total max
	two three-bladed propellers
Speed	42.7 knots max
Range	5,030nm at 20 knots
Electricity	DC 115V circuit
	two 120kVh turbo-generators
	two 75kVh, one 18kVh diesel generators
Firefighting pumps	two turbine-driven 160t/h total at 16kg/cm^2
	one electrically-driven 40t/h at 16kg/cm^2
Salvage and bilge pumps	eight 10t/h, one 30t/h, three 80t/h, four 100t/h, four 150t/h
Navigation equipment	Kurs-2 gyrocompass, four 5in ZMI magnetic compasses, GO-3 log,
	EMS-2 echosounder, Gradus-K direction finder
Main guns	six 130mm/50 B-13 in three B-2LM-I twin turrets, 900 rounds
Main fire control	Galileo Duplex DCT with two OG-3 rangefinders and APG director sight
	Galileo CT (aft) with one OG-3 rangefinder
	two Galileo DM-1.5 rangefinders (bridge)
	Galileo Centrale fire control computer
HA guns	two 3in/55 3-K in one 39-K twin mount
HA fire control	Galileo DCT with Galileo Centrale FC computer
Light AA guns	six 37mm/67.5 70-K in six single mounts
	six 0.5in DShK in six single mounts
Searchlights	two Galileo 90cm
Torpedo tubes	three 39-Iu triple mountings: 9 21in torpedoes (+ 9 spare)
Torpedo fire control	calculator with Galileo Centrale computer
A/S weapons	two DC chutes: ten M-1 charges, four B-1 charges
A/S detection	none
Sonic station	Arktur
Mines	76 KB type
Paravanes	1 K-1
W/T	Shtorm-M, Briz-MK, Okyn' transmitters
	Groza-M, Purga, two Vikhr, Metel' receivers
	Rejd, two RB transceivers
Complement	250 (15 officers, 45 POs, 190 men)

Army) and approved by Viktorov on 13 December 1937, two weeks before he was relieved from his post. There was some difficulty in choosing a contractor for the sketch design, however. Despite the reference in the *Sovnarkom* decree to the advantages of the *Leningrad* class, the experienced design team of Vladimir Nikitin had already been broken up: Pavel Trakhtenberg[10] was executed in May 1937, Aleksandr Speranskij[11] was in prison awaiting trial, while others were completely occupied with the design of the capital ships of the blue-water navy programme. The Design Bureau of Works No 198 (Marti Yard) at Nikolaev was therefore selected, despite a lack of experience in running such a complex project.

From the Bureau workforce of 400, a team of 30 designers was allocated to the project under the leadership of V Rybalko as Chief Constructor, who was relatively new to this work. This meant that the experience acquired over the previous decade with great difficulty and at enormous cost was largely wasted. Instead, experience freshly gained from insights into design and construction of *Tashkent* was applied to the maximum, and many of the solutions were simply copied to the extent that Soviet technology allowed.

The design process coincided with the reorganisation of the naval forces, which from 1 January 1938 regained independent status, becoming the *Voenno-morskoj flot*

Tashkent in 1941 with her designed armament of six 130mm guns in B-2-LM twin mountings. (Vitaliy Kostrichenko Collection)

Tashkent late in 1941 with her final outfit of twin 130mm guns and the enclosed mounting for the twin 3in HA (barrels elevated at 45°) at the stern abaft 'Y' mounting. The bridge of the battleship *Parizhskaia kommuna* can be seen to the left of the 3in turret. (Boris Lemachko collection)

Tashkent in 1942. (Vitaliy Kostrichenko collection)

SSSR (Soviet Navy) under a separate ministry (Peoples' Commissariat). Frequent changes in the post of Peoples' Commissar did not help to preserve experience, nor to maintain a stable view of the tactical and operational tasks of the newly designed ships.

The sketch design was formally approved on 19 July 1938 by Flagman of the Fleet 1st Rank Ivan Isakov,[12] with significant modifications requested. The major shortcomings noted in comparison to the *Leningrad* class

were larger size (2,220 tons standard as compared with 2,030 tons for the *Leningrad* class) and inferior stability; the others were insufficient range and the excessive variety of gun calibres: 130mm, 3in, 45mm, 0.5in, 0.3in. Recommendations included a reduction in topweight, a speed not less than that of the *Leningrad* class, a range of 3,500nm at 15 knots, the replacement of the two triple torpedo tubes by quintuple mountings, and the removal of the 45mm guns and their 1.5-metre rangefinders, the 0.3in MG and the seaplane facilities. A change in electricity voltage to 220V was advised, and Isakov also wanted protection for the wheelhouse with 8mm plating and a change from single DShK MG to twin DShKM-2B mountings. The modified design was recommended on 7 September 1938 by *Sovnarkom* as the one generally fulfilling the requirements outlined by the Defence Committee the previous year. However, further changes in personnel put the flotilla leader on hold. The following day Frinovskij became the People's Commissar of the Soviet Navy. In the meantime, the struggle for seats in the People's Commissariat of the Defence Industry began, with the result that it was divided into four separate bodies from January 1939.

At the same time the Design Bureau of Works No 198 was struggling with the technical design of *Proekt 48*,

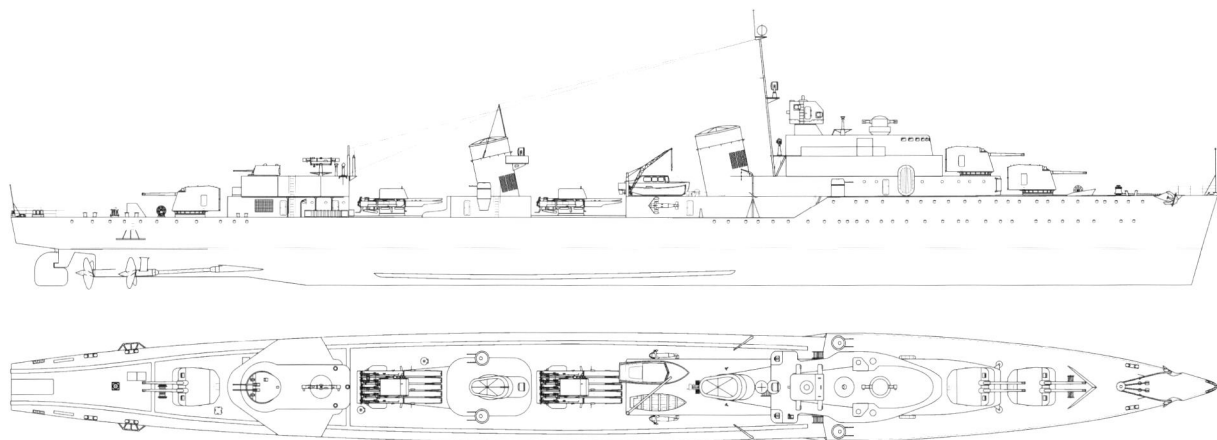

Proekt 48 design as approved on 21 November 1939. (Drawn by Tomasz Grotnik)

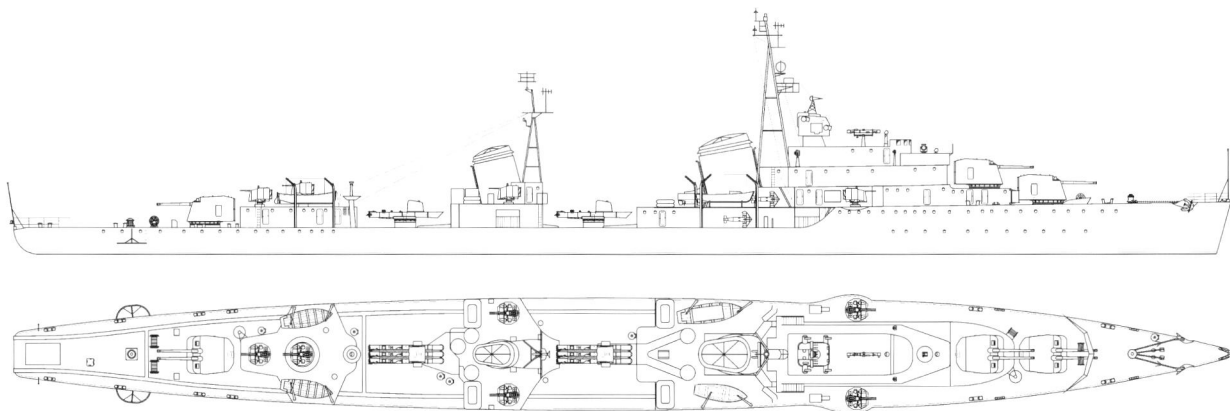

Proekt 48K 1948 design of Works No 444. (Drawn by Tomasz Grotnik)

which in March 1939 was reviewed by the Naval Scientific-Technical Committee and rejected as incomplete and inferior to the *Leningrad* class. The design reverted to the sketch stage with the following changes:

– standard displacement increased to 2,350 tons
– new, fuller body lines
– a range of 3,200 miles
– a change of geared turbines from the three-stator type of Works No 190 (Zhdanov Yard) to the

two-stator type of the Kirov Works
– a change of 130mm turrets from the B-31 type (abandoned) to the B-2-LM type.

The technical design was ready at the end of September 1939, and the following month was presented to Stalin by the new People's Commissar Flagman of the Fleet 2nd Rank Nikolai Kuznetsov.[13] It received the go-ahead from the former, and *Proekt 48* was finally approved on 21 November 1939 by the Defence Committee of

Table 8: **Characteristics of *Proekt 48*, 21 November 1939**
(*Proekt 48K* version 1948 by Works No 444 in square brackets)

Displacement	2,350 tons standard; 2,697 tonnes normal; 3,045 tonnes full load [2,722 tons standard]
Dimensions	length oa 127.8m; beam (max) 11.7m; draught (full load) 4.2m [n/a]
Boilers	three three-drum watertube, 27kg/cm², 350ºC 65t feed water, ? evaporators [two-120t/h]
Fuel	Marine Oil Fuel: 600t normal, 750t full load [600t normal, 625t full load]
Main machinery	three geared turbines, 90,000shp total
-	three three-bladed propellers
Speed	39.5 knots max [37.5kts max]
Range	3,000nm at 20 knots
Electricity	DC 120V circuit two 165kVh turbo-generators two 50kVh diesel generators
Navigation equipment	Kurs-2 gyrocompass, four 5in ZMI magnetic compasses, Gauss-50 log, EMS-2 sounder, Gradus-K direction finder [Kurs-3 gyrocompass, four 5in ZMI magnetic compasses, Gauss-50 log, NEL-2 sounder, Burun-K direction finder]
Main guns	six 130mm/50 B-13 in three B-2LM-I twin turrets, 900 rounds
Main fire control	Mina-48 system: KDP2-4L-I DCT with two DM-4 rangefinders, two 1-N night sights, director sight; DM-3 rangefinder (bridge); 2 – ZD rangefinders [none] TsAS-2M fire control computer
HA guns	two 3in/55 3-K in 39-K one twin mount, 300 rounds
HA fire control	Soiuz system: SVP-29 DCT with DM-3 rangefinder, VN-2 director sight [none]
Light AA guns	twelve 37mm/67.5 70-K in six V-11 twin mountings
Machine guns:	eight 0.5in DShK in four DShKM-2B twin mountings
Searchlights:	one MPE-e 990cm, one MPE-e6 60cm [none]
Torpedo tubes:	two 2-N quintuple mountings [two quintuple PTA-53-47 TT]: ten 21in torpedoes (none spare)
Torpedo fire control	Mina II system: 2 PMR-21 torpedo sights [n/a] TAS-1 computer [n/a]
A/S weapons	two DC chutes: 20 M-1 charges, 10 B-1 charges [4 BMB-2 DC throwers: 48 B-1 charges]
A/S detection	none [sonar Tamir 5N]
Sonic station	Poliaris [none]
Mines	86 Model 1926 [n/a]
Paravanes	two [n/a]
Radars	none [Giujs-2 air search, Rif surface search, Redan-2 FC]
W/T	Shtorm-M, Briz-MK, Okyn'transmitters Groza-M, Purga, two Vikhr, Metel' receivers Rejd, two RB transceivers [n/a]
Complement	264 (16 officers, 5 staff officers, 50 NCOs, 198 ratings) [350]

Tashkent steaming at high speed into a gale in March 1942. (A/D/E Author's collection, B Vitaliy Kostrichenko collection, C Boris Lemachko collection)

Sovnarkom, 27 months after the decision to embark on the project.

Of the twelve flotilla leaders to be built under the blue-water navy programme, eight hulls were laid down before cancellation of most of the programme on 19 October 1940. The speed of Germany's victory over France (considered by Stalin to be Europe's greatest land power) shattered Soviet plans, which were based on the premise of a protracted war between Germany and the Allies. The German domination of Western Europe prompted fear of an imminent German attack on the Soviet Union. Armament priorities were rapidly revised, and the shipbuilding programme found itself on a collision course with other requirements, in particular the modernisation of the armoured forces of the Red Army.

The development of Proekt 35 (*Udaloj* class), which had a comparable displacement and armament, but with B-2-U dual-purpose mountings with an elevation to 85 degrees, effectively put a nail in the coffin of the *Kiev*s. This in turn threw into question the entire flotilla leader concept and with it the outdated *Proekt 48* design, which had its roots in the *Proekt 1* design of the *Leningrad* class, which in turn had been based on a tactical concept that dated back to early 1920s. This led to the cancellation of all except the two most advanced hulls of the *Kiev*

class (see Table 14). Two ships, *Kiev* and *Erevan* (48.9% and 25.4% complete respectively), were launched at Nikolaev before the fall of that town in August 1941; the hulls were evacuated first to Sevastopol then, in October 1941, to Poti (*Kiev*) and Kerch (*Erevan*); in October 1942 they would be towed to Yeysk, and later to Poti.

Both survived the raids of the *Luftwaffe*, and during the closing stages of the war these two undamaged hulls had considerable value when set against the serious

Another photo of *Tashkent* steaming at high speed into a gale in March 1942, but taken on board. The twin mounting closest to the camera is trained to port, the forward mounting to starboard. Note the ice on the forecastle deck. (US Naval History & Heritage Command, NH 100191)

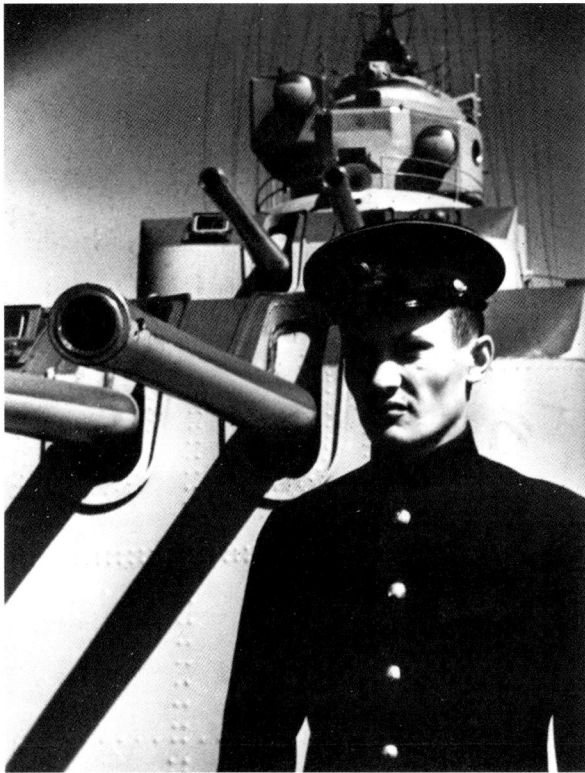

Close-up of the forward 130mm guns. Note the distinctive paint scheme. (Boris Lemachko collection)

The hulls of *Kiev* and *Erevan* on the stocks at Works No 198 in Nikolaev. (Author's collection)

The hull of *Erevan* being towed from Nikolaev to Sevastopol, August 1941. (Author's collection)

depletion of the Soviet destroyer force. In September 1944 Admiral Kuznetsov therefore ordered an update of the *Proekt 48* design to include the latest types of weaponry and sensors. The design was entrusted to the Design Bureau of Works No 444 (formerly No 198, but organised from scratch due to war damage following its liberation by the Soviets). Despite a substantial increase in displacement and the resulting loss of two knots of speed, in 1947 the completion of both ships was inserted into the 1945 Programme, which in effect re-activated the prewar Great Ocean and Seagoing Navy plan. At that time Kuznetsov fell out of favour with Stalin, and the new variant of the design prepared in 1948, which received the code-name *Proekt 48K* ('K' for *korrektirovannyj* = Corrected), lived to see a fifth Head of the Navy during its ten-year history: Admiral Ivan

Iumashev.[14] There were several variants of the design, including one with a two-shaft power plant, but finally the one outlined in Table 8 was approved for implementation. Both ships were re-rated as destroyers; they were finally cancelled in 1952, and three *Proekt 30-bis* destroyers ordered instead.

The Armoured Flotilla Leaders

Proekt 24
The idea of a flotilla leader with armour protection was planted into the heads of Soviet shipwrights in 1933

Table 9: **Characteristics of the Armoured Flotilla Leader Venture Design, 1935**

Project	TsNIIVK (mid 1935)	TsKBS-1 (3rd version) (Dec 1935)
Standard displacement	3,750 tons	3,350 tons
Machinery	186,000shp = 50 knots	140,000shp = 41.5+ knots
Range	2,100nm	2,500nm
Armament	10–130mm DP, 2–130mm LA	8–130mm B-31
	6–45mm AA, 4–0.5in MG	10–45mm AA
	6–21in TT (2 x III)	8–21in TT (2 x IV)
Protection	machinery/magazine sides 50mm	sides 45mm
	deck 40mm	deck 25mm

Tashkent in the port of Batum, with the submarine *Shch-212* in the foreground. (Author's collection by courtesy of Marek Twardowski)

when Sıvkov, during his stay in Italy, obtained a portfolio of sketch plans of warships from Ansaldo which included fast, lightly armoured ships. The concept of an armoured flotilla leader was based on the assumption that the battle line, in addition to destroyers, could utilise fast light cruisers (like the British *Arethusa* class) for scouting and screening. The flotilla leaders might therefore face a more powerful opponent than had hitherto been supposed. The idea of a ship of extraordinary technical and tactical qualities fired the imagination of the Russian designers, who had traditionally had a penchant for outlandish ideas such as the round ships of Admiral Popov in which technical constraints, the laws of physics and the practicality of the application counted for little.

The first design studies were undertaken in 1935 on a venture basis by TsNIIVK (Central Scientific-Research Institute of Naval Shipbuilding) and TsKBS-1 (Central Special Construction Bureau, 'special' being a euphemism for 'warships'). The study from TsNIIVK was completely detached from reality in that it assumed machinery with a weight to power ratio of 6.53kg/hp. The proposal from TsKBS-1, prepared by the team of Pavel Trakhtenberg, assumed a more realistic figure of 8.93kg/hp, based on the use of a high-pressure power plant with straight-flow Wagner-type boilers, which had been incorporated into the *Proekt 45* design of experimental destroyer already laid down in June 1936 at the Zhdanov Yard (later Works No 190) at Leningrad. Instead of the futuristic 130mm DP guns, they assumed lightweight B-31 type mountings, which at the time appeared to be a more practical proposition. TsKBS-1 prepared five variants of their design with different side armour thicknesses that ranged from no protection at all for Variant 1 up to 60mm for Variant 5.

The idea seemed to fall on fertile ground. At that time, for the purposes of the Great Seagoing and Ocean Navy programme, plans for the construction of battleships and aircraft carriers, until recently considered unrealistic, were being worked on. Flagman 2nd Rank Pyotr Svetlovskij,[15] at that time Head of the Naval Inspectorate, was well aware of both the characteristics of *Tashkent* and the scope of the studies at TsKBS-1. He now presented their design studies to Orlov, stating a preference for Variant 3. Orlov duly passed the documentation to Muklevich (then Head of the Shipbuilding Directorate) with an accompanying note stating that he likewise favoured Variant 3. This was understood to be one more order to start design work among a host of projects. On 10 February 1936 the design order was

Table 10: Characteristics of *Proekt 24*, 18 February 1936

Displacement	3,400 tons standard, 3,650 tonnes trial
Boilers	six straight-flow MPN 70/75 type, 75kg/cm², 450°C
Main machinery	2-shaft geared turbines, 140,000shp = 47 knots max
Main guns	8–130mm/50 B-13 in four twin armoured turrets, 150rpg
Light AA guns	8–37mm/67.5 (2 x IV), 1,500rpg
Torpedo tubes	10–21in (2 x V) mounted abreast at sides
Mines	50
Paravanes	two
Protection	citadel of 50mm sides/bulkheads with 25mm deck wheelhouse 35–65mm turrets 35–60mm

Table 11: **Characteristics of the Armoured Flotilla Leader, 17 January 1940**

Standard displacement	4,500 tons
Speed	36 knots
Range	8,000nm
Armament	10–130mm B-2-U (5 x II)
	8–37mm 66-K (4 x II), 4–0.5in MG
	10–21in TT (2 x V)
Aircraft	one seaplane, no catapult
Protection	machinery/magazine sides 70mm (45°)
	forward bulkhead 70mm
	deck 25mm

Table 12: **Characteristics of *Proekt 47*, Autumn 1940**

Displacement	5,400 tons standard, 5,900 tonnes normal
Dimensions	length oa 160.0m; beam (max) 15.0m; draught (mean) 5.0m
Boilers	two
Main machinery	2-shaft geared turbines, 75,000shp = 34.0 knots max
Range	6,000nm
Main guns	10–130mm/55 in five twin B-2U turrets, 900 rounds
Main fire control	Smena system
Light AA guns	8–37mm/67.5 in two quad 66-K mountings
	8–0.5in DShK in four twin DShKM-2B mountings
Torpedo tubes	10–21in tubes in two quintuple 2-N torpedo mountings
A/S weapons	two DC chutes: 24 M-1 charges, 12 B-1 charges
Mines	60
Aircraft	one seaplane
Protection	sides 70mm, forward bulkhead 70mm, deck 25mm, turrets 25mm

given the code-name *Proekt 24* and was returned to TsKBS-1 with the order to proceed; the preliminary work had already been completed and the characteristics submitted only eight days later.

After receiving these, the Directorate of the Naval Forces put *Proekt 24* to one side, as the priority was the big ships in which Stalin himself was interested. The project number '24' was later allocated to the follow-on design of the battleships of the *Sovetskij Soiuz* (*Proekt 23*) class.

Proekt 47 of 1940

The large destroyers of the *Udaloj* class, which featured a dual-purpose main armament and which effectively killed off *Proekt 48*, contributed to the revival of the idea of an armoured leader. A heavier armament in combination with armour was to provide the new flotilla leader with a qualitative advantage, enabling it to lead destroyers in an attack on the enemy battle line. The characteristics of a flotilla leader with protection comparable to that of the *Proekt 26-bis* cruisers (Improved *Kirov*) were prepared by a commission of naval and industrial experts and issued on 17 January 1940.

Following a review by the Shipbuilding Directorate of the Soviet Navy, a speed of 40 knots was requested; the cost was a reduction in armour thickness to 50mm with only two triple banks of torpedo tubes. The final definition of the characteristics was delayed by a request by the Chief of the General Naval Staff Flagman 2nd Rank Lev Galler,[16] who was apparently inspired by the German *Zerstörer 1937* and *Zerstörer 1938* designs presented to him during the visit of Tevosian's commission to Germany in 1939.[17] This request was countered by Isakov, who was higher in the command structure at that time, and apparently influenced by American destroyer construction practice thanks to his own visit in the USA.[18] However, time was pressing, and in June 1940 the characteristics were confirmed: the ships would have triple torpedo tubes and a 50mm belt with a 25mm deck.

The sketch design was entrusted to TsKB-17 (Central Design Bureau) – as TsKBS-1 was renamed after yet another reorganisation – but the work proceeded only slowly as a consequence of the bloody purges which reduced the size of the design team, the survivors being put to work on the big ships. In the event the *Proekt 47* sketch design was ready only in the Autumn of 1940, when the blue-water navy programme was mostly cancelled.

Design work proceeded in parallel on a competing concept of a long-range light cruiser (*Proekt 94*), of which the *MK-6* variant displaced only 500 tons more than the *Proekt 47* flotilla leader but had double the range, a speed of 38 knots and belt reduced by 20mm (forward bulkhead: 100mm). The design team worked in a *sharashka*,[19] an elite labour camp which was in effect the design and construction bureau of the NKVD and operated in closed and guarded accommodation within Works No 196 (Sudomekh, known in Tsarist times as the New Admiralty Yard).[20] Despite the closed nature of this establishment, the information leaked anyway, and TsKB-17 responded with a new version of *Proekt 47*, with standard displacement reduced to 4,000 tons and the thickness of the armoured forward bulkhead increased to 120mm; this was achieved by confining the armoured citadel to the forward part of the ship only.

Based on an amendment to the shipbuilding plan dated 9 January 1940, it was planned to lay down at least six ships of the *Proekt 47* type with completion scheduled for 1942. The unrealistic and uncoordinated nature of these plans is evidenced by the fact that preparation of the design was scheduled for either TsKB-32 or the Design Bureau of Works No 198 (Andre Marti in Nikolaev) with an end date in October 1941. It is, however, indicative of the chaotic nature of Soviet planning in general, and it is unsurprising that the project failed to bear fruit.

After the cancellation of the blue-water navy programme on 19 October 1940 development came to a

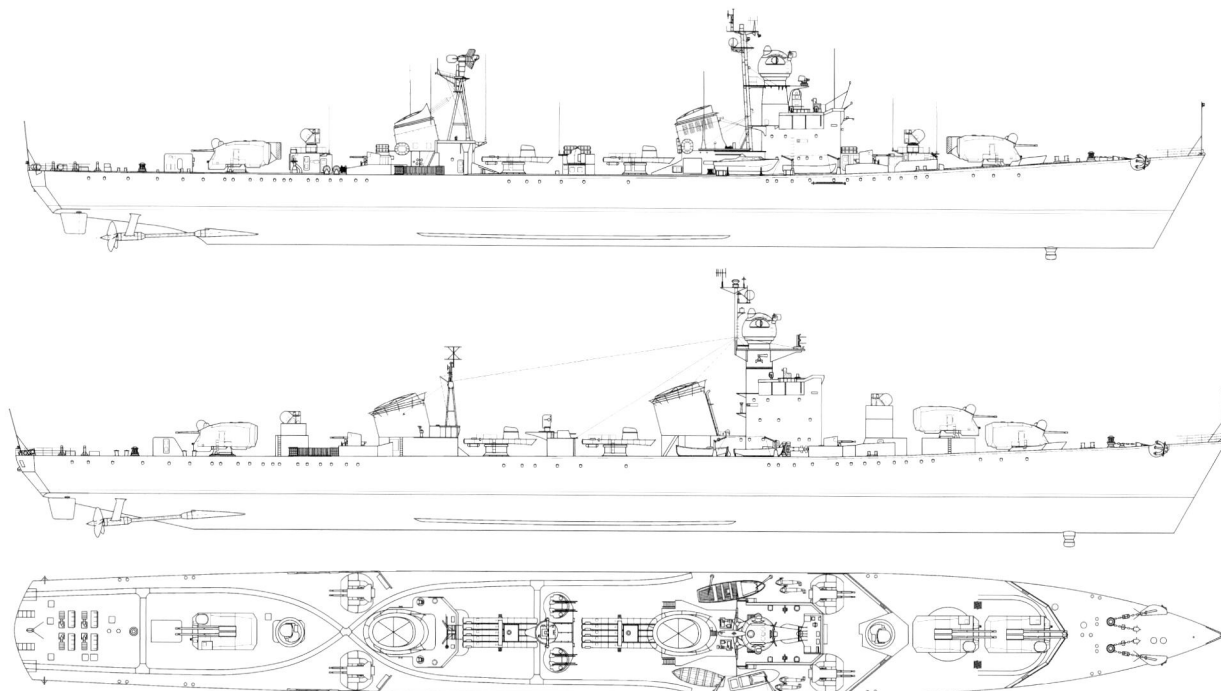

Proekt 41 technical design (top) and *Proekt 47* three-turret version preliminary design (bottom). Note that general arrangements of both ships were identical, except one turret less in the *Proekt 41*. (Drawn by Tomasz Grotnik)

halt. In April 1941 both the *Proekt 47* and *Proekt 94* designs were cancelled, and it was stated that:

> ... results of all the designs confirmed that in the current state of shipbuilding and engineering technology, it is not possible to build an appropriately armoured flotilla leader without a significant reduction in speed or an increase in displacement and size, bringing it close to the cruiser type.

Proekt 47 of 1948

The project of armoured flotilla leader returned in 1947,

again through the back door. In January 1945, when victory was no longer in doubt, Stalin decided to reactivate the blue-water navy programme. The '10-year Plan of Naval Shipbuilding' was approved by *Sovnarkom* on 27 November 1945; it included 188 destroyers, requiring series production based on a standard design. A dispute between the fleet command and the managers of the shipbuilding industry then began regarding the choice of a 'standard'. The latter were able to convince Stalin that the launch of construction of a radically new project would require a considerable length of time not only to

Table 13: **Characteristics of *Proekt 47* (3-turret variant), 1948**

Displacement	3,688 tonnes normal
Dimensions	length oa 138.0m, wl 132.0m; beam wl 3.6m; draught (mean) 4.25m
Boilers	four KV-41, 64kg/cm^2, 450°C
Main machinery	four sets of TV-8 geared turbines, 2 shafts, 70,000shp = 36.0 knots max
Range	5,500nm
Main guns:	four 130mm/58 in two SN-2 twin mountings
Main fire controls	D-2 system: SPN-500 DCT with ZDMS-4 optical 4-metre rangefinder
	Shtag-B radar rangefinder in each turret
	Zenit-41 calculator, 1-N night sight
Light AA guns:	eight 45mm/78 in SM-16 twin mountings
Light AA fire controls	DU-Shtil' system, Kliuz radar
Automatic AA guns	eight 25mm/80 in two 4M-120 quad mountings
Torpedo tubes	ten 21in in two PTA-53-47 quintuple mountings
Torpedo fire control	Stalingrad T-41
A/S weapons	four BMB-1 DCTs, four DC chutes (2 internal), 48 DCs
Mines	48 KB-Krab
Radar	Giujs-1
Sonar	Pega

The Italian *esploratore oceanico Scipione Africano* shortly after her completion in 1943. She belonged to the *Capitani romani* class, the design of which was based on *Tashkent*. (Leo van Ginderen collection)

Proekt 41 destroyer *Neustrashimyj* in mid-1958 presenting her high freeboard flush-deck hull, which was one of the differentiators of the *Proekt 47* armoured flotilla leader. Note the depth charge chutes beneath the quarterdeck. (US Naval History & Heritage Command, NH 79994)

Neustrashimyj in silhouette in 1958; this is how the *Proekt 47* two-turret version would have looked if built. (US Naval History & Heritage Command, NH 80023)

prepare documentation and instrumentation in the shipyards, but also to design and start production of new fittings and armament. The last prewar project to reach the production phase was *Proekt 30* (*Ognevoj* class), which since the end of the war had been revived in a slightly improved version as *Proekt 30K*. The best way to start series production immediately was therefore to adopt the existing project with modifications. Stalin, who in his late sixties was conscious of the limited time that remained to realise his imperial ambitions, decided that all the wartime designs (*Proekt 40*, *Proekt 41*) were to be suspended, and all the resources of TsKB-53 (former TsKB-17) were to be devoted to an upgrade of the *Ognevoj* class, *Proekt 30-bis* (future *Smelyj* class, NATO 'Skory').

When in the second half of 1947 the *Proekt 30-bis* design work came to an end, in order to keep the team busy Nikitin (then Chief Constructor of the bureau) decided to revive the armoured flotilla leader concept and to re-use the code-name *Proekt 47*. This was the best way for him to evade the ban issued by Stalin on new studies for large destroyers while enabling him to access the funding allocated in the budget to studies of the future destroyer (theme SP-46). The reasoning was simple: the ban applied to the destroyer, not the flotilla leader type, particularly an armoured variant.[21] Of course, no one intended to revive the pre-war concept – on the contrary, it was planned to implement the latest (and if possible futuristic) solutions. As armour provided an excuse to design a large destroyer, Nikitin envisaged a vessel with 14mm side protection with a 10mm deck over the magazines and boiler uptakes. Moreover, the armour steel plates were to be part of the hull structure, so could practically be discounted when considering the other characteristics of the design. It should be noted that the Soviets had no experience of using armour plating as a structural element of a large ship, which complicated strength and stability calculations, and had no reliable welding technology.[22]

This breakthrough design was created because of the freedom accorded to the engineers working on it: they were able to employ new types (many of them prototypes) of fittings and armament without the close surveillance of conservative naval officers and without the size of the ship being limited by budgetary constraints. In March 1948 two variants were ready: one with two twin and one with three twin 130mm DP mountings. The flush-deck hull, which featured a high degree of rake and sheer, allowed minimal structures above the weather deck thanks to spacious internal volume. Anti-aircraft guns were mounted directly on the weather deck behind breakwaters. The high-pressure steam plant, which featured two independent machinery rooms each combining two boilers, a single set of turbines and a full set of auxiliaries, was 100 tonnes lighter than that of

USS *Mitscher* photographed on 8 February 1957, illustrates the post-war Western tactical approach: no torpedo tubes but significant A/S weaponry instead. (US National Archives & Records Administration, USN 1015458)

Table 14: **Building Data for the Second Generation of Soviet Flotilla Leaders**

Name	Builder	Yard No	Laid down	Launched	Accepted
Proekt 20I					
Tashkent	OTO, Livorno	216	11 Feb 1937	28 Nov 1937	19 Apr 1939
Proekt 20					
[*Baku*][1]	190, Leningrad	5112	–	–	–
	190, Leningrad	5122	–	–	–
	198, Nikolaev	–[2]	–	–	–
Proekt 48					
Kiev	198, Nikolaev	357	29 Dec 1939	11 Dec 1940	–
Erevan	198, Nikolaev	358	30 Dec 1939	30 June 1941	–
Stalinabad	190, Leningrad	5423	27 Dec 1939	–	–
	190, Leningrad	5433	–	–	–
	190, Leningrad	5443	–	–	–
Ashkhabad	190, Leningrad	5453	n/a	–	–
Alma-Ata	190, Leningrad	5463	n/a	–	–
Pertozavodsk	198, Nikolaev	3593	n/a	–	–
Ochakov[4]	198, Nikolaev	n/a[3]	n/a	–	–
Perekop[4]	198, Nikolaev	n/a[3]	n/a	–	–
Arkhangel'sk[4]	402, Molotovsk	n/a[3]	–	–	–
Murmansk[4]	402, Molotovsk	n/a[3]	–	–	–

Notes:

1 Name *Baku* proposed, but not accepted by Voroshilov and deleted with pencil on his order dated 29 April 1937.
2 Cancelled in September 1938.
3 Cancelled 19 Oct 1940.
4 Names unconfirmed.

Fates of the Proekt 48 ships:

Kiev: August 1946 sunk for experimental purposes off Odessa; 1947 raised and hulked at Works No 444 at Nikolaev; stricken 8 August 1952; 1955 transferred via the Volga-Don Canal to the Caspian Sea and listed in the Caspian Flotilla as target vessel for missile tests; 1961 transferred to the Black Sea; 1962 sunk by P-6 missile (minus live warhead) fired from the experimental vessel *OS-15* (former freighter *Ilet'*).

Erevan: 1944 accommodation hulk; 1953 transferred via the Volga-Don Canal to the Caspian Sea and listed in the Caspian Flotilla; stricken 8 May 1954 and rated as target vessel for missile tests; 16 February 1957 sunk by KSShch test missile, raised and BU.

Ashkhabad, Alma-Ata, Stalinabad: put into conservation for a while, later dismantled on the slipway.

Arkhangel'sk, Murmansk, YN-543, YN-544: material expended.

Ochakov, Perekop, Petrozavodsk: probably damaged on slipways, material scrapped by the Germans.

Proekt 30-bis. An AC electrical circuit was adopted for the first time in the Soviet Navy. To reduce topweight, the forward turret was placed in a recess in the deck. The three-turret variant was considered the more balanced design, but the two-turret variant had greater potential for subsequent modernisation – a feature unheard of in Soviet practice (both earlier and later).

Although extremely 'modern' by Soviet standards, the project lagged behind Western developments: in essence it was still based on prewar concepts. It was a heavily gunned ship (albeit with DP mountings) with an impressive complement of torpedo tubes, but was seriously lacking in underwater sensors and A/S weaponry.

From a technical point of view it could reasonably be claimed that the design was the closest to Western standards in the entire history of the Soviet Navy. The Soviets had benefited to an extraordinary degree from the influx of modern Western technology during the Second World War as a result of Lend-Lease deliveries that featured modern weaponry, engines and sensors. To this should be added the treasure trove of modern equipment, armaments, entire factories and scientific and engineering staff seized from Germany. All this provided a boost to the development of Soviet military technology on the eve of the Cold War.

Proekt 47 was intended to be a typical 'paper' design, with no prospect of being built. However, the Soviet system, dependent on the whims of one man, proved capable of another *volte-face*. It turned out that the *Smelyj* class of standard destroyer, which lacked size, adequate seaworthiness, and was armed with low-angle guns, proved to be unsuitable for oceanic service as escorts for the planned 39,500-tonne *Stalingrad*-class cruisers. This was the basis of Stalin's decision to unfreeze a destroyer project other than *Proekt 30-bis*.

Design work on the *Proekt 41* destroyer was resumed, and this time Nikitin took over the role of Chief Constructor. The *Proekt 47* design was well-developed, and it was sufficient to shorten the hull of the two-turret version by 4 metres to prepare the preliminary design of

Proekt 41, which was ready by August 1948. And in this way the last Soviet design of flotilla leader[23] was transformed into a thoroughbred destroyer design. When built *Neustrashimyj* (NATO 'Tallin' class) attracted considerable attention in the West – in contrast to the Soviet Union, where she was considered something of a white elephant. Nevertheless, she was a trendsetter for Soviet naval shipbuilding for two decades to come.

Endnotes:

1   Zof was a former locksmith, a communist activist from 1910, a political commissar, then *Namorsi* 1924–26. Following his dismissal for incompetence, he was steadily demoted to the position of factory director. He would be executed in June 1937 for alleged participation in a Trotskyist conspiracy.

2   A Communist activist from 1906, from 1912, then a mechanician in the Black Sea Fleet and an NCO from 1915, Muklevich was *Namorsi* from 1926 to June 1931, then Surveyor of the Navy. In 1934 he was appointed Chief of the Naval Shipbuilding Directorate, and in 1936 deputy Peoples' Commissar of the Defence Industry. He was executed February 1938 for alleged espionage.

3   Sivkov was an engineer of landowner origin, and a member of the Communist Party from 1920. Following the Civil War he was appointed CO of the destroyer *Stalin* and the battleship *Oktiabr'skaia revoliutsiia*, and in 1930 he became Head of the Technical Branch of the Directorate of Naval Forces. From 1935 he was Chief of the Naval Staff, and in 1937 he became C-in-C of the Baltic Fleet, being promoted Flagman 1st Rank – Stalin re-introduced naval ranks in November 1935. He was executed in February 1937 for alleged 'wrecking'.

4   Of Estonian origin, Oras was a naval engineer, becoming a Communist activist in 1917. He was naval attaché in Stockholm 1926 to 1928, then Chief of the NTK (Naval Scientific-Technical Committee). In 1933 he was again naval attaché, first in Rome then in Washington. From 1936 he occupied various positions in the Peoples' Commissariat of the Defence Industry. Arrested in 1937 for alleged anti-Soviet conspiracy, he worked in the *sharashka* of OKB-196 and died in prison 1943 – see also *Warship 2020*.

5   From a noble family, Nikitin had served with the Artillery Corps of the Imperial Army, and in 1925 graduated from the Shipbuilding Faculty of the Leningrad Technical Institute. He would become Chief Designer of the escort ships of the *Uragan* class, the flotilla leaders of the *Leningrad* class, the destroyers of the *Gnevnyj* class (responsible for the preliminary design), and the heavy cruisers of the *Kronshtadt* class. From 1953 he was head of TsKB-53, which was responsible for the design and construction of the destroyers of the *Proekt-30K*, *Proekt 30-bis*, *Proekt 41* types. He was also Chief Designer for the missile cruisers of the *Groznyj* class (*Proekt 61* – NATO 'Kynda'), Following retirement, he worked as the Lead Designer of TSKB-53 until 1976; he died in 1977.

6   Orlov, a would-be lawyer, was a warrant officer in the Imperial Navy in 1917, and from 1919 served in various *politruk* (political officer/commissar) functions. From 1926 to 1931 he was C-in-C of the Naval Forces of the Black Sea, and in July 1931 was appointed *Namorsi*, with the rank of Flagman of the Fleet 1st Rank. He was arrested in July 1937 and executed in 1938 for alleged espionage.

7   Son of an officer, Viktorov was a *Gardemarin* in the Imperial Navy, and a naval officer during the Great War. From 1918 he was CO of the destroyer *Vsadnik*, battleships *Andrej Pervozvannyj* and *Gangut*. In 1921 he became C-in-C of the Naval Forces of Baltic Sea, then in 1924 of the Black Sea Fleet. He was Chief of the Hydrographic Service from 1924 to 1926, then C-in-C of the Naval Forces of the Baltic Sea. In 1932 he was transferred to the Far East, becoming C-in-C of the Pacific Fleet. *Namorsi* from Aug 1937 and promoted Flagman of the Fleet 1st Rank, he was removed from office in December 1937, arrested and executed in 1938, allegedly for participation in an anti-Soviet plot.

8   A carpenter by trade, Smirnov served in the Communist Party from 1917 in various *politruk* posts. In 1937 he became Head of the Political Department of the Red Army, then from December 1937 the People's Commissar of the Soviet Navy. Having led a purge of naval officers and *politruk*s, he was himself arrested in June 1938 and executed in 1939, allegedly for participation in the 'fascist plot' of the military.

9   Frinovkiy was a seminarist, then a cavalryman from 1916. He was appointed Chief of the secret police section of Budenny's 1st Cavalry Army in the Cheka 1919. In 1934 he was appointed Chief of the Border Guard of the NKVD, in 1936 was Deputy Peoples' Commissar of the NKVD, and in September 1938 was the Peoples' Commissar of the Soviet Navy, being promoted Komandarm 1st Rank. He was arrested in April 1939 and executed 1940, allegedly for organising a fascist-Trotskyist plot in the NKVD.

10 Trakhtenberg was an engineer. Following his graduation in 1929 from Leningrad Technical University, he was appointed in 1931 to posts in the TsKBS and its successor design bureaux, and was part of the design team of the escort ships of the *Uragan* class, the flotilla leaders of the *Leningrad* class and the experimental destroyer *Sergo Ordzhonikidze* (later *Opytnyj*). He was arrested for a time for concealing his non-proletarian origin but released shortly afterwards. Trakhtenberg was general designer of the *Proekt 7* destroyers, and after the design was considered by Stalin to be 'wrecking', he was executed in 1937, allegedly for participation in a terrorist Trotskyist sabotage organisation.

11 Speranskiy graduated from the St Petersburg Technological Institute in 1906 as a mechanical engineer. In 1914 he was appointed constructor at the Pulilov Works. After the Revolution he would be in charge of the design of the propulsion plant of the *Uragan*-class escort ships, the *Leningrad*-class flotilla leaders, the experimental destroyer *Sergo Ordzhonikidze* (later *Opytnyj*) and the *Proekt 7* destroyers. After the destroyer design was considered by Stalin as 'wrecking', he was executed in 1938 for alleged espionage.

12 Isakov, of Armenian descent, was a former *Gardemarin* in the Imperial Navy. In 1919 he became CO of the escort ship *Kobchik*, followed by the destroyers *Deiatel'nyj* and *Petrovskij* and the minesweeper *Iakor'*. He was then executive officer of various ships, and from 1922 was mostly involved in staff work. In 1932 he became a lecturer in the Voroshilov Naval Academy and in 1933 was Chief of Staff of the Naval Forces in the Baltic. Between 1935 and 1937 he reverted to academic work following the loss of the submarine *B-3* during exercises. In 1937 he was appointed C-in-C of the Baltic Fleet, and between 1938 and 1939 served as Deputy Peoples' Commissar of the Soviet Fleet. He became a member of the Communist party only from 1939. He was Chief of the Naval Staff 1940–1942, and occupied various military and, after the war, civilian senior positions, and rose to the rank of Admiral of the Fleet of the Soviet Union. An

academic author, he is known in the West for his book *The Red Fleet in the Second World War* published by Hutchinson in 1947. He died in 1967.

13 Kuznetsov was a *protégé* of Stalin. Following two years as CO of the cruiser *Chervona Ukraina* and one year as naval advisor to Republican Spain, he was given command of the Pacific Fleet in January 1938, replacing Flagman Kireev (under arrest and later executed), who had replaced Viktorov when he was promoted to *Namorsi* in August 1937 in place of the arrested Orlov. In April 1939, at the age of 34, Kuznetsov became People's Commissar of the Navy and was to hold this post throughout the entire war. When, due to inter-service rivalry, the separate Navy Ministry was abolished in 1946, he became C-in-C of the Soviet Navy with the rank of Admiral of the Fleet. In 1947 he fell out of favour with Stalin because of a disagreement about the cruiser building programme and became head of the Naval Training Command. In 1948 he was downgraded to the rank of Rear Admiral, accused of passing classified information to the Allies during the war. Returned to the C-in-C position in 1951, in 1953 (following Stalin's death) he was promoted to the rank of Admiral of the Fleet of the Soviet Union. The loss of the battleship *Novorossijsk* (ex-Italian *Guilio Cesare*) was a good pretext for Khrushchev to replace Kuznetsov by Gorshkov in 1956, and the former was again downgraded to the rank of Vice Admiral. Later attempts to return him to higher rank were blocked by Gorshkov, and Kuznetsov was only restored to the rank of Admiral of the Fleet of the Soviet Union posthumously in 1988, one month after Gorshkov's death.

14 Yumashev had been a petty officer in Imperial Navy. After the Revolution he had served as executive officer on various ships. In 1927 he commanded the destroyer *Dzerzhinskij*, then in 1932 the cruiser *Profintern*. In 1934 he became senior office of the destroyer flotilla, and from 1935 the cruiser brigade. In 1937 he was appointed Chief of the Naval Staff, then in 1938 C-in-C of the Black Sea Fleet and in 1939 C-in-C of the Pacific Fleet. In 1947 he replaced Kuznetsov as head of the Soviet Navy, and in 1951 he was replaced by Kuznetsov and became head of the Voroshilov Naval Academy with the rank of Admiral. He retired from the service in 1957 and died in 1972.

15 Smirnov-Svetlovskij was of Cossack origin. From 1916 he was a student at Petrograd Technical University and became a Bolshevik activist. In 1919 he was appointed senior officer of the Volga Flotilla, then the Dnieper Flotilla. In 1927 he was captain of the destroyer *Novik*, then senior officer of the destroyer flotilla, and in 1930 he was part of the Soviet delegation to Germany that was seeking assistance in submarine construction. In 1934 he became Inspector of the Navy, becoming C-in-C of the Black Sea Fleet in 1937, then Deputy Peoples' Commissar of the Soviet Navy, with the rank of Flagman of the Fleet 2nd Rank. He was arrested in 1939 and executed in 1940, allegedly for participation in counter-revolutionary activity and 'wrecking'.

16 Of noble origin with German roots (von Haller), Galler was a *Gardemarin* in the Imperial Navy, then from 1915 gunnery officer of the battleship squadron of the Baltic Fleet. From 1917 he served successively as the CO of the destroyers *Turkmenets Stavropol'skij* and *Mecheslav*, the cruiser *Baian*, and the battleship *Andrej Pervozvannyj*. Following staff work, in 1932 he became C-in-C of the Baltic Fleet, then deputy *Narkom* in 1937. From 1938 he became Chief of the Naval Staff, and from 1940 Deputy Peoples' Commissar of the Soviet Navy for shipbuilding affairs. From 1947 he was head of the Kirov Naval Academy. In the same year he was accused of handing over materials to the Allies in 1942–1944 using a high-altitude parachute torpedo, samples of this weapon, and maps of two islands and the southern coast of Kamchatka. Sentenced in 1948 to four years in prison and demotion from Admiral, he died in prison in 1950. (See also Note 22).

17 The Commission for Placing Orders in Germany, led by Ivan Tevosyan, paid two visits to Germany, in September 1939 and March 1940, so Galler must have obtained this information in 1939. This was a honeymoon period in Soviet-German relations, and the Germans were very open concerning their armaments industry. In order to appear in Germany in decent clothing, delegates were specially kitted out by a Soviet textile factory: they wore identical suits and grey hats with a black band. In his memoirs Nikitin recalled that German taxi drivers quickly learned to recognise them.

18 The Commission for Negotiations with the Gibbs Company led by Isakov was in the USA from February to May 1939 and, in addition to the widely known attempts to purchase design documentation of a battleship, aimed to purchase two destroyer designs and to evaluate the possibility of a purchase of decommissioned US warships.

19 The word appeared in the USSR at the beginning of the 20th century, and was derived from the dialect word *Sharan*, which meant 'crook' or 'deceit'. The *sharashka* followed the communist rule of 'no work – no food', which was applied to intellectual creativity supervised and evaluated by prison guards. Many prominent Russian scientists and engineers worked in such conditions. The most famous of them was aircraft constructor Andrei Tupolev, who also designed the first Soviet MTBs (see *Warship* No 8).

20 The team was led by Valerian Bzhezinskij who was jailed there from 1937 – he would be released only in 1947; he worked on projects auch as the submersible MTB *Blokha* and the midget submarine *M-400*. Jailed again in 1949, he was exiled to the town of Yeniseysk in mid-Siberia. He was ultimately released early from exile in 1952, as he had developed a device which was of use on warships.

21 This was a dangerous game, as in early 1948 five admirals (including Kuznetsov) were prosecuted under the pretext of spying for the British during the war. In the best style of the NKVD, the aged Vice Admiral Goncharov, who died after 17 days of interrogation, was arrested earlier to prepare the charges. The whole affair was intended to distract the attention of Stalin from the low technical level of the newly built ships.

22 When it was no longer possible to hide the problems with welding the hull of the new cruiser *Dzerzhinskij*, Stalin asked his usual question: 'Who is to blame?', and after receiving the answer that it was ignorance, he replied 'Ignorance is not a plea. Fix it.'

23 On 12 January 1949, the term 'flotilla leader' was deleted from the official classification list of ships of the Soviet Navy.

Note: The US Library of Congress transliteration protocol (slightly modified to avoid the use of special characters) has been employed throughout for Russian script. The names of well-known historical personages and geographical places are given in their standard English renderings.

ACTION OFF THE BOSPHORUS, 10 MAY 1915

When German and Ottoman warships shelled several Russian ports on 29 October 1914 and took the Ottoman Empire into the First World War, Russia's Black Sea Fleet was significantly larger, more powerful and more efficient than the Ottoman naval forces. However, making the balance of forces less uneven were two newly-arrived modern German vessels, *Goeben* and *Breslau*, much faster than any large Russian ship, though weaker than their combined Russian opponents. **Toby Ewin** looks at the second major encounter between these ships.

On 14 August 1914 Grand Admiral von Tirpitz, State Secretary of the German Imperial Navy Office and overall architect of the Kaiser's navy, wrote to Vice Admiral Souchon, then commanding the 'battle cruiser' *Goeben* and light cruiser *Breslau* at Constantinople, that:

… the Russians in the Black Sea can be poorly estimated … The top speed of the big Russians is not over 18 [knots]. In reality, lower. Marksmanship is bad as well.[1]

And an August 1914 Turco-German conference involving Ottoman War Minister Enver Pasha, Marshal Liman von Sanders, Vice Admiral Souchon, Germany's military and naval attachés, and other senior officers, had a low estimate of the Russian fleet's efficiency.[2] Yet after the war the German official history judged that the Black Sea Fleet 'was well-trained – much better than the forces in the Baltic. The Russians shot well at long ranges, they were often underway, always in concentrated force …'[3] The following study helps explain why the German Navy's evaluation of their Russian opponents underwent such a transformation.

The Black Sea Fleet by 1914

Among the conclusions the Russian Navy drew from the 1904–05 naval war with Japan were: that their heavy guns needed to fire faster; that future major naval engagements might be at very long range; and that most of the numerous 37mm to 75mm guns festooning its

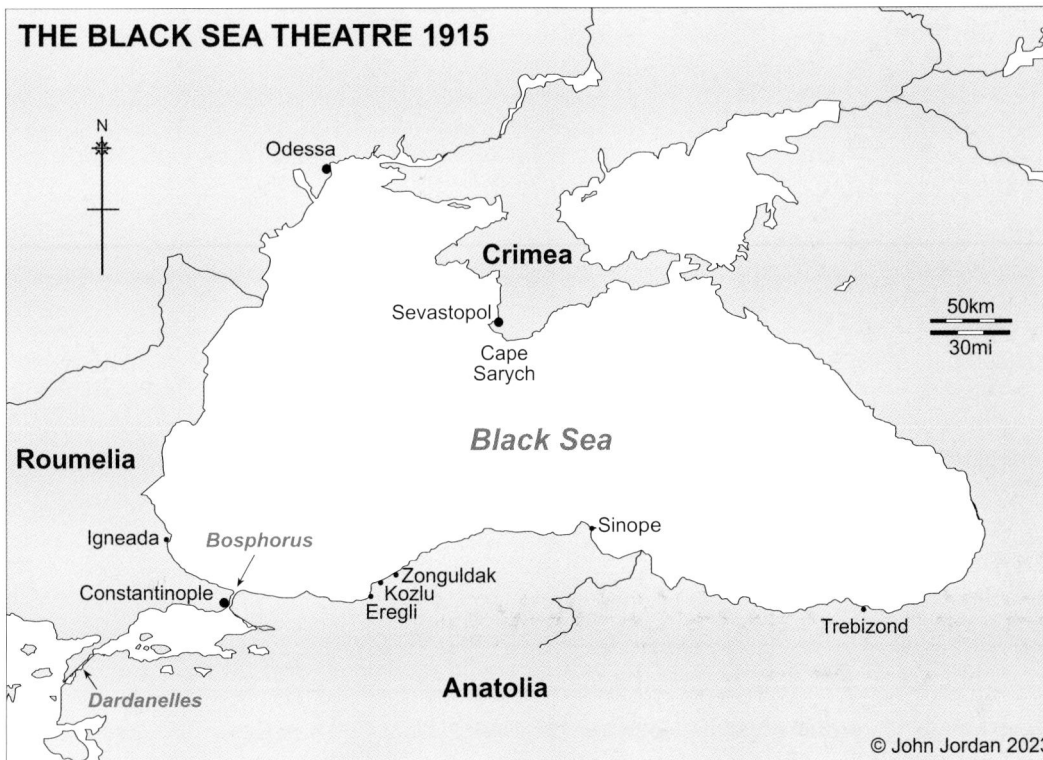

THE BLACK SEA THEATRE 1915

N

Odessa

Crimea

Sevastopol

Cape
Sarych

50km

30mi

Black Sea

Roumelia

Igneada

Bosphorus

Sinope

Zonguldak
Kozlu
Eregli

Constantinople

Trebizond

Anatolia

Dardanelles

© John Jordan 2023

The Black Sea, showing significant places mentioned in the text (spelled as was common at the time).

Table 1: **Major Black Sea Warships 1914–1915**

Name	Displacement[1]	Main armament	Maximum armour thicknesses (mm)[2]				Max speed[3]
			Belt	Deck	Main guns	CT	
Russian							
Sviatoi Evstafi & *Ioann Zlatoust*	13,780 tons	4–12in/40, 4–8in/50, 12–6in/45	229	76	254	229	16 knots
Panteleimon	~13,000 tons	4–12in/40, 16–6in/45	229	76	254	229	16.9 knots
Tri Svititelia	~13,800 tons	4–12in/40, 14–6in/45	457 (379)	76 (63)	406 (337)	305 (253)	~16 knots
Rostislav	10,520 tons	4–10in/45, 8–6in/45	356 (295)	76 (63)	254 (210)	152 (126)	15.37 knots
Pamiat Merkuria & *Kagul*	7,667 tons	16–6in/45, up to 290 mines	–	70	25–127[4]	140	~20 knots[5]
Almaz	3,285 tons	7–4.7in/45, up to 4 seaplanes	–	?76[6]	–	–	19 knots
Imperator Alexander I	9,240 grt	6–4.7in/45; up to 8 seaplanes	–	–	–	–	15 knots
Imperator Nikolai I	9,230 grt	6–4.7in/45; up to 7 seaplanes	–	–	–	–	15 knots
Turco-German							
Goeben (›Yavuz)	~25,000 tons	10–28cm/50, 12–15cm/45	270	75	230	350	~26 knots
Barbaros Hayreddin & *Turgut Reis*	~10,500 tons	4–28cm/40, 2–28cm/35, 6–10.5cm/35	400 (332)	60 (50)	120 (100)	300 (249)	~10 knots
Mesudiye (sunk *B11* 13.12.14)	8,980 tons	12–6in/45[7]	305 (127)	127 (53)	230	200	~12 knots
Breslau (›Midilli)	5,499 tons	12–10.5cm/45; up to 120 mines	60	60	50	100	~27 knots
Hamidiye	3,805 tons	2–6in/45, 8–4.7in	–	102	51	?	~18 knots
Mecidiye (mined 3.4.15)	3,693 tons	2–6in/45, 8–4.7in	–	102	51	?	~18 knots

Sources: McLaughlin, *Russian and Soviet Battleships*; Dodson/Nottelmann, *The Kaiser's Battlefleet* and *The Kaiser's Cruisers*; Mielnikow, *Imperatrica Maria*; Kuznetsov, *Ioann Zlatoust*; Layman, *Before the Aircraft Carrier*; *CAWFS* 1906-1921; Navypedia; *Jane's Fighting Ships* 1914. [Losses up to May 1915 are noted.]

Notes:

[1] Actual or estimated displacements are full load in 'long' tons, except the converted liners, which are Gross Registered Tons.

[2] Armour was Krupp, or Krupp Cemented (KC) steel except for *Tri Sviatitelia* and *Rostislav* (Harvey steel) and *Mesudiye* (iron belt and deck). KC equivalent thicknesses are in parentheses, calculated via conversions based on based on 1901 Target Book on Russia, ADM 231/33.

[3] Speeds are maxima in knots in 1914–15 (though on 10 May *Panteleimon* may have reached 17.5 knots).

[4] Twin 6in turrets had 127mm armour; four guns were in 80mm casemates; the rest had 25mm shields.

[5] Le Page's 18 February 1915 report (ADM 137/754, 12–19): 'The speed of the *Mercury* and *Kagoul* [sic] is estimated at 22½ knots, but I venture to say that now it is very little over 20 … When 22½ knots were obtained, the ship had been refitted, [and had] Welsh coal, clean hull and clean boilers. At present the boilers have not been cleaned internally … also tubes are weak and need renewing … several cases of burst tubes being reported from *Kagoul*. Coal very bituminous … enormous quantities of very dense smoke, when travelling at any speed. The hulls also covered with weed, ships not docked now 7–8 months.'

[6] Only *Jane's Fighting Ships* suggests *Almaz* had an armoured deck.

[7] Two 9.2in/40 guns were not mounted.

The 'battle' cruiser *Goeben* in the Sea of Marmara in 1915. (Courtesy of Dirk Nottelmann)

ships were of little value. Modifications were made to ships still under construction – accepting consequent delays in their completion – and to existing ships judged worth the effort. Budgetary constraints after the 1904–05 war, and the necessary priority of first rebuilding Russia's army, meant it was several years before significant naval spending came on stream. But by 1914 there had been important changes in the Black Sea Fleet's major units.

For instance, additional armour was fitted to the most modern battleships, *Sviatoi Evstafi*, *Ioann Zlatoust* and *Panteleimon*, near the waterline forward and aft. New (probably 12-foot) rangefinders were fitted in *Evstafi* and *Ioann Zlatoust*, while *Panteleimon* received new telescopic sights for her main and secondary guns as well as rangefinders. Six 3in guns were removed from both *Evstafi* and *Ioann Zlatoust*, although a plan to mount a

Table 2: Heavy & Medium Guns in Black Sea Ships 1914–1917

Model	Gun	Shell wt (pounds)	Max elevation[1]	Max range (yards)	Max rpm[2]
Russian					
Pattern 1907[3]	12in/52	1,038	25°	~26,000 at 25°	1.5–3
Pattern 1895	12in/40	1038 (Mod 1911) *or* 731 (older shells)	25° *or* 35°	26,600 at 35°,23,000 at 25°	1.5
Pattern 1891	10in/45	496	15°	?17,000 at 15°	1.5–2
Pattern 1905	8in/50	307 *or* 246	20°	19,230 at 19.8°	3–4
Pattern 1892	6in/45	91	20° (*Tri Sviatitelia* 25°)	12,600/15,400 at 20° 17,000 at 25°	7–10
Pattern 1913[3]	5.1in/55	81	20°	16,800 at 20°	5–8
Pattern 1892	4.7in/45	63 *or* 45	20° *or* 25°	13,000 at 20° 14,000 at 25°	12–15
Pattern 1911	4in/60	38	17–22° *or* 30°	12,600/16,200/17,600[4] at 30°	12
Turco-German					
28cm C/09	11in/50	660	13.5°	19,790 at 13.5°	3
28cm MRK	11in/40 *or* 11in /35	529	25°	17,400 at 25° (40-cal) 15,800 at 25° (35-cal)	0.5
15cm SK	5.9in/45	99.8	20°	16,300 at 20°	7
6in Armstrong[5]	6in/45	100	15° (later 20°)	14,600 at 20°	5–7
10.5cm SK	4.1in/45	38.4	30°	13,890 at 30°	15
10.5cm SK	4.1in/35	30.9	30.3°	11,810 at 30.3°	7.5

Sources: Campbell, *Battle Cruisers*; Friedman, *Naval Weapons of WWI*; McLaughlin, *Russian and Soviet Battleships*, 'Predreadnoughts vs a Dreadnought'; Ley, 'New Information'; NavWeaps. [Rounded ranges are – or are converted to – yards.]

Notes:

[1] *Evstafi* and *Ioann Zlatoust*'s 12in guns elevated to 35°, *Tri Sviatitelia*'s to 25°. For *Panteleimon*'s 12in guns maximum elevation was 25° or 35° – a contemporary report quoted below suggests the latter.

[2] Rounds per minute are the reported maximum practical.

[3] These were fitted in Russia's dreadnoughts, and thus not present on 10 May 1915.

[4] Varying reported ranges may (and alternative shell weights do) reflect different shell types.

[5] These were on *Hamidiye* (later replaced by German 15cm/45); the author has not located data for the 4.7-inch guns carried by Ottoman cruisers.

single extra 8in/50 each side in their place was not carried through. Modifications to breeches and loading mechanisms of the 12in/40 guns on *Panteleimon* and *Tri Svititelia* raised their rate of fire from less than one round every four minutes (in *Panteleimon*'s case) to one round every 40 seconds. The 12in guns of ships then under construction were given 35-degree elevation; and the elevation of guns already in service was increased to 25 degrees (in *Panteleimon*'s case perhaps 35 degrees). *Panteleimon*'s bow torpedo tube was removed, her bridge and bridgework reduced and foremast fighting top removed, and some boilers refurbished. In a major 1911 refit, one deck of *Tri Sviatitelia*'s superstructure, her military masts and most smaller guns were removed.

Limited funding meant some ideas for modifying the smallest battleship, *Rostislav*, were not pursued, but most of her 47mm and 37mm guns, together with her above-water torpedo tubes, were removed; 15-foot rangefinders were fitted for her main guns; engines and boilers were overhauled, turrets upgraded and new telescopic sights fitted.[4]

Twelve 3in guns were also removed from each of the protected cruisers *Kagul* and *Pamiat Merkuria*. In 1914–15 each received four more 6in guns, raising their total to 16 apiece.[5] (During the war, some thought was given to replacing the two twin 6in turrets on each ship with single new 8in/50 guns.[6]) The four destroyers of the *Leitenant Shestakov* class were re-armed, two 4.7in replacing their former 3in guns.[7]

Russia also developed triple 12in gun turrets for her dreadnoughts, and twin and triple torpedo tube mountings for her new large destroyers.

There were important developments in the fleet's training, too. While visibility in the Baltic was often limited, that in the Black Sea was frequently excellent. The Black Sea Fleet therefore rightly anticipated the need for battleships to engage at 16–20,000 yards,[8] reflected in a demonstration on 20 October 1914 in which *Tri Sviatitelia* made several hits on a target from twenty rounds at 16,000 yards, and *Panteleimon* and *Rostislav* also fired at this range.[9] During the war, the Black Sea Fleet also set up a training school for destroyer officers.[10]

Moreover, realising that it might have to face one of the dreadnought battleships the Ottomans had ordered from

Britain before its own first dreadnought came into service in 1915, the Black Sea Fleet also developed a method of concentrating the fire from the twelve 12in guns of three predreadnoughts: a 'master ship' – the middle ship of the three when they were in line ahead – would co-ordinate their firing via a dedicated radio network. The three ships would ideally be the Black Sea Fleet's most modern at the time – *Sviatoi Evstafi*, *Ioann Zlatoust* and *Panteleimon* – although if one of these was unavailable, the older *Tri Sviatitelia* (with the same model of 12in guns) could substitute.[11] This technique was employed at the brief skirmish off Cape Sarych in November 1914 with the battlecruiser *Goeben*, but as poor visibility led the 'master

ship' *Ioann Zlatoust* to overestimate the range it was not successful there: the only Russian hit was scored by the fleet flagship, *Evstafi*, whose officers realised their sister-ship's error and used their own (correct) estimate of range.

The years immediately before 1914 saw several other significant developments in the Imperial Russian Navy, including the establishment of a Naval General Staff, the beginnings of structured naval intelligence activity, and the development of naval aviation,[12] though most of these are beyond the scope of this article.

By spring 1915, the Black Sea Fleet included:[13]

– Five predreadnoughts: *Sviatoi Evstafi* (flag of the

Table 3: Russian & Ottoman Black Sea Destroyers 1914–1917

Name	Displace-ment[1]	Main armament			Max speed[2]	Comments[3]
		Guns	17.7in torpedoes	Mines		
Russian large destroyers						
Kerch, Gadzhibey, Fidonisi, Kaliakriya (+*Tserigo, Zante, Korfu, Levkas* building)	1,555 tons	4–4in/60	four triple TT	60	~30 knots[4]	Four built 1915–17; rest still building 1917
Shchastlivyi, Bystryi, Gromkiy, Pospeshnyi, Pylkiy	1,309 tons	3–4in/60	five twin TT[5]	80	30 knots[6]	Built 1913–15
Derzkiy, Bespokoiny, Gnevnyi, Pronzitelnyi	1,220 tons	3–4in/60	five twin TT[5]	80	32.7 knots	Built 1912–Oct 1914
Russian destroyers						
Leitenant Shestakov, Kapitan Saken, Kapitan-Lt Baranov, Leitenant Zatsarenni	836 tons	2–4.7in/45	three single TT	40	~24 knots	Built 1906–09
Zavetnyi, Zavidnyi, Zharkiy, Zhivoy, Zhivuchiy, Zhutkiy, Zorkiy, Zvonkiy, Leitenant Pushchin	~520 tons	2–3in/50	two single TT	18	23-24 knots	Built 1901–07
Strogiy, Smetlivyi, Stremitelnyi, Svirepyi	~300 tons	2–3in/50	two single TT	12	22 knots[7]	Built 1899–1902
Ottoman destroyers						
Muavenet-i Milliye, Yadigar-i Millet, Numune-i Hamiyet, Gayret-i Vataniye	753 tons	2–3in/50	three single TT	–	~26 knots	Built Germany 1908–10
Samsun, Yarhisar, Tafloz, Basra	~300 tons	1–65mm	two single TT	–	~17 knots	Built France 1906–07
Ottoman torpedo gunboats						
Peyk-i Şevket, Berk-i Satvet (hit Russian mine 2.1.15; under repair until 1917)	763 tons	2–10.5cm/40 6–57mm	three single TT		~18 knots	Built Germany 1906–07
Peleng-i Derya (torpedoed & sunk by HM Submarine *E11* 23.5.15)	900 tons	2–12cm/40, 2–88mm/30[8]	three single 14in TT		~14 knots	Built Germany 1889–96

Sources: Afonin, *Leitenant Shestakov*; Zablotsky & Levitsky, *Dzerzkiy*; Zablotsky, *Shchastliviyi*; Güleryüz, *Torpedoboats & Destroyers*; Le Page's reports; Navypedia. The table omits both navies' old ~100-ton torpedo boats; losses up to May 1915 are noted.

Notes:
1 Full load displacements (except mines) in 'long' tons; mines could add 150–200 tons to a large Russian destroyer.
2 Top speeds are the most recent reported, in knots.
3 Russian destroyers were all built in Russia.
4 Le Page's report 22/17 (23 June 1917) noted that, early in her service, *Fidonisi* made 29 knots.
5 Phillimore's report 28 of May 1916 notes that these classes carried 150 rounds of 4in and 13 torpedoes (ten in the tubes + three spares).
6 *Shchastlivyi*; Le Page's report 3/15 stated that *Gromkiy* also made 30 knots on trials.
7 *Bespokoiniy*; Le Page's reports 10/15 and 6/16 recorded that she made 32 knots on trials and in February 1916; in a letter of 20 May 1916 Admiral Phillimore's Flag Lieutenant, Parsons, described *Gnevnyi* as being capable of 34 knots.
8 Le Page's report 7/16 (26 February 1916) stated that *Smyetlivyi* could make 22 knots for a short period.
9 In early 1915 all these guns were replaced by three 3in/40.

C-in-C, Admiral Andrei Augustovich Ebergard[14]), *Ioann Zlatoust* (flag of Vice Admiral Pavel Ivanovich Novitsky), *Panteleimon*, *Tri Sviatitelia* (flag of Rear Admiral Prince N S Putiatin), and *Rostislav*.

– Two heavily-armed protected cruisers, *Pamiat Merkuria* (flag of Rear Admiral Andrei Georgievich Pokrovsky) and *Kagul*, and the small cruiser-yacht (and seaplane carrier} *Almaz*.

– Two liners converted to serve as seaplane carriers: *Imperator Alexander I* and *Imperator Nikolai I*. As they were about as fast as the fleet's battleships, they could be incorporated into fleet sorties without compromising the warships' speed.

– Seven new fast and powerfully-armed large destroyers (with more under construction), comprising the First and Second Destroyer Divisions.[15] The Third Destroyer Division comprised the four *Leitenant Shestakov*s, the Fourth and Fifth Divisions the nine small 'Zh'-class destroyers, and the Sixth Division the four smaller *Stremitelny* class, all under the command of Captain First Rank Mikhail Pavlovich Sablin

– Three gunboats, built in the late 1880s and modernised 1911–12 and given modern 6in guns. One of these was sunk in Odessa harbour in the Ottoman surprise attack of 29 October 1914, but was now being repaired.

– To supplement the small and not very capable submarines available in 1914, a class of larger vessels entered service in 1915, as did the submarine minelayer *Krab*.[16]

– About a dozen steamers capable of 10–12 knots were fitted as minesweepers for operations against the Bosphorus. A similar number of small steamers were used as minesweepers around Sevastopol. More steamers were prepared as transports, and two had already been converted into hospital ships. (The transport flotilla already mustered over 100 freighters, and would later grow to around 140.)

Four dreadnoughts and more large destroyers were under construction, many of which would enter service in 1915–1917. The war years also saw the completion of purpose-built landing craft and several submarines. Four fast light cruisers were building, too, although these had lower priority and none were completed by 1917.

Turco-German Naval Forces in the Black Sea

The only first-class ships available to the Ottomans and their German allies were *Goeben* and *Breslau*, notionally purchased from Germany in August 1914, but which only truly came under Turkish command after the 1918 armistice. As the German official history put it:

Breslau was superior in speed to all the Russian ships [except the new large destroyers] but her armament [was] weak in comparison with the Russian cruisers, which were also more heavily armoured. The two Turkish cruisers *Hamidiye* and *Mecidiye* were inferior in every respect. The two Turkish battleships could not strengthen the *Goeben* when fighting together with her against the Russians. Their armament, extreme range and speed were all so far

The battleship *Ioann Zlatoust* before the war. The three port-side 3in guns are still evident, immediately above the 6in casemates. Sister-ship *Sviatoi Evstafi* was identical. (Wikimedia Commons)

The battleship *Tri Sviatitelia* in Sevastopol on 20 July 1914. (Wikimedia Commons)

inferior to the Russians that the Fleet Commander abandoned any thought of active employment of these vessels against the Russian Fleet and assigned to them the role of protecting the Straits ... The Russian torpedo-boats were superior to the Turkish in every respect ...[17]

The Ottoman Navy also had three torpedo gunboats (a slow type of early destroyer), a dozen small destroyers and torpedo boats, and about a dozen gunboats, all smaller and weaker than their Russian counterparts. Only the four largest destroyers (purchased new from Germany in 1910) and the two cruisers had the speed and range to be useful much beyond the Bosphorus and Dardanelles. In mid-1914 those destroyers were among the many Ottoman warships in the process of being refitted by British engineers, who noted that they were:

All in bad condition. Two of these vessels are under extensive repair including new coal bunkers, part renewal of decks, new boiler tubes &c. These repairs will take at least one month if not more. The turbines of these boats were removed some time ago by the Turks for overhaul, but reports say they have lost considerably in speed, and since the Germans have been in charge they have been overhauling the turbines.[18]

The Germans organised further work on these four ships from October 1914 to January to February 1915.[19] One, *Numune*, had a small part to play in the events of 10 May 1915.

The Bosphorus Defences, 1914–15

The Ottomans' capital – then generally known in the West as Constantinople – was by a considerable margin their most significant centre of government, industry, and population, and their most important transport hub. It is therefore no surprise that the only two sites in the empire with massive artillery (and in wartime, minefield) coastal

defences were the two straits leading to Constantinople: the Dardanelles to the south and the Bosphorus to the north. In 1873–75 and 1885–86, the Ottomans purchased at least 134 large Krupp guns, ranging in calibre from 21cm to 35.5cm (8.2in–14in). The first batch were 22 calibres long; the later guns were 35-calibre.[20] Most were sited in forts and batteries defending the Dardanelles or Bosphorus. Immediately before, or during the First World War a handful of modern German ships' guns were landed to reinforce the defences, mainly 10.5 or 15cm.[21] In 1914–15 the Bosphorus defences amounted to 41–46 guns of 21cm or larger, 78 smaller guns and five mortars.[22]

The work of actually deploying these guns in forts and batteries at the two straits seems to have taken many years, but was complete by 1914. British observers in the 1890s and 1900s noted that most of the Bosphorus forts and batteries were vulnerable to attack by land from the rear, but this was a hazard only if an enemy could land enough troops to make such attacks practical. Some additional mobile artillery pieces, initially sited at the Bosphorus, were sent to the Dardanelles in September 1914: the 8th Heavy Artillery Regiment with 22 (later 32) 15cm (5.9in) howitzers, and eventually also fourteen 12cm (4.7in) howitzers.[23]

By comparison with the main armament of major

Rear Admiral Richard Fortescue Phillimore (the Royal Navy's representative at the Tsar's HQ in 1915–16), Engineer Lt Cdr George Wilfred Le Page, RN, and Phillimore's Flag Lieutenant John Randal Parsons, RNVR, on 27 October 1915, aboard *Panteleimon*. (Admiral Phillimore's album, IWM)

warships from the mid-1890s onwards – guns of 40 (and, by 1914, 50 or more) calibres – those defending the Bosphorus and Dardanelles were almost all obsolescent, lacking the range and penetrating power of later weapons. However, as the Dardanelles' defences proved in 1915, at the short ranges involved once attackers had to enter the straits, such old guns were more than capable of dealing severe blows.[24] And as the vast majority of the Bosphorus coastal defence batteries, like those at Gallipoli, were situated along the sides of the channel rather than at its mouth, many were vulnerable *only* to attackers who entered the straits, and any ships that did so laid themselves open not only to short-range gunfire but also to the danger of minefields which the batteries defended.[25] The Russians contemplated building some monitors to attack the Ottoman defences, and did some preparatory work towards converting their smallest predreadnought battleship and the guardship *Sinop* into monitors,[26] but the greater priority of other shipbuilding projects and other naval operations meant that these plans made only limited progress.

British and French capital ships could risk close engagement with the Dardanelles defences as both navies had plenty of predreadnoughts that were surplus to their other requirements, and thus expendable. In contrast, the Black Sea Fleet could not risk its four 12in-gunned predreadnoughts, certainly not until its first dreadnought came into service later in 1915. So while Russian ships could bombard the outer Bosphorus defences, the Russians were not in a position to emulate the intensity and risk of the Anglo-French attack at the Dardanelles.

Black Sea Naval Operations 28 March–3 May 1915

From late March to early May 1915 was a busy period in the Black Sea, partly prompted by the start of the Gallipoli campaign on land. *Goeben*, having struck two mines off the Bosphorus on 26 December 1914, was slowly repaired with the use of specially-made coffer-dams. (Her port side hole took from 11 February to 28 March 1915 to repair; that on her starboard side from 5 April to 1 May.[27]) *Breslau* completed an overhaul on 2 March 1915, and was thus also ready for renewed operations.[28] Meanwhile, after naval bombardments had failed to suppress the Dardanelles defences, the first Allied troop landings on the Gallipoli peninsula took place on 25 April.

Apart from continued patrols to disrupt the coal trade along the north Anatolian coast, during this period the Black Sea Fleet undertook a series of bombardments of the batteries and lighthouses at the mouth of the Bosphorus and, as *Goeben* became fit for renewed service, Turco-German forces likewise mounted a number of sorties. The Russians also laid minefields outside the straits: more than a thousand mines had been laid in that area by April 1915. And as they came into service, the submarines *Tyulen*, *Nerpa* and *Morzh* also sortied to the straits, though their endurance was insuffi-

The submarine *Morzh* in dock in Sevastopol; spaces for 'drop-collars' along the hull are very evident. (I V Alexeev, A S Goncharov, and V P Zablotskiy, *Submarines of the 'Morzh' type*, via Wikimedia Commons)

cient to permit a continuous presence.[29] We can follow the next few weeks' events though the detailed reports of the RN officer attached to the Black Sea Fleet, Engineer Lt-Cdr GW Le Page, and the 1928 German and 1964 Soviet official histories, supplemented by modern research.

The Russian fleet of five battleships, three cruisers (with a seaplane on *Almaz*), the seaplane carrier *Imperator Nikolai I* (with five aircraft), three large and seven small destroyers, and five minesweepers, arrived at the Bosphorus on 28 March. They saw four Ottoman destroyers patrolling the mouth of the straits. Four Russian seaplanes were sent to reconnoitre the Bosphorus and found no large Turco-German warships as far south as the Golden Horn (an inlet at Constantinople, near the southern end of the straits). Two of the seaplanes then assisted the bombarding ships by bombing batteries at the mouth of the straits, while the other two observed the entrance to the Bosphorus. Between 10.30 and 11.30, *Tri Svititelia* and *Rostislav* under Prince Putiatin bombarded the battery at Riva

The cruiser/seaplane carrier *Almaz*, seen here on 13 March 1916, looking forward, with a Grigorovich M-5 seaplane carried on a platform abaft her funnels. (Boris Drashpil collection, US Naval History & Heritage Command, NH 94394)

Russian battleships bombarding the Bosphorus defences on 28 March 1915: *Rostislav* is nearest the camera; beyond her is *Tri Sviatitelia*. The photo was taken from the 'Zh'-class destroyer *Leitenant Pushchin*. (Wikimedia Commons)

from about 14,000 yards, and later the Panas battery on the European side, using altogether thirty-four 12in, twenty-one 10in and fifty 6in rounds, while the main fleet covered the operation from some 16 miles offshore and the cruisers *Pamiat Merkuria* (with *Nikolai I*) and *Kagul* (with the large destroyer *Derzkiy*) kept watch on either flank of the bombarding ships. The Riva battery briefly opened fire at 10.35 but the Russian ships were out of its range. The Russians sank a large steamer trying to enter the Bosphorus, and two Ottoman destroyers, briefly emerging from the straits, were engaged by *Rostislav* and retired.[30]

The seaplanes reported an explosion in Fort Panas on the European side of the straits, that a possible barracks behind the fort had been set on fire, and that one seaplane's small bomb had scored a near-miss on an Ottoman destroyer. Lewis Einstein, at the time a senior diplomat at the US Embassy in Constantinople, recorded:

> The Russians bombarded the Bosphorus forts this morning. Shells fell as far as Beicos and Buyukdere, and an Italian steamer lying off the quay had a narrow escape. The bombardment was distinctly heard here and the windows rattled in the houses near the German Embassy.[31]

The following day the Russians deployed in the same way, apparently with a view to engaging the batteries at the mouth of the straits with 6in guns, while using their heavier guns against batteries within the straits themselves. The bombardment began at 09.15, but soon had to be abandoned due to growing heavy mist – though a sailing ship that loomed out of the mist was engaged and sunk by 09.45. Weather conditions also meant that only one seaplane was able to take off for reconnaissance. At 10.45 the seaplane reported that *Goeben*, *Breslau* and some Ottoman warships were on the move, and the Russian fleet remained about 30,000 yards off the mouth of the Bosphorus until 13.30, and farther offshore until evening, presumably in case the Turco-German forces

ventured out; however, apart from one Ottoman destroyer emerging at 13.15 and 14.15, no enemy warships were sighted. The fleet therefore proceeded eastwards to the 'coal coast', six small destroyers being sent in to attack Eregli harbour – they reported that they sank eleven sailing barges and two schooners, and destroyed three cranes and a jetty. *Kagul* and *Pamiat Merkuria* bombarded Zonguldak between 11.00 and 12.20, where the coastal guns (noted by a seaplane) did not, however, return fire, and where a combination of Russian 6in shells and small bombs from seaplanes was reported to have damaged various coal-mining facilities and a local electricity station.[32]

The German official history claims no serious damage was caused at Eregli or Zonguldak. It also states that the bombardment of 28 April targeted lighthouses at the mouth of the straits, not batteries, and identifies the target of the Russian seaplane's near-miss as the small destroyer *Samsun*.[33] And we now know that the various bombardments of the Bosphorus defences, including those in spring 1915, caused only a few Ottoman casualties.[34]

Before the Russians could return to the Bosphorus, and as the partially-repaired *Goeben* could now make 20 knots, on 1 April Ottoman and German ships sortied towards Odessa to attack Russian troop transports which they thought were there. A force comprising the cruisers *Hamidiye* and *Mecidiye* and four destroyers, under the command of *Mecidiye*'s senior German officer Korvettenkapitän Ernst Büchsel, was covered from a considerable distance by *Goeben* and *Breslau*, which themselves drew near to Sevastopol. But as the Ottoman ships approached Odessa on the 3rd, *Mecidiye* struck a mine, sank in shallow water and had to be abandoned. (The Russians soon raised, repaired and rearmed her and took her into naval service as *Prut*.[35])

As the Turco-German ships then retreated, the Black Sea Fleet emerged from Sevastopol, and a Russian seaplane spotted *Goeben*. However, as *Goeben* and *Breslau* were making 20 knots and the Russian battleships were capable of no more than 15–16 knots, the Germans could not be brought within range. At one point *Goeben* fired an 11in salvo at the pursuing *Pamiat Merkuria*, 26,000 yards away. Given the maximum range of her guns, *Goeben*'s shells unsurprisingly fell at least 6,000 yards short. The gap between the Russian battleships and their quarry progressively widened. Despite this, the Russians continued to pursue, the German official history noting that 'as usual in the Black Sea, the visibility was extraordinarily great and the smoke of individual ships could be seen over thirty miles.'

Breslau, having been detached at 09.20 to identify a pursuing Russian cruiser (in fact *Kagul*), was briefly engaged by two Russian battleships at 16,500–18,500 yards, and some of the ten Russian broadsides fell so close as to leave splinters on *Breslau*'s deck. (The Russians later released their destroyers to try to close with the German ships, and by 20.41 three large First Division ships were within 4,500 yards of *Breslau*.

Gnevnyi fired some torpedoes but made no hits. The Germans mistakenly thought they made some hits on a Russian destroyer.[36]) In the process of evading the Russian fleet, the German ships encountered and sank two Russian freighters. The Russian submarine *Nerpa*, proceeding towards the Bosphorus, sighted *Goeben* and *Breslau* but did not get an opportunity to torpedo them.[37] *Goeben*'s presence on 3–4 April – and her evident speed – warned the Russians that she remained a potent threat.[38]

Lt-Cdr Le Page was concerned at the apparent lack of recent Russian battleship gunnery practice, at least amongst the most modern ships:

> The 1st Division [Brigade] battleships, with the exception of a few rounds at Junguldak on March 7th last, have carried out no firing since the action of November 8th, 1914. Two new 12-inch guns have been fitted to the ... [*Evstafi*], forward, and these have not been calibrated. A meeting of the gunnery officers of the fleet was held to consider as to whether practice with aiming tubes, half charges, etc should be carried out, it was finally decided that in view of the extensive firing practice carried out last summer, when excellent results were obtained, and also the same gun-layers, etc still being available, no further practice was necessary, but with the opportunities we have of carrying out aiming tube practice when cruising, with destroyers scouting well ahead, I should have thought that every opportunity would have been taken to still further increase the efficiency of the firing.[39]

Page's reference to the First Battleship Brigade indicates that it was the Second Brigade (*Tri Svitatelia* and *Rostislav*) which had fired on *Breslau* on 3 April.

The Black Sea Fleet sortied again 10–12 April towards Odessa, following news that some Turco-German ships were at sea, lest these intended to disrupt the salvage of *Mecidiye*. But nothing was seen of the enemy.[40]

On 14 April Le Page met Admiral Ebergard, and learned that the latter was awaiting news of events in the Dardanelles before attacking the Bosphorus forts again. In the meantime some of the large Russian destroyers made a sweep along the Anatolian coast, sinking a steamer, two schooners and three sailing barges at Eregli, two steamers and a coal lighter at Kozlu, and a steamer at Zonguldak, returning to Sevastopol on the 16th. These destroyers sortied again on the 17th for another sweep of the same coast, and on the 18th reported having sunk twelve sailing ships near Kerrasund. From 17–19 April the whole fleet cruised up to 50 miles from Sevastopol.[41]

On 21 April the Russian fleet – all five battleships and three cruisers, plus three minesweepers and *Imperator Nikolai I* with five seaplanes – sailed for the Bosphorus. On this occasion Le Page was on Prince Putiatin's flagship *Tri Sviatitelia*, rather than *Almaz* (where he was normally accommodated). Three large and seven small destroyers provided escort. When they arrived on the morning of the 24th, dense fog inhibited further operations – except that the submarine *Nerpa* reported sinking a sailing ship.[42]

There was still mist on 25 April, but a Russian

The seaplane carrier *Imperator Nikolai I* c1915–16. She was renamed *Aviator* in May 1917 and sold into French merchant service in 1921. (Boris Drashpil collection, US Naval History & Heritage Command, NH 94397)

The Ottoman merchant ship *Amalia* when captured by *Kagul* on 4 May 1915 off the Roumelian coast; see note 48. (Boris Drashpil collection, US Naval History & Heritage Command, NH 94798)

seaplane was able to reconnoitre the Bosphorus and found the old battleship *Turgut Reis*, cruiser *Hamidiye* and a supposed submarine at Buyukdere (some six nautical miles within the Bosphorus), and a torpedo boat and seaplane near the mouth of the straits. The minesweepers and six small destroyers led the way for *Tri Sviatitelia* and *Panteleimon* to approach, followed by the three large destroyers, while the submarines *Tyulen* and *Nerpa* remained near the mouth of the straits. The Russians sighted an Ottoman torpedo boat, but she retreated rapidly into the Bosphorus. At 07.55 *Tri Sviatitelia*, followed by *Panteleimon*, fired at the batteries on the Asiatic shore, at 14,800 yards. They gradually closed to 13,200 yards, *Tri Sviatitelia* shifting fire to batteries on the European shore while *Panteleimon* fired at the Anadolu Kavak and Roumeli Kavak batteries well within the straits. *Tri Sviatitelia*'s shells were seen falling near Roumeli lighthouse, but the mist obscured *Panteleimon*'s fall of shot. The only Ottoman reply was a couple of shells that fell 3,000 yards short of *Tri Sviatitelia*. The Russians saw what appeared to be an explosion in one Ottoman battery about 09.10. Russian fire ceased at 09.20: *Tri Sviatitelia* had fired seventy-five 12in rounds and seven 6in, while *Panteleimon* fired sixty-seven 12in. The fleet then returned to Sevastopol, arriving on the 26th.

On 30 April two new large destroyers, *Bystriy* and *Schastlivyi* 'hoisted the war flag' and were officially considered fit for service, although in acceptance trials, with boilers not at full capacity, they made 28 knots rather than the designed 33. At a meeting before the fleet sailed from Sevastopol again on the morning of 1 May,[43] Admiral Ebergard stated that now Allied troops had landed at the Dardanelles, the Russian fleet's duty was to continually worry the Turks from the Bosphorus side. On this occasion the fleet's five battleships and three cruisers were accompanied by *Imperator Nikolai I*, two large and six small destroyers, and four minesweepers. At 06.00 on the 2nd, as planned, the fleet arrived off the Bosphorus, while the cruisers *Kagul* and *Pamiat Merkuria* covered the Romanian and Bulgarian coast, and Zonguldak, respectively. Four seaplanes explored the Bosphorus, and

found no warships except a submarine at Beicos Bay (on the Asiatic side of the Bosphorus, a couple of miles south-east of Buyukdere). On this occasion *Tri Sviatitelia* and *Panteleimon* sailed twice across the mouth of the Bosphorus, some 7,000 yards from the coast and perhaps 15,000 from the major forts at Anadolu Kavak and Roumeli Kavak (either side of the strait, but well inside it). Le Page noted that they meant to strike 'Elmas, Filbournou and Anatol Kavak on the Asiatic side; Roumeli Kavak, Kdiljeh, Fanaraki and Kilia on the European side.' The Roumeli Kavak and Anadolu Kavak batteries replied with a few rounds, but their fire was very slow and their shells fell short. The battleships fired for 2¼ hours and rejoined the main fleet at 14.00, as did *Kagul* and *Pamiat Merkuria*.[44] *Panteleimon* fired altogether a hundred and ten 12in and three hundred 6in rounds, *Tri Sviatitelia* fifty-six and 195 respectively.[45] The destroyers coaled from the larger ships, while *Pamiat Merkuria* and *Kagul* proceeded along the coast to Zonguldak, sinking a small steamer near Kozlu and also a sailing barge.

The following day, fog precluded reconnaissance flights by Russian seaplanes. Preceded by minesweepers and the cruiser *Kagul*, *Tri Sviatitelia* and *Rostislav* approached the European shore and bombarded batteries at Kara Burnu and the northern end of the Chataldja Lines (fortifications protecting Constantinople and the Bosphorus from the west). *Rostislav* fired thirty-nine 10in rounds, *Tri Sviatitelia* 132 6in. These ships rejoined the main fleet at 18.00, at which point the smaller destroyers coaled from larger ships, and the two large destroyers returned to Sevastopol to refuel with oil. On 4 May *Rostislav*, *Pamiat Merkuria* and *Almaz*, preceded by minesweepers, sailed to Igneada (three miles from the Bulgarian border), where seaplanes identified four gun batteries; *Rostislav* and *Pamiat Merkuria* fired about sixty 6in shells. Le Page understood that this site had been considered a possible landing place for Russian troops, given its long beach. (*Kagul* also sank two sailing ships smuggling kerosene to Constantinople.) The Russians had received reports that the area was mined, but minesweeping up to 6,000 yards from the shore revealed no sign of this. On 5 May the large destroyers *Gnevnyi* and *Pronzitelnyi* sank two steamers at Eregli, shelled another and destroyed a sailing ship between Eregli and Kozlu. The whole fleet returned to Sevastopol on 6 May.[46]

8–9 May 1915

Much of the Black Sea Fleet put to sea again at 06.00 on 8 May: the three First Brigade battleships and the three cruisers, this time accompanied by seaplane carrier *Imperator Alexander I* and large destroyer *Derzkiy*; and instead of the minesweeping trawlers of the previous sortie, minelayer *Xenia* and minesweeper *Sviatoi Nikolai*. *Goeben*, *Breslau* and *Hamidiye* were reported only 20 miles south-west of Sevastopol at the time (having sortied from the Bosphorus on the 7th),[47] and *Pamiat Merkuria* reported *Goeben* and *Hamidiye* to be heading south-west

Two of the large Russian destroyers in Sevastopol during the war. The striking camouflage on one ship (apparently in various shades of blue) is known to have been used at least on *Schastlivyi* and *Gromkiy*. (Wikimedia Commons)

at 18 knots, with *Breslau* turning east; but the Russian main body (cruising at 12 knots) did not sight the enemy.

At 16.00 the Russian units rendezvoused at Khersonese Point with the Second Battleship Brigade of *Tri Sviatitelia* and *Rostislav* (the latter under Captain First Rank Kazimir A Porembsky), five large and six small destroyers and four minesweepers. At 18.00, four of the large destroyers returned to Sevastopol. On the 9th, with the main fleet 40–50 miles off the Ottoman 'coal coast', Russian destroyers sank two colliers at Eregli and forced a third to run ashore, her crew escaping in lifeboats. The large destroyer *Bespokoiniy* sent in a motor boat to blow up the grounded ship, but rifle fire from ashore obliged the Russian boat to retire.[48] (*Pamiat Merkuria* also reportedly destroyed two steamers and 27 sailing ships in the Eregli area.[49])

Having dispersed to sweep the coast, the Russian cruisers and destroyers rejoined the main fleet at 15.00, and an hour later the small destroyers coaled from larger ships. The whole fleet headed towards the Bosphorus at 18.00.

The Russian 'Landing'

Bespokoiniy's motor boat at Eregli was instrumental in drawing *Goeben* out to sea and thus in occasioning the action of 10 May, because initial reporting overstated its significance. At 10.00 on the 9th, the authorities at Eregli informed Constantinople that there had been a Russian landing, which their troops had repulsed. Ottoman

General Headquarters understandably feared a large-scale Russian landing to destroy the coal mines and related facilities. Thus at 10.25 *Goeben* was ordered:

> Proceed at once to Bender Eregli where Russian cruiser is attempting a landing. Report departure. The Fleet Commander will not accompany the *Goeben*.[50]

(Fleet commander Souchon was absent at a conference with Ottoman War Minister Enver Pasha and the most senior German military adviser, Field Marshal Colmar von der Goltz.) While *Goeben* raised steam her captain, Richard Ackermann, received further information in a radio message at 10.50:

> At 10:00am the *Kagul* bombarded Eregli continuing until 10:30am. The landing force is withdrawing. Other warships are coming in sight headed for Eregli.

Goeben put to sea at 13.00 and steered north at 23 knots, with a view to arriving at Eregli early the next morning. At 14.25 'fleet HQ' radioed Captain Ackerman that:

> The bombardment of Eregli ended at 10:40am. The cruiser and torpedo-boats are joining forces with the battleship now in sight *Joan Slatust* [sic]. Course north, *Goeben* will attempt to damage enemy.[51]

At 14.38 *Goeben* sighted Russian submarine *Nerpa*,

which then dived.[52] And on the night of 9–10 May the Russian fleet probably came close to *Goeben*, as both were near Eregli.

While they could not ignore the news from Eregli, the Germans may not have been certain of its accuracy. Their official history noted of such alarming reports:

> … the precision and accuracy of the reports received were detrimentally influenced by the vivid imagination of these eastern children of nature. … [e.g.] The Commandant of Police at Makrikoi near Constantinople reported … on 2 September: 'a submarine ran ashore between Kutschuf Tschekmedsche and the match factory on the strand. The officers and men went ashore. But they were surrounded by our outposts …' This cheerful news was soon corrected … : 'A submarine passed Tschekmedscke this morning early. Trading steamer No.20 on seeing the submarine approach, ran aground on the strand. The crew of six men then went ashore and the submarine disappeared. General HQ was notified that a submarine had stranded but in reality it was the small steamer and not the submarine which grounded.[53]

A flavour of the Ottoman army's understandable concern is also apparent in some modern accounts. Gary Staff claims 'some Russian troops were put ashore at Eregli to destroy the power station … and a fight developed with Turkish troops ashore', while George Nekrasov reports that the power station was destroyed.[54] (This mention of a power station perhaps conflates the events of 9 May with those of 29 March described above, where a power station at Eregli was bombarded.) More cautiously, Vincent O'Hara and Leonard Heinz note that 'Russian raiders landed at Eregli'.[55] Even if a few of *Bespokoiniy*'s men briefly set foot ashore, their actual target was the grounded Ottoman steamer, and there does not seem to have been any attempt at a landing of the sort the Ottomans feared.

10 May 1915[56]

At 04.00 the Ottoman destroyer *Numune* under Leutnant Otto Sommer[57] stood out to reinforce a gunboat patrol off the Bosphorus, and at 05.05 she sighted heavy smoke to the north and steamed towards it at 15 knots. At 05.40 she sighted the Russian fleet and reported by radio: 'Seven Russian battleships square 228 on course SE', following this at 06.35 with: 'Three enemy ships one cruiser in square 229. Twelve torpedo-boats and minesweepers are approaching the Bosphorus.'[58]

And as the Black Sea Fleet drew closer to the entrance to the Bosphorus, at 05.40 Admiral Ebergard detached *Tri Sviatitelia* (Rear Admiral Prince Putiatin; Captain First Rank V K Lukin) and *Panteleimon* (Captain First Rank M I Kaskov) to shell the fortifications, accompanied by seaplane carrier *Imperator Alexander* and cruiser/carrier *Almaz*, and preceded by minesweepers and some destroyers. The rest of the fleet covered the operation from 20–25 miles off shore and, as on previous occasions, *Kagul* and *Pamiat Merkuria* scouted to east and west of the fleet.

The weather was calm and clear, with a light mist over the coast; *Alexander* launched a seaplane to reconnoitre, which the Ottomans spotted at 06.50. *Numune* fired at the Russian minesweepers at 07.10, at an estimated range of about 7,700 yards, but withdrew by 07.20 under the fire of *Panteleimon*. *Panteleimon* also fired seven 12in shells at a large ship seen in the straits. Meanwhile, the cruiser *Pamiat Merkuria* sank an Ottoman coal schooner, then spotted smoke to the east and identified *Goeben* approaching: Rear Admiral Pokrovsky reported this to Ebergard and his ship made speed to rejoin the main fleet. For her part, *Goeben* sighted what turned out to be *Pamiat Merkuria*'s smoke at 06.30, and by 07.15 had closed sufficiently to see the Russian battleships, in separate groups of three and two ships.

Numune's sighting of the Russian fleet prompted Vice

The Bosphorus seen from the Black Sea; perhaps *c*1920. In good visibility it was clearly possible to see well into the straits, but individual gun batteries would be hard to spot; and some were concealed by promontories. (US Naval History & Heritage Command, NH 122308)

Admiral Souchon to order *Breslau* to raise steam to be ready to support *Goeben*, while available torpedo craft were ordered to sea to guard the Bosphorus against hostile submarines. But only the torpedo gunboat *Peik-i Shevket* (under the flotilla commander, *Fregattenkapitän* Adolf Pfeiffer) and the destroyer *Yadigar* (*Oberleutnant zur See* Otto von Schrader) were immediately available; other torpedo craft were convoying transports to the Dardanelles or conducting anti-submarine patrols in the Sea of Marmara.[59]

On learning that *Goeben* had been sighted, Admiral Ebergard moved the main fleet towards Putiatin's detachment, and at 07.05 ordered *Tri Svititelia* and *Panteleimon* to re-join the fleet. It took 18 minutes for Putiatin's minesweepers to re-stow their sweeps and for his detachment to reassemble and retrace their course via their swept channel. And while the German official history implies that the Russians had reunited by the time the two sides' capital ships opened fire, in fact Putiatin's battleships were still about two miles behind the Russian main body at that point. Captain Ackermann had doubtless hoped to exploit this division and engage just part of the Russian fleet. However, while not all the Russian battleships were immediately able to fire, *Goeben* would have only a few minutes before Putiatin's ships rejoined their consorts. The main responsibility now devolved upon the senior gunnery officers – Lieutenants A M Nevinski in *Evstafi*, V M Smirnov in *Ioann Zlatoust*, V G Malchikovski in *Panteleimon*, and *Korvettenkapitän* Arnold Knispel in *Goeben* – and their subordinates. The Russian acting Fleet gunnery officer, Senior Lieutenant M Iu Giubner, was on board *Pamiat Merkuria* (which took position 2,000 yards ahead of *Evstafi* in the coming battle), and observed and took careful notes, as did the gunnery officers of the other unengaged Russian ships.[60]

Evstafi and *Ioann Zlatoust* opened fire on *Goeben* at 07.53 at 18,800 yards as she drew closer, the two Russian ships co-ordinating fire according to the range measured by *Ioann Zlatoust*. When the Germans returned fire, aiming at *Evstafi*, they recorded the range as 17,500 yards and the Russians as 17,400. At this point *Goeben* was turning onto a course almost parallel with the Russian battleships, but gradually closing, and she was initially slightly abaft the Russians' beam but soon drew level with, and then overtook them, given her higher speed. George Nekrasov understood – presumably from an *émigré* source – that *Evstafi* and her immediate consorts briefly slowed to 5 knots to allow Prince Putiatin's detachment to catch up, but *Evstafi*'s captain reported their speed was in fact 10 knots; the Russians' speed increased to about 15.5 knots once the battleships had reunited. *Panteleimon* and *Tri Sviatitelia* made their very best speed to rejoin the C-in-C, and Nekrasov (again presumably drawing on an *émigré* account) reports that at this point *Panteleimon* reached 17.5 knots, over half a knot faster than she had managed even on trials. Before long, this allowed her to overtake both *Tri Sviatitelia* and *Rostislav* to take her normal place in the Russian battle line behind *Ioann Zlatoust*. (*Rostislav* did not fire on

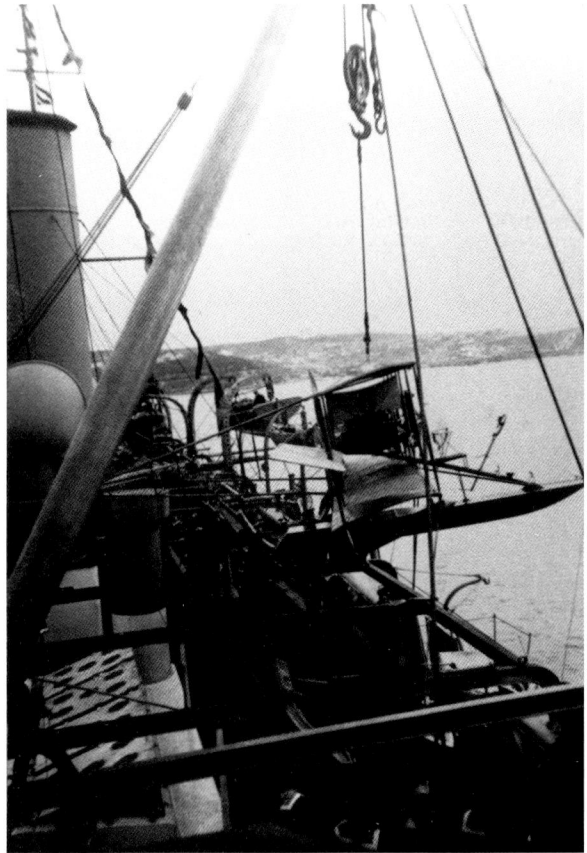

On some sorties, the cruiser *Kagul* carried an aircraft. Images survive from February 1915 (NH 100152, shown here, with Curtiss floatplane no 6) and 28 March (NH 100153, with Curtiss F-type flying boat no 15). Problems with the Curtiss aircraft meant that the Black Sea Fleet more often used Grigorovich seaplanes. (Boris Drashpil collection, US Naval History & Heritage Command)

Goeben, presumably because the action began beyond the range of her 10in guns, and to avoid her fall of shot confusing others' gunnery.)

From their third or fourth salvo, *Evstafi* and *Ioann Zlatoust* shifted to rapid fire from their 12in guns. The first Russian salvoes fell short of their target. *Goeben*'s initial shots were closely-grouped but also short, and then a few fell close enough that some splinters fell on *Evstafi*'s deck; one near-miss caused a small leak which was however quickly repaired. Admiral Ebergard ordered his flagship to zig-zag, and she varied her speed between 10 and 12 knots. *Panteleimon* and *Tri Sviatitelia* joined in the firing by about 08.08 (Le Page, on *Tri Sviatitelia*, gives 08.03 and 08.07 respectively), when the former was 20,800 yards from *Goeben*. By the time *Panteleimon* rejoined the other Russian battleships the range had fallen to 16,000 yards, and as the action continued, the two sides' slightly converging courses meant that the range shortened still further, until it was about 14,600 yards.[61] Thereafter *Goeben* turned away to starboard and quickly opened the range.

Goeben sought to draw the Russians away from the

POSITIONS AT 7AM ON 10 MAY 1915

42°N

Note: The maps are based on Le Page (ADM 137/754).

Black Sea

N

Imperator Alexandr I

Kagul

1st Div

Alma

2nd Div

GOEBEN

TBDs

Pamiat Merkuria

10 5 0 5 10
Scale in nm

Bosphorus

41°N

28°E

Sea of Marmara

30°E

© John Jordan 2023

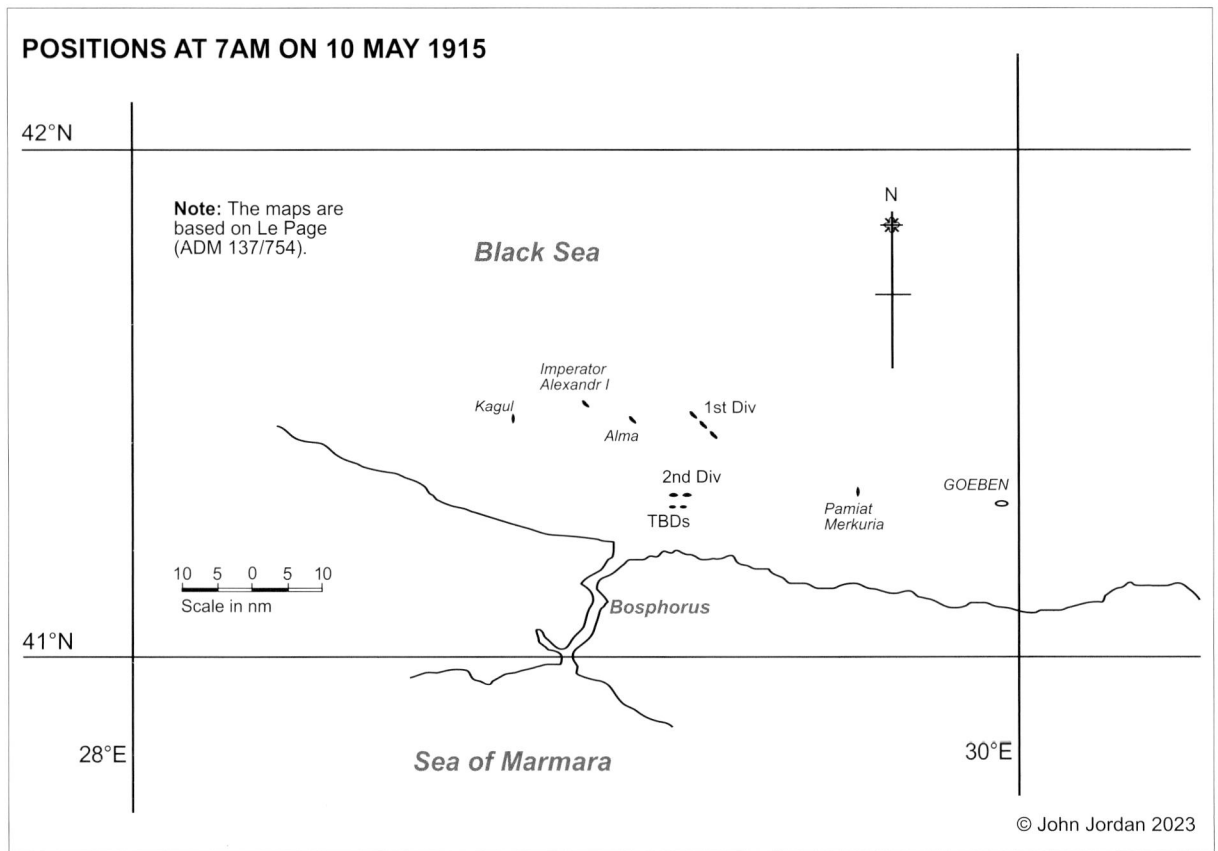

Bosphorus and headed northwards, and at 08.18 avoided what her crew thought were two torpedo tracks. She initially limited her speed to keep the Russian fleet in sight, and the Russians followed her until 14.10. At about that time *Goeben* turned south towards the Bosphorus at her best speed, and lost sight of the Russians by 14.50; at 16.20 she joined forces with *Breslau* and stood into the Bosphorus under cover of Ottoman destroyers. For their part, the Russians lost sight of *Goeben* at 15.00; the submarine *Tyulen* reported *Goeben*'s arrival in the Bosphorus, and the fleet returned to Sevastopol, arriving on 11 May.

German sources acknowledge that *Goeben* sustained two major hits. One high-explosive shell struck the forecastle and penetrated to the main deck. The second was underwater, striking (or near-missing) the lower edge of the armour belt and exploding next to frames 97–103 on the port side, thus damaging the outer skin below the armoured belt for about 5–6 metres, and damaging the sighting telescope and elevating mechanism of No 2 15cm gun to port, putting it temporarily out of action. Torpedo-net spars were also damaged. There were no casualties. The Russians thought they had made two to four hits in all, at least one being by *Panteleimon* at 20,000 yards. Russians and Germans judged each other's salvoes to be well grouped; the German official history acknowledges that they scored no hits. The Russians noted that while some of *Goeben*'s early salvoes included near misses, later salvoes were 3,000 yards short.

As the range opened, *Goeben* ceased fire at 08.12 and the Russians at 08.16, by which time the range was about 22,000 yards from *Panteleimon* and *Ioann Zlatoust*. In all the Russian ships fired 169 12in (*Evstafi* 58–60, *Ioann Zlatoust* 75, *Panteleimon* 23, *Tri Sviatitelia* 13) and thirty-six 8in rounds (*Evstafi* 32, *Ioann Zlatoust* 4). A fault in the obturator cushion of the right-hand 12in gun in *Evstafi*'s forward turret meant that turret fired 25 shots, while the aft turret fired 35. *Goeben* fired 124 or 128 11in rounds – though Russian sources claim it was 160.

The entire engagement had lasted just 22–23 minutes – a short action compared to the January 1915 battle of Dogger Bank, which took more than four hours, but almost twice as long as the November 1914 fight off Cape Sarych. The action off Cape Sarych took place in poor visibility and at short range (6,800–7,700 yards), but that of 10 May was in excellent visibility at up to 22,000 yards. The greater odds of hitting provided by the flatter trajectory of *Goeben*'s main armament was *much* less significant at extreme range,[62] while the Russians' 12in guns had significantly longer range, somewhat longer-base rangefinders, and their gunnery personnel were arguably more familiar with very long-range firing. *Goeben* began by firing five-gun salvoes from her main guns, but by the end of the action the Russians noted that the salvoes were smaller, and it was their impression that *Goeben*'s foremost turret had ceased firing. They seem to have assumed this diminution reflected battle damage,

SHIPS' MOVEMENTS BETWEEN 7AM AND 8.25AM

N

08.25

08.15

08.25

08.10

07.55

Imperator
Alexandr I

Almaz

08.10

07.55

Kagul

Sviatoi Evstafi

Ioann Zlatoust

Rostislav

07.55

07.40

Pamiat
Merkuria

Tri Sviatitelia

Panteleimon

GOEBEN

Mine
Sweepers

Derzhkiy

Bespokoiniy

07.00

07.00

© John Jordan 2023

POSITION OF SHIPS AT 8.15AM

Imperator Alexandr I

Sviatoi Evstafi

Ioann Zlatoust

Almaz

Rostislav

GOEBEN

Tri Sviatitelia

Panteleimon

Kagul

N

N

**POSITION OF SHIPS
SOME TIME LATER**

N

Derzhkiy

Bespokoiniy

Sviatoi Evstafi

Ioann Zlatoust

Rostislav

Tri Sviatitelia

Panteleimon

Imperator Aleksandr I

Almaz

Small
TBDs

4 Mine
Sweepers

Pamiat Merkuria

Kagul

© John Jordan 2023

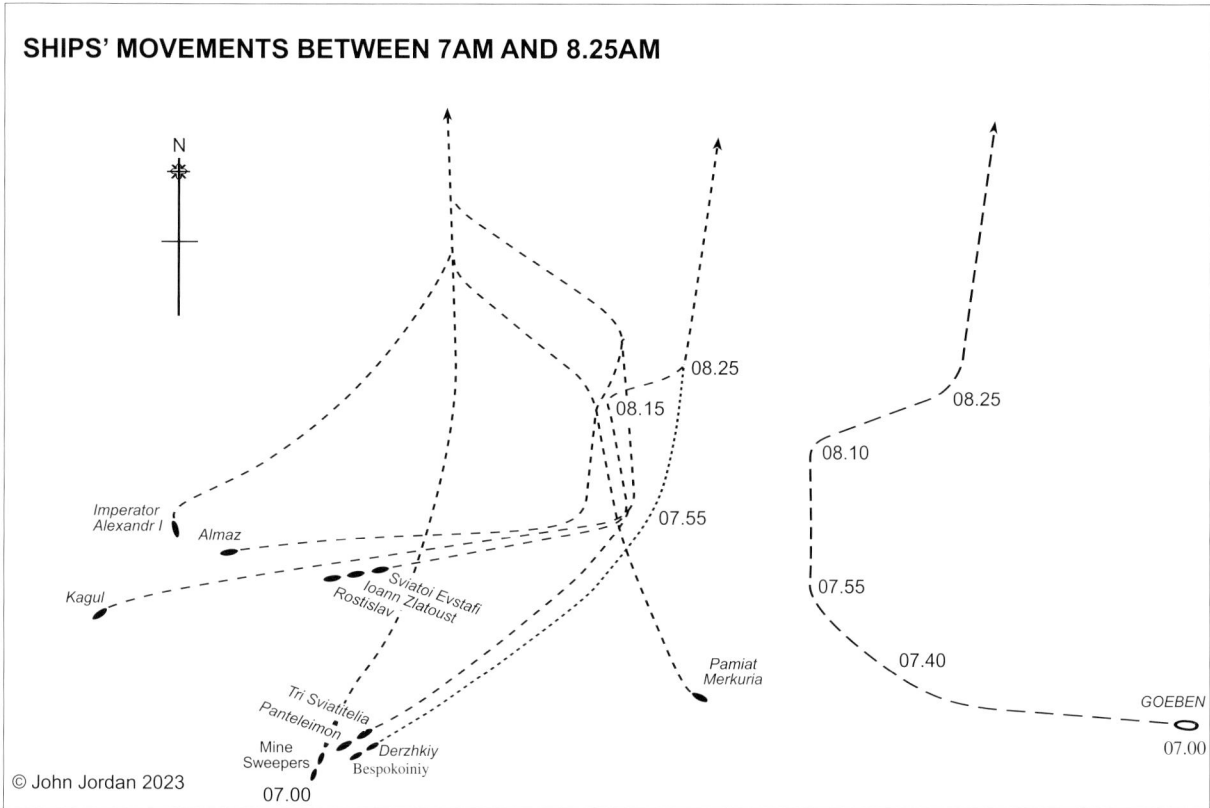

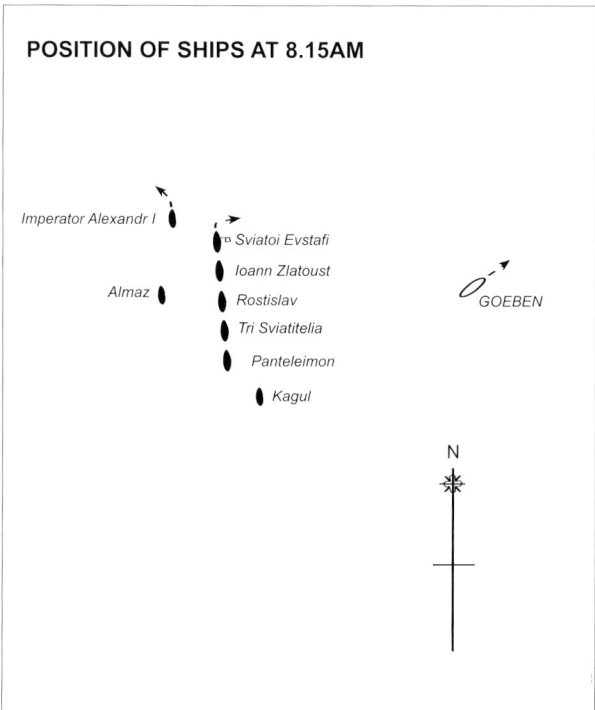

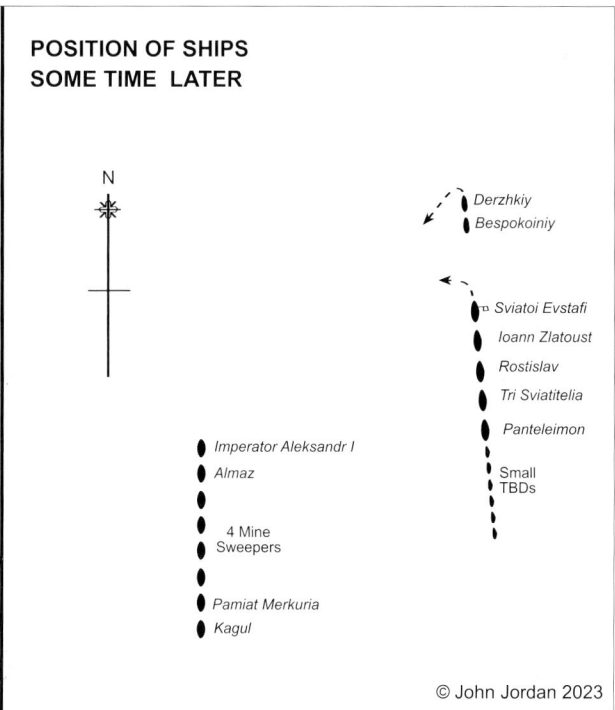

but it may simply be that, as *Goeben* moved ahead of the Russian battle line and then turned away, the angle to fire at *Evstafi* limited how many guns would bear.[63]

On the Russian side there were different opinions as to the gunnery on 10 May. On the one hand, observers thought they had seen two to four hits on *Goeben*, while the Germans made none. Le Page understood Admiral Ebergard was disappointed by the Russian gunnery, though the two officers' impressions likely reflected the greater visibility on 10 May. (It was poor visibility – and

The battleship *Panteleimon*, which secured at least one hit on *Goeben*. The photo was taken at Sevastopol in 1911–14. (Sergei Trubitsyn collection)

later reports from informers in Constantinople, overstating German casualties – that meant that the number of hits at Cape Sarych was much overestimated.) *Panteleimon*'s executive officer, G K Leman, criticised the centralised shooting by *Evstafi* and *Ioann Zlatoust* as too inflexible compared to the 'decentralised' method employed by *Panteleimon*, under which each ship simply observed a pre-agreed *sequence* of firing but did not use shared range data. (At a different range from *Goeben* when firing began, *Panteleimon* was not in a position to use data from *Ioann Zlatoust* anyway.) The German official history noted that '*Goeben* was not equipped with the fire-control installation later available on the other German battlecruisers'. Indeed, it was only over the

The forward hit on *Goeben*, 10 May 1915. (Bundesarchiv, 134-B2083)

period July–November 1916 that 'direction indicator' equipment was installed on *Goeben*, while work on increasing the maximum elevation of her 11in guns from 13½ to 22½ degrees did not begin until June 1917.[64]

In any event, the Russian battleships clearly out-fought *Goeben*, and this was reflected in a number of awards. These included: Captain Fedorovich of *Evstafi*, awarded the St George Order; Lieutenant Malchikovski, *Panteleimon*'s gunnery officer, awarded the Order of St Anne 3rd Class, Order of St Stanislas 2nd Class, and promoted to Senior Lieutenant; and Andrei Zhukov, *Panteleimon*'s long-serving senior seaman gunner, awarded the Order of St George 4th Class. *Panteleimon*'s captain, executive officer, and a number of other officers and men were also among those decorated after the battle (including Captain Second Rank E N Kvashnin-Samarin, head of the Naval General Staff's Historical Section, who participated in several of her wartime sorties), and *Panteleimon*'s good shooting was mentioned in an order by Admiral Ebergard.

Aftermath

As with many naval battles, the first news to reach the public was garbled and inaccurate. US diplomat Lewis Einstein, in Constantinople, noted that:

> In the afternoon came a report that the *Goeben* had sunk the Russian flagship. The Grand Vizier announced it officially to his callers, and said the news had been confirmed, but at the Club [Germany's Ambassador, Freiherr von] Wangenheim knew nothing, and it turns out to be the usual canard.[65]

and that *Breslau* left Constantinople early on 10 May – presumably heading north to support *Goeben* – with her funnels painted white.[66] A couple of days later, Einstein added:

> *Goeben* is back at Stenia, injured. She is said to have been struck four times, and one of her big guns put out of action.[67]

The hit that penetrated *Goeben*'s deck took nine days to repair. The waterline hit could be repaired only slowly, as it was hard to access the damaged area.[68]

One undeniable sequel to the 10 May engagement was that *Goeben* never again sought or accepted engagement with the Russian predreadnoughts – let alone with the Black Sea dreadnoughts when they came into service later in, and after 1915. The mine damage of December 1914 (and January 1918; fully repaired only 1927–30) and inferior local coal meant she could probably not exceed 24–25 knots for the rest of the war, and on 8 January 1916 had difficulty escaping from the dreadnought *Imperatritsa Ekaterina Velikaya*, whose trials speed was 21.55 knots.

Acknowledgements:

This article could not have been written without the research

and publications of Stephen McLaughlin – and in particular his invaluable translations of Russian sources. His researches and those of Prof Marek Herma also provided useful biographical information about Russian naval officers.

Principal Sources:

National Archives, Kew: ADM 137/754 ('Russian Black Sea Fleet: reports of Eng Lt Cdr G W Le Page RN, February 1915–September 1916'); ADM 137/881 ('Dardanelles, reports of proceedings and other papers, September–December 1914'); ADM 137/1146 ('Dardanelles: operations: December 1915, January 1916'); ADM 137/1389 ('Russia: Phillimore reports: October 1915–November 1916'); ADM 137/4068 ('Naval operations against Turkey: papers received in NID 1914–16'); ADM 231 series (naval intelligence reports).

Russian State Naval Archive (*Rossiiskii gosudarstvennyi arkhiv Voenno-Morskogo Flota*): file Ф.696 Оп.1 д.87.

Einstein, Lewis, *Inside Constantinople during the Dardanelles expedition, April–September 1915*, E P Dutton (1918).

Ewin, Toby, 'British account of the action off Cape Sarych, 1914', *The Mariner's Mirror* 102:2, May 2016.

———, 'Bosphorus Defences – and a Russian Minefield – in 1915', *Magna* 30:2 (magazine of the Friends of the National Archives), November 2019.

———, 'Naval Interrogations of PoWs in the Black Sea War, 1914 and 1916', *The Mariner's Mirror* 108:3, August 2022.

Grechanyuk N M, Lyakhovich A A, Shlomin V S, 'Russian Naval Operations in the Black Sea 1914–1917', in Rear Admiral Prof Nikolai Bronislavovich Pavlovich (ed), *The Fleet in the First World War, Volume I: Operations of the Russian Fleet*, Amerind Publishing (New Delhi, 1979).

Greger, René (trans Jill Gearing), *The Russian Fleet 1914–1917*, Ian Allan (London, 1972).

Lorey, Hermann, *Der Krieg in den türkischen Gewässern. Erster Band: Die Mittelmeer-Division*, Mittler & Sohn (1928); translated by Cdr HS Babbitt of the US Naval War College as *The War at Sea 1914–1918, published from the Naval Archives: The war in Turkish waters, prepared by Herman Lorey, Rear-Admiral retired. 1st volume: the Mediterranean Division*.

McLaughlin, Stephen, 'Predreadnoughts versus a Dreadnought: the Action off Cape Sarych, 18 November 1914', *Warship 2001–2002*.

———, *Russian and Soviet Battleships*, US Naval Institute Press (Annapolis, 2003).

———, 'Coast Defence and Coast Offence: Russian Monitor designs of the First World War Era', *Warship 2018*.

Nekrasov, George, *North of Gallipoli: the Black Sea Fleet at War 1914–1917*, East European Monographs (1992).

O'Hara, Vincent and Heinz, Leonard R, *Clash of Fleets: Naval Battles of the Great War, 1914–18*, US Naval Institute Press (Annapolis, 2017).

Phillimore, Adm Sir Richard, 'Some Russian experiences 1915–1916', *The Naval Review* 23:2 (May 1935) to 25:3 (August 1937); ten-part series drawing on his wartime correspondence.

Staff, Gary, *Battle on the Seven Seas: German Cruiser Battles 1914–1918*, Pen & Sword Maritime (Barnsley, 2011).

————, *German Battlecruisers of World War One: their design, construction and operations*, Seaforth Publishing (Barnsley, 2014).

Editor's Note: A full bibliography for this article is available from the Editor.

Endnotes:

1 Klaus Wolf, *Victory at Gallipoli*, 38.

2 Liman von Sanders, *Five years in Turkey*, 25.

3 Lorey, 84–85; the emphasis is the author's. Quotations and page numbers are from Cdr Babbitt's translation.

4 The above paragraphs summarise data from McLaughlin, *Russian and Soviet battleships*, 76, 95, 119, 149, 287–289, 294–295 and his 'Predreadnoughts versus a dreadnought', 119–120 and 138.

5 Le Page's 19 November 1914 report notes that *Pamiat Merkuria* already had sixteen 6in guns, but *Kagul* still had twelve (ADM 137/881, 388–395). His 1/15 of 8 March 1915 says *Kagul* now had 16 guns (ADM 137/754, 21–24). Engineer Lt-Cdr George Wilfred Le Page was the Royal Navy's observer with the Black Sea Fleet; his career is summarised in Ewin, 'A British account of the action off Cape Sarych'.

6 Le Page's report of 18 February 1915 and 1/15 of 8 March 1915 (ADM 137/754, 12–19 and 21–24).

7 N N Alfonin, *Leitenant Shestakov*.

8 McLaughlin, 'Russia: *Rossiiskii imperatorskii flot*', 232; 'Predreadnoughts versus a Dreadnought', 121; Herma, '*Na służbie w marynarce wojennej Imperium Rosyjskiego Kontradmirał Kazimierz Kietliński*'.

9 The demonstration is described in Ewin, 'A British account of the action off Cape Sarych', 216–217.

10 Rear Admiral RF Phillimore, report 28 of May 1916, 'Transport and disembarkation of troops, Black Sea', ADM 137/1389, 271–281.

11 Details are in McLaughlin, 'Predreadnoughts versus a Dreadnought', 120–121; see also Norman Friedman, *Naval Firepower*, 272–273. Phillimore's observation of the system in action, in a shore bombardment, are in his report of 30 October 1915, ADM 137/1389 35–41.

12 See Peter Gatrell, 'After Tsushima', *Economic History Review* 43:2, 1990; Evgenii Podsoblyaev, 'Russian Naval General Staff', *Journal of Military History* 66:1, 2002; S McLaughlin, 'Russia: *Rossiiskii imperatorski flot*'; Herma, '*Zbrojenia morskie Rosji w okresie pocuszimskim*'. On aviation, see R D Layman , 'Euxine wings', *Before the Aircraft Carrier* (95–102) and *Naval Aviation in the First World War* (46–47, 57, 95, 107–08, 208); also Timothy Wilson, 'Broken wings', *Journal of Military History* 66:4, 2002.

13 This paragraph is mainly drawn from Le Page's report 2/15 of 22 March 1915 (ADM 137/754 26–31). He estimated that the two converted seaplane carriers *displaced* about 5,000 tons.

14 Ebergard's earlier commands included that of the battleship *Panteleimon* after her 1905 mutiny which, as Stephen McLaughlin has observed, suggests he was considered an especially capable officer.

15 Four large destroyers were in service by 1914. Five slightly larger ships were built in 1913–15; eight more (larger again) were begun, of which four were completed by or during 1917. In March 1915 *Bespokoiny* was having a turbine repaired, while *Gromkiy* and *Bystryi* were undergoing acceptance trials.

16 The submarines were *Tyulen*, *Nerpa* and *Morzh*, displacing *c*760 tons submerged, armed with four 17.7in torpedo tubes

and up to eight more torpedoes in external Drzewiecki 'drop-collars' (though the latter were unreliable and their use was later abandoned). *Krab*, also entering service in 1915, displaced *c*740 tons submerged, had two 17.7in torpedo tubes and could carry up to 60 mines.

17 Lorey, 84–85.

18 James Stewart, 'State and condition of Turkish warships, October 1914', in ADM 137/881, 259–270.

19 Güleryüz, *Torpedoboats and Destroyers*, 98–99.

20 Michael Forrest, *The Defence of the Dardanelles*, 12 and 228; Naval Intelligence Department report 838 of May 1908 (ADM 231/49).

21 Three 15cm guns came from an old Ottoman ironclad lost in the First Balkan War (Forrest, 24). Other medium-calibre guns came from *Breslau* and *Goeben*, including two 15cm/45 guns from the latter on 15 May 1915 (Staff, *German Battlecruisers*, 124).

22 Ewin, 'Bosphorus defences'. Two 35.5cm/35, two 28cm/24, sixteen to nineteen 24cm/35, ten to twelve 24cm/22, nine 21cm/22 guns, and two 24cm howitzers. Details are from Naval Intelligence reports in ADM 231/14, /19, /26, /34, /45 and /49 (1889, 1891, 1896, 1901, 1905 and 1908), Volume 2 of Lorey's German official history (covering the Dardanelles; published 1938), and a modern Turkish official history.

23 Erickson, *Gallipoli: Command Under Fire*, 50.

24 Ayhan Aktar, 'Who sank the battleship *Bouvet*', *War & Society* 36:3, 2017. Also Stephen S Roberts, *French Warships in the Age of Steam 1859–1914* (Seaforth Publishing, 2021), 198.

25 From a report in Admiral Limpus' papers we know that, about May 1914, the Ottomans had some 700 sea mines, some of which were being refurbished. There were also a number of ex-Army 'observation mines' and some 70 captured Bulgarian mines. Friedman notes there may have been as many as 378 observation mines, and that the Ottomans acquired some of the hundred M1906 sea mines Russia provided to Bulgaria in 1912. By March 1915 some 576 mines were laid in the Dardanelles, and 93–112 in the Bosphorus. This left few for deployment elsewhere compared to those available to the Russian Navy, which laid almost 13,500 in the Black Sea in 1914–17. ('Paper Number 3' in LIM/9; Friedman, *Naval weapons*, 383–384 and 386; Forrest, 232–233; Greger, 70.) (LIM/9, National Maritime Museum),

26 McLaughlin, 'Coast Defence and Coast Offence'. Planned modifications to *Rostislav* and *Sinop* are described in Le Page's reports 31/15 and 32/15 of 15 and 23 December 1915, and 23/16 and 30/16 of 9 August and 5 October 1916 (ADM 137/754).

27 Staff, *German Battlecruisers*, 120; Lorey, 103, 119–120, 130–131, 498–501.

28 Lorey, 125.

29 Le Page's report 6/15 of 10 April 1915. In November 1915 the Russian Navy shared a map with the Admiralty, showing lines of Ottoman mines across the Bosphorus, and a line of mines laid further out by the submarine *Krab* (ADM 137/1146, 93; see Ewin, 'Bosphorus Defences'). Grechanyuk *et al*, 320, 323 and 332, note many dates in spring/summer 1915 when submarines sortied to the Bosphorus: *Nerpa* 5–8 March, 3 April (return not stated), 27 May–5 June, 19–25 June, 8–16 July, 22–29 July and 18–24 August; *Tyulen* 29 March–4 April, 7–14 June, 25 June–2 July, 8–16 July, 8–16 August and 27 August–3 September; *Morzh* 8–16 July and 6–13 August. *Krab* sortied to the Bosphorus 8–12 July. Le Page's reports 8/15 and 9/15 note that *Tyulen* and *Nerpa* also sailed to the Bosphorus on 21 April and returned on the

28th. However, the transit to and from the Bosphorus took altogether some three days for the submarines, thus eating into the time any vessel could remain in her patrol area.

30 Grechanyuk *et al*, 321, notes that the Ottoman steamer *Rostislav* sank was *Sabbakh*, and adds that the battleships' targets included Cape Elmas, Antoli-Fener, Feneraki and later Panas-Burnu on the Roumelian coast.

31 Lewis Einstein, *Inside Constantinople During the Dardanelles Expedition*, E P Dutton (Boston, Ma, 1918), 4.

32 Le Page's report 3/15 of 31 March 1915 (ADM 137/754 33–39); he states that there were six small destroyers. Grechanyuk *et al*, 320, states that the minesweepers were *Veliky Kniagina Xenia*, *Alexei* and *Veliky Kniaz Konstantin*, plus the trawlers *Tsar* and *Ruslan*, and that there were seven small destroyers. Langensiepen & Güleryüz, *Ottoman Steam Navy*, 47, note that the steamers sunk included *Sadiç* (765grt) at Zonguldak and *Dafni* (1321grt) at Kozlu.

33 Lorey, 129 and 131.

34 An official Ottoman report at the time stated that the March 1915 bombardment caused no casualties, and later reports only 'a few'. This is consistent with surviving Ottoman hospital records (Ozdemir, *The Ottoman Army 1914–1918*, 78 and 124).

35 The author hopes to cover *Mecidiye*'s story in a separate article.

36 Grechanyuk *et al*, 323. Lorey, 137.

37 This paragraph is based on Lorey, 130–137 (who recorded that *Gnevnyi* had fired three torpedoes), and Le Page's report 5/15 of 5 April 1915 (ADM 137/754 46–50). Langensiepen & Güleryüz, 47, and Greger, 49, note that the Russian steamers lost were *Vostochnaia Zvezda* (844 or 944grt) and *Providence* (748grt).

38 Le Page's report 6/15 of 10 April 1915 (ADM 137/754 52–54).

39 Le Page's report 6/15 of 10 April.

40 Le Page's report 7/15 of 20 April 1915 (ADM 137/754 56–58).

41 Le Page's report 7/15 of 20 April; Grechanyuk *et al*, 324. Nekrasov, 53–54, and Greger, 49, note that the large Russian destroyers also sank 30 Ottoman sailing ships. Langensiepen & Güleryüz, 47, say the Russian destroyers sank two ships on the 15th, one being *Despina* (714grt – later refloated). Greger, 68, records Ottoman losses as *Neva* (549grt) *Dafni* (1,321rt), *Morna* (1,496grt) and ?*Alkistis* (1,213grt).

42 Le Page's report 8/15 of 26 April 1915: ADM 137/754 60–63.

43 It is not clear from Le Page's report exactly what type of meeting he meant; presumably it was of Black Sea Fleet officers.

44 Le Page's report 9/15 of 6 May (ADM 137/754 65–68); Grechanyuk *et al*, 325. Le Page added that on 2 May *Pamiat Merkuria* sank a steamer at Kozlu and a large sailing-ship collier under the Persian flag; *Kagul* sank a sailing ship off the Bulgarian coast. On 4 May *Kagul* captured a steamer and sent her in to Sevastopol with a prize crew. And on 5 May two large Russian destroyers reported sinking four Ottoman colliers off the 'coal coast'. Langensiepen & Güleryüz, 47, state that the captured steamer was the Italian *Amalia* (413grt), on charter to the Ottomans, while the sailing ship was of 950grt; ships sunk on 5 May were *Yesilirmak* (865grt), *Güzel Girit* (1,232grt) and *Morna* (1,495grt). Greger, 68, records the steamer *Pamiat Merkuria* sank on the 2nd as ?*Sahir* (1,215grt), and adds that she also sank *Iskondar* of 2,900grt off Kozlu on the 4th, while the large destroyers sank *Millet* (817grt) and damaged *Kizil Irmak*

(2,794grt) on the 5th. *Iskondar* was the former Craig Line *Craigforth*, detained in the Bosphorus in August 1914 and taken over by the Ottomans when they entered the war.

45 Grechanyuk *et al*, 326.

46 Le Page's report 9/15 of 6 May; Grechanyuk *et al*, 326.

47 Langensiepen & Güleryüz, 47.

48 Le Page's report 10/15 of 10 May 1915 (ADM 137/754 70–74). Lorey, 175, confirms the loss of three colliers. Langensiepen & Güleryüz, 47, name the losses as *Selanik* (1,127grt) at Kozlu, *Millet* and *Sadic* (each 817grt) at Eregli.

49 Grechanyuk *et al*, 326.

50 Lorey, 175; Langensiepen & Güleryüz, 47; Staff, *Seven Seas*, 53.

51 This and the previous quoted radio message are from Lorey, 175–176. 'Fleet HQ' was presumably the German accommodation ship (and former Ostafrika Line steamer) SS *General*, moored in Constantinople.

52 Lorey, 75; Staff, *Seven Seas*, 53.

53 Lorey, 156–157.

54 Staff, *Seven Seas*, 53; Nekrasov, 55.

55 O'Hara and Heinz, 140.

56 For details of the 10 May engagement see Le Page's 10/15 and 11/15 of 10 and 18 May 1915 (ADM 137/754 70–80); Lorey 177–179; Grechanyuk *et al*, 328; O'Hara and Heinz, 140–142; Staff, *German Battlecruisers*, 123–124; Staff, *Seven Seas* 55–57; R M Melnikov, '*Panteleimon*', V Iu Gribovski, 'Chernomorskii flot v boiakh s Gebenom', L A Kuznetsov, *Linienie korabli typa 'Ioann Zlatoust' 1906–1918*, 46–48 and 70 (transcribes Russian senior officers' reports); Nekrasov 56.

57 Wolf, 310.

58 Lorey, 176. *Numune*'s senior Ottoman officer was Uskudarli Nezir Abdullah (Güleryüz, *Torpedoboats & Destroyers*, 118). Square 228 on the wartime German charts was at least 20 miles north of the Bosphorus, judging by the maps in the German Official History.

59 Lorey, 179. Pfeiffer commanded the Ottoman torpedo flotilla from January 1915 to June 1917, when he was succeeded by then-*Korvettenkapitän* Hermann Lorey (Lorey, 78).

60 Melnikov, 18 (drawing on Russian State Naval Archives file Ф.418 Оп.1 д.874).

61 Russian ranges in 'cables' (of 200 yards) and German ranges in metres are here translated into yards.

62 For consideration of a ship's 'danger space' – the distance a shell might be due to fall *beyond* its target, yet still hit it because of having a relatively flat trajectory – see McLaughlin, 'Predreadnoughts vs a dreadnought', 133. For examples of danger spaces for British 12in and 13.5in guns – from 157–227 yards at a range of 4,000 yards to only 28–33 yards at 12,000 yards – see Friedman, *Naval Firepower*, 18.

63 Naval Attaché Petrograd's telegram of 12 May 1915 and NID reports of 14 and 25 May 1915, in ADM 137/4068.

64 Lorey, 178–179; November 2016 e-mail from Gary Staff. For German rangefinding and fire control, see Friedman, *Naval firepower*, 158–164, and McLaughlin, 'Predreadnoughts vs a dreadnought', 117–118.

65 Einstein, 34 (entry for 10 May 1915).

66 By early 1916 *Breslau* had the foremost of her four funnels painted white (Lorey, 286), presumably to make her less distinguishable from the Russian *Pamiat Merkuria* and *Kagul* and the Ottoman cruisers *Hamidiye* and *Mecidiye*, all of which had three funnels. When *Breslau* was first so 'disguised' is not clear.

67 Einstein, 38–39 (entry for 12 May 1915).

68 Staff, *German Battlecruisers*, 124.

SUFFREN & DUQUESNE: FRANCE'S FIRST MODERN CARRIER ESCORTS

During the late 1950s, French ambitions to operate carrier task forces centred around three aircraft carriers of modern design required the development of suitable large escorts capable of both area air defence and anti-submarine warfare. For a suitable design they looked to the 'missile frigates' currently being built for the US Navy; the result was the *frégate lance-engins*, or FLE. **Jean Moulin**, and **John Jordan** present a detailed study of the first (and only) two ships of the programme, *Suffren* and her sister *Duquesne*.

By the mid-1950s the first of a series of eighteen 'fleet escorts' (*escorteurs d'escadre* or EE) were entering service with the *Marine Nationale*, the 9,400-tonne 'anti-aircraft cruiser' (*croiseur anti-aérien*) *De Grasse*, a reconstruction of a prewar design, was fitting out and a second, purpose-built AA cruiser, *Colbert*, had been laid down at Brest. The first of three planned new fleet carriers of French design, *Clemenceau*, had been authorised under the 1954 Estimates and would be laid down in October 1955. These were the first fruits of an ambitious programme to build a fleet modelled on the Royal and US Navies that could intervene in support of French interests abroad, and in particular for intervention in the French overseas territories (*l'Union Française*), a fleet that could act independently of naval forces dedicated to NATO ASW missions in the North Atlantic and Mediterranean and largely funded by the Americans under the Mutual Defense Assistance Program (MDAP).

The newly-completed fleet escorts and the AA cruisers were armed with conventional weapons: high-performance dual-purpose 127mm (5in) guns, 57mm Bofors twin AA guns, and (in the case of the fleet escorts) long-range 'heavyweight' 550mm anti-ship and anti-submarine homing torpedoes. However, the French Navy was conscious of the need to keep abreast of the latest technological developments. The US Navy was currently working on a series of missiles for fleet air defence, the so-called '3 Ts' (the long-range Talos, the medium-range Terrier and the short-range Tartar), and the British were not far behind with their Seaslug missile.

In March 1955 the then-Chief of Naval Staff Henri Nomy, a former naval aviator and a key advocate of the carrier task force as an instrument of power projection, published a document entitled *Directives pour un programme naval* which included two cruisers equipped with missiles to serve as carrier escorts alongside *Colbert* and *De Grasse*. Nomy envisaged a ship of 4,000–5,000 tonnes, designated *croiseur léger d'Union française*. The

deployment of missiles with an area defence capability would permit a reduction in the number of fleet escorts, so the ships could be larger and carry both surface-to-air and anti-submarine missiles. Studies began in 1956. The project was then put on hold to await technical developments, as it was unclear what the space and weight requirements of the new missile systems would be.

On 1 November 1955 the US cruiser *Boston* (CAG-1) completed with the new Terrier SAM, and it became clear that a carrier escort armed for both AAW and ASW would have to be a ship of cruiser size. On 27 Nov 1956 the *Commission de la défense nationale* stated that a missile cruiser displacing 5,000 tonnes was required, to be funded under the 1958 programme; it proposed the addition of surface-to-surface missiles (SSMs) to enable hostile surface ships and land targets to be engaged. At the end of the same year (1956) Navy Minister Paul Anxionnaz drew up and submitted construction programmes for 1957 and 1958. The second of these was to include a missile cruiser (still designated *croiseur léger d'Union français* in order to make clear that the project was independent of NATO). Studies began immediately, and by early 1957 the characteristics for the new *croiseur-escorteur* had been firmed up: 5,000 tonnes, one Masalca (French counterpart to the long-range Talos) and two Masurca SAM systems, three twin 100mm guns, a 305mm A/S mortar, Malaface surface-to-surface missiles, 2–3 helicopters, and the LOFAR submarine detection system. The ship was to be integrated with follow-on units with nuclear propulsion to be completed around 1970. The 1958 Estimates were to include a third fleet carrier plus minor craft. However, pressure on budgets reduced the funding available for new construction in 1958–59, and both the carrier and the cruiser were put on hold; the carrier would never materialise, a victim of de Gaulle's decision to pursue an independent nuclear deterrent, but carrier escorts would still be needed for *Clemenceau* and *Foch*, which would complete in 1960 and 1963 respectively.

The inspiration for the French missile frigates were the US Navy carrier escorts of the *Farragut* (DLG-6) class, which were armed with the RIM-2 Terrier SAM. The *Farragut*s were originally to have been conventional air defence escorts armed with 5in guns. Advanced anti-submarine capabilities in the form of ASROC and a large SQS-23 low-frequency sonar were an afterthought; the 8-cell 'pepperbox' launcher for ASROC was simply 'bolted on' (only the name-ship had a reload capability), replacing No 2 gun mounting. This is the name-ship of the class in the North Atlantic on 22 June 1967. (Naval History and Heritage Command, NH 106801)

In late 1959 new requirements were drawn up for a 5,000-tonne missile cruiser with anti-submarine helicopters. The ship was to be able to engage supersonic aircraft or missiles up to 60,000ft (*sic*! = 18,300m), all types of submarine (including boats with nuclear propulsion), and small missile boats. It proved impossible to achieve this on 5,000 tonnes, and the studies concluded on 4 November 1959 with a 5,700-tonne ship armed with a twin-arm launcher for the Masurca SAM, a single launcher for Malafon, two single 100mm D-P guns, two single 30mm fully-automatic AA guns, A/S torpedoes launched by catapult, and a helicopter platform. The project was first christened F 60 (F for *frégate*), then FLE 60 (*frégate lance-engins*); the adoption of the term *frégate* was undoubtedly influenced by developments in the US Navy, which from 1958 classified its new DLGs of the *Farragut* class as 'frigates' to distinguish them from the smaller missile-armed 'destroyers' (DDG). With the exception of Masurca and its associated tracking and guidance radars, which were currently under development, the weapons and electronics were those currently being fitted in the modified fleet escort *La Galissonnière*, which would begin her trials in July 1962. It was initially envisaged that six ships would be ordered, but the programme was subsequently reduced to five due to rising costs.

The *Loi-programme* adopted on 6 December 1960 envisaged that three frigates would be ordered 1960–65. *Suffren* was duly authorised under the 1960 Estimates and a sister *Duquesne* under 1962; however, the third ship was cancelled, again due to inflationary pressures; the estimated cost of a ship without missiles had risen from 18 billion old francs to 25 billion. The amount saved by cancellation of the third ship was used to purchase the US Tartar missile system to modernise four of the T 47-type fleet escorts, and to order 42 F-8E Crusader carrier-borne interceptors (see the authors' article on *Clemenceau* and *Foch* in *Warship 2023*) – US funding under MDAP had now ended.

The first of the two frigates was ordered from Lorient Naval Dockyard on 14 May; the steam propulsion machinery was to be built by the state establishment at Indret. The second ship would be built by the Arsenal de Brest.

Construction

Lorient had recently completed no fewer than seven of the 18 *escorteurs d'escadre*, including *La Galissonnière*, which trialled the anti-surface and anti-submarine armament of the new ships. *Suffren* was laid down in the Lanester ship hall following the floating out of the *aviso-*

Duquesne on 14 February 1966, following her 'floating out' at Brest. (Jean Moulin collection)

escorteur Protet and the logistics ship *Rhône*. The *aviso-escorteur Enseigne de Vaisseau Henry* was built alongside her from September 1962 to 14 December 1963, followed by the logistics vessels *Garonne* and *Rance*. In 1964 her hull number was changed from D 601 to D 602 (freed up by the decommissioning of the destroyer *Hoche*, ex-Z 25).

Suffren was floated out on 15 May 1965 in the presence of Defence Minister Pierre Messmer and VAE Meynier, representing the Chief of Naval Staff Admiral Cabanier; *Rance* was floated out one hour later. Fitting-out threw up numerous problems typical of a first of class, but trials began in October 1965. The distinctive radome housing the DRBI 23 target designation and tracking radar was painted a darker grey during this period, but would be repainted to conform to the rest of the ship in 1966. Following two years of exhaustive trials and testing, *Suffren* was commissioned (*armement définitif*) on 1 Oct 1967 and was transferred to Toulon during the same month. She underwent her initial acceptance cruise (*Traversée de longue durée*) in the Atlantic November–December, returning to Lorient, and there was a second cruise to northern waters in early 1968. She returned to Toulon on 19 June and was admitted to active service. She was then transferred to the Atlantic

Squadron, operating with the new carriers 1969–70.

Her sister *Duquesne* was built at Brest in the Laninon no 9 dock, following on from *Clemenceau*, the helicopter carrier *La Résolue*, and the transport dock (TCD) *Ouragan*. She took her D 603 pendant number from the decommissioned destroyer *Kléber* (ex-Z 6). Laid down in February 1965, she was floated out on 11 Feb 1966 in the presence of Messmer, Admiral Cabanier and the head of the DTCN. Trials took place from July 1968–1970; *Duquesne* then undertook her acceptance cruise in the Atlantic from 19 Feb to 30 March 1970, returning to Brest. She was admitted into active service 1 Apr 1970, serving with the Atlantic Squadron.

Hull & Sea-keeping

Suffren and *Duquesne* introduced a completely new hull-form that would characterise French surface ship construction for the next two decades. A long forecastle with pronounced sheer and flare terminated in a short, low quarterdeck on which the missile launcher for Masurca and the towing apparatus for the large variable depth sonar were located. The bow, which housed a centreline anchor, was given considerable overhang to protect the large fixed sonar dome, and had negative sheer

Building Data

Pendant No	Name	Builder	Laid down	Floated out	Trials	In service
D 602	*Suffren*	Arsenal de Lorient	21 Dec 1962	15 May 1965	23 Oct 1965	20 Jul 1968
D 603	*Duquesne*	Arsenal de Brest	1 Feb 1965	11 Feb 1966	27 Jul 1968	1 Apr 1970

Suffren being fitted out at Lorient on 23 July 1966. The twin launcher for Masurca SAMs and the superimposed DRBR 51 tracking/guidance radars are in the foreground. (Jean Guiglini, courtesy of Jean Moulin)

at its forward end to enable the 100mm DP guns to fire at low angles of elevation. A high length to beam ratio (10:1) was adopted to secure high speed; the performance of *Suffren* and *Duquesne* was similar to the earlier fleet escorts, despite a 2,000-tonne increase in displacement and only a 10,000shp increase in horsepower.

The two missile frigates were fully stabilised to ensure a steady platform for the DRBI 23 radar: three pairs of non-retractable fins, controlled from two gyro centres, limited roll to 5 degrees, the system being supplied by Brown Bros of the UK. The fins were effective from 15 knots and attained maximum efficiency at 24 knots. Each gyroscopic centre had two gyros: one for angular speed,

the other for roll; the data they supplied was used to action a hydraulic electro-pump. All three pairs of fins could be controlled from a single gyro centre. The only caveat was that the non-retractable fins projected from the hull, and care was needed when coming alongside. *Suffren* and *Duquesne* had exceptional sea-keeping qualities with moderate roll and pitch, and were able to steam at full speed in Sea State 4 with a head sea.

Suffren and *Duquesne* had only light protection in the form of some splinterproof plating. *Duquesne* was completed with an air-tight NBC citadel, but this was never retrofitted in *Suffren*. The thickness of the plating for the hull and the decks was as follows: 6mm for the

Main Deck, 12mm for the 1st Deck and the roof of the Masurca magazine, and 14mm sides, increasing to 18mm at the bilges and 24mm at the keel.

Machinery

The ships had an advanced high-pressure steam propulsion plant built by Indret; it represented a major advance on the classical installation in the earlier fleet escorts and featured a high degree of automation, particularly for the control of boiler combustion. The four Indret asymmetrical boilers were rated at $45kg/cm^2$ with superheating at $450°$ C. They still used heavy furnace fuel oil (*mazout*), so the fuel had to be pre-heated.

There were two large machinery spaces (Sections G and I) each 16.5 metres long and housing two boilers, a set of geared turbines, a turbo-alternator and an evaporator, separated by a 10-metre compartment for auxiliary machinery that included three diesel generators and a third evaporator (Section H). The two main machinery spaces were completely independent for damage control purposes. The boilers were paired within air-tight boxes, and were controlled from a sound-proofed, air-conditioned *PC machines* located at the upper level in each compartment (see GA Plans). In order to conserve centreline space, the exhaust uptakes for the four boilers were led up into a single tall 'mack' (US: 'mast' + 'stack') amidships.

The turbines were of the Rateau impulse type: each set comprised a high-pressure (HP) and a low-pressure (LP) turbine. The HP turbine was of the 'parallel series' type: it was complex but was designed to reduce mass and for prolonged use (it was dismantled for maintenance every six years). The LP turbine was of the double flux type and had the astern turbine incorporated in the casing. Designed horsepower was 72,500CV for a maximum speed of 34 knots – *Duquesne* made 30.7 knots even after her 1990 refit. The condenser cooling used natural circulation above 18 knots and a turbo-pump below this speed. Each of the two main condensers was aligned athwartships with the LP turbine casing welded on top. Double reduction gearing reduced the shaft revolutions to a maximum 300rpm. The turbines in the forward engine room drove the starboard shaft and those in the after engine room the port shaft. One consequence of this was that all four of the boilers had to be offset to port to allow the longer starboard shaft to run outboard of them (see GA Plans). Each of the two four-bladed propellers had a diameter of 4.3 metres.

The various electrical circuits were fed by two Breguet SW turbo-alternators each rated at 1,000kW; one was located in each of main machinery spaces together with an evaporator to supply fresh water for the crew and for the boilers. Three MGO V12 ASR diesel alternators, each rated at 480kW, were located in the auxiliary machinery room, which also housed three air-conditioning units to

Suffren at Cannes on 4 July 1976. The Masurca missiles and the large variable depth sonar are prominent in this stern quarter view. Note the original TACAN radome atop the short lattice mainmast, and the 30mm Hispano-Suiza cannon to starboard. (Pradignac & Leo)

provide fresh water for air cooling systems and an auxiliary boiler for use when alongside.

There were two 'stabilisation centres' (*centrales de cap et de verticale*, or CCV), each equipped with two gyroscopes and an electro-hydraulic pump. These were supplied and fitted out by SAGEM; they provided perfect stabilisation for the weapons and electronics.

Masurca

French studies for cruise missiles and guided rockets using German wartime technology began in 1948, but it was some ten years before these early projects came to fruition. Prototypes of the US Navy's RIM-2 Terrier medium-range surface-to-air missile (SAM) had been trialled in the old battleship *Mississippi* from early 1953, leading to a first installation of a production model in the converted heavy cruiser *Boston* (CAG-1), which recommissioned on 1 November 1955. French developments lagged some way behind: *Boston* entered service the same day as the first of the new generation of fleet escorts, *Surcouf*, which was armed with conventional 127mm DP guns.

The first French prototype SAM was christened Maruca; the name was an acronym indicating the armed service, the manufacturer and the role of the missile (MA = Marine, RU = RUelle, CA = Contre Avion). Trials of Maruca, a subsonic missile with liquid propellant derived from the German HS 117 Schmetterling, began as early as 1951, and an improved A6 version was developed in 1954, with shipboard trials taking place using the converted ex-German freighter and hospital ship *Ile d'Oléron* throughout 1959. However, the trials demonstrated that a liquid-fuelled missile was not a practical proposition for operational naval use.

Maruca was to have been followed by a solid-fuel missile, Masalca, developed by the aviation company Latécoère. A long-range missile similar to the US Talos intended for cruisers, Masalca was likewise tested in 1959 but subsequently abandoned. However, a parallel development of Maruca as the supersonic Masurca (MArine SUpersonique Ruelle Contre Avion), which like Masalca employed solid fuel, showed greater promise; it would be the only surface-to-air missile retained by the French Navy.

Masurca prototypes underwent land-based trials at CERES (*Commission d'Etudes et Recerches sur Engins Spéciaux*), then on board *Ile d'Oléron* from May to October 1960. On 13 July 1968, CERES was superseded by the CEM (*Centre des Essais de la Méditerranée*) on the Ile du Levant; the centre would be responsible for the development of SLBMs, Malafon (see below), the SS 12 helicopter-launched missile, and later models of Masurca. No fewer than 50 missiles would be test-fired from *Ile d'Oléron* over a period of fifteen years, including prototypes of the Mk 1, Mk 2 Mod 2 and Mk 2 Mod 3 variants embarked in *Suffren* and *Duquesne*. Operational service with *Suffren* was achieved 1967–68.

The missile and the launcher were built at ECAN Ruelle, the state-owned weapons development establishment, from 1968 onwards. A third system intended for the helicopter carrier *Jeanne d'Arc* would be retrofitted in the AA cruiser *Colbert* in 1970–72 in place of the after 100mm and 57mm gun mountings, but development was plagued by the limited number of installations; no new missiles were manufactured after 1986 and upgrades ceased in 1990. The installation was too bulky for installation on the modernised fleet escorts, for which the US Tartar missile was purchased, or in the air defence variant of the C 70 design.

Masurca was similar in conception to the US Navy's

A Masurca surface-to-air missile on the starboard arm of the twin launcher on board *Suffren*. The later Mk 2 Mod 3 variant of Masurca used semi-active guidance. (Bertrand Magueur)

Suffren: Profile & Plan

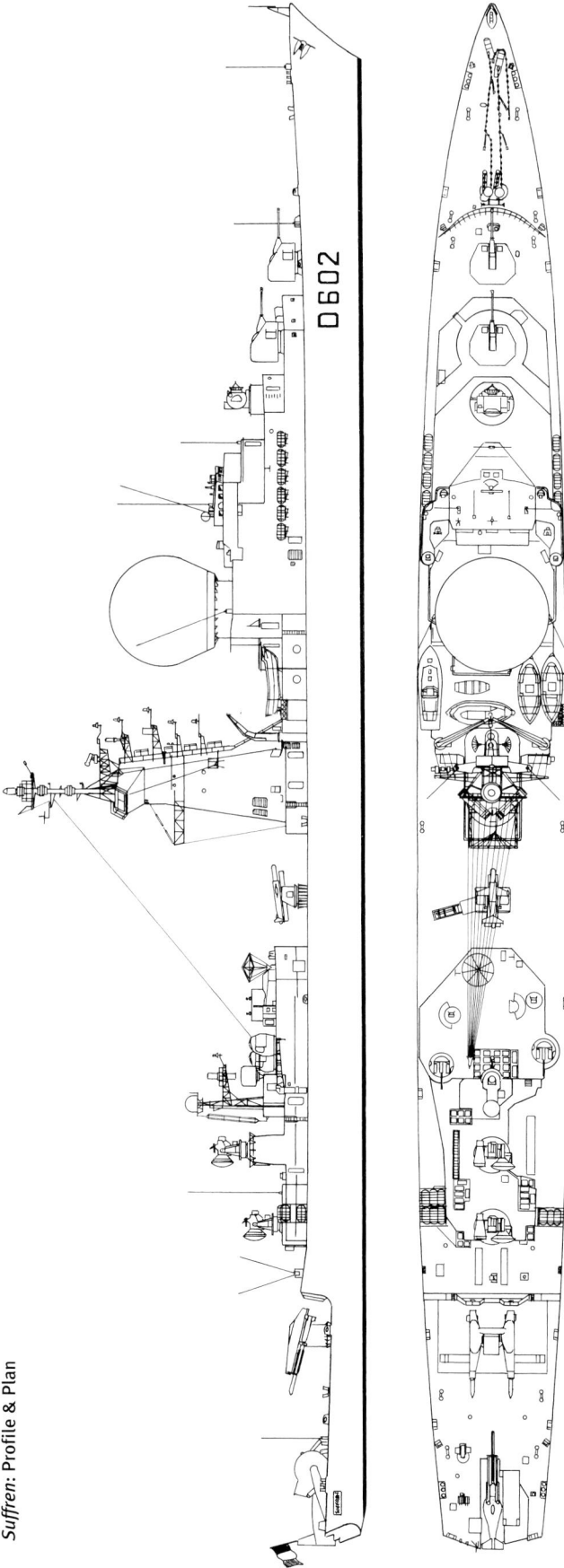

D602

Characteristics: *Suffren* as Completed

Displacement:	5,090 tons standard
	6,090 tonnes full load
Dimensions:	
length	148m pp; 157.62m oa
beam	15.54m
draught	6.10m
Machinery:	
boilers	four Indret asymmetrical Sural boilers,
	45kg/cm², 450°C
engines	two-shaft Rateau geared steam turbines
horsepower	72,500CV
speed	34 knots (trials)
endurance	5,100nm at 18 knots
electricity	two Breguet turbo-alternators each 1,000kW
	three MGO diesel alternators each 480kW
Armament:	
AAW	Masurca SAM system (44 combat missiles)
ASW	Malafon A/S system (13 missiles)
	four KD 59 catapults for ten L3 A/S torpedoes
DP guns	two 100mm Mle 1953 DP
light guns	two 30mm Hispano-Suiza
Electronics:	
tactical data	SENIT 1 system
radars	DRBI 23 3D air surveillance & target designation
	DRBV 50 low-altitude air/surface surveillance
	DRBR 51 missile tracking/guidance (x 2)
	DRBN 32 navigation
	DRBC 32A gun fire control
sonar	DUBV 23 LF in bow dome
	DUBV 43 LF variable depth
ESM	ARBR 10E
ECM	ARBB 31 & 32
Complement:	38 officers, 118 petty officers, 270 men

Suffren: GA Plans

Inboard Profile

© John Jordan 2021

Command Spaces

Radio Office
Weapons
Transmissions Centre
Cypher Room
Operations Centre
Sonar Room
Radar Room
ASW
ECM

deck store
paint store
WTC
WTC
DUBV 23 transducer
143 FP
seamen's mess
DUBV 23 sonar rm
general store
annex
136
cable locker
capstan machinery
seamen's mess
seamen's mess
general store
130
electr'l spares
mag
100
120
p/w
passageway
seamen's mess
seamen's mess
mag
100
p/w
POs' mess
passageway
mag
stores
ops office
108.1
passageway
cafeteria
cool room
wine store
100mm calculator position + annex
converters
DRBI converters
DRBI radar room
Comms Office
Enclosed Bridge
passageway
flour store
gyro room
machinery control room
central passageway
sick bay
96.5
Ops Centre
Radar Room
DRBI transmitter room
DRBI radar room
DRBI radome
DRBV 50 radar room
air intakes
FORWARD MACHINERY ROOM
80
RFW
engineering workshop
p/w
diesel
FFO
AUXILIARY MACHINERY ROOM
70
air intakes
central passageway
RFW
53.5
AFTER MACHINERY ROOM
machinery control room
Malafon launch ramp
Malafon hangar
central passageway
gyro room
Masurca control room annex
FFO
41
converters for TRE nº 1
converters for TRE nº 2
hydraulics for rammers
30mm gun p&s
ESM Office
p/w
officer cabins
reserve magazine (stbd)
33.5
Masurca repair shop
converters
FFO
DRBR 51 missile tracker radars
hydraulics for rams
Masurca handling room
Masurca magazine ring p&s
21.5
WTC
Masurca RPC
12.8
seamen's mess
store
30
converters
WTC
twin launcher for Masurca
VDS winch room
DUBV 43 towed sonar
5.6
AP
steering gear compartment

1st Deck

100mm gun Nº 1
officer cabin [x1]
officer cabins [x1]
officer cabins [x2]
p/w
passageway
p/w
Wardroom & pantry
officer cabins [x1]
Flag Staff office
office
Sick Bay
torpedo room
trainable launcher for Malafon
launcher commander's office
Malafon hangar
strike-down hatch for Masurca missiles
converters for TRE nº 2
blow-out hatches
twin-arm launcher for Masurca
DUBV 43 variable depth sonar

1st Platform Deck

paint store
WTB 143
DUBV 23 transm'r
WTB 136
seamen's mess nº 2
seamen's mess nº 3
WTB 130
POs' mess
POs' mess
WTB 120
POs' mess
POs' mess
WTB 108.1
provision issue room
cold room
wine store
sonar converter rm
potato store
flour store
cool room
WTB 96.5
fwd machinery control room
turbo-alternator
boilers nºs 11 & 12
refrigeration machinery p&s
evaporator
WTB 80
engineering workshop
evaporator
three diesel generators
boilers nºs 21 & 22
evaporator
WTB 70
HP turbine
LP turbine
reduction gearing
evaporator
aft machinery control room
turbo-alternator
CPO cabins
gyro room
POs' mess
WTB 53.5
officer & CPO cabins
cabins
seamen's mess nº 6
reserve magazine for Masurca
magazine rings to 63.5
Masurca missiles
WTB 41
prison
store
Masurca spares
DUBV 43 transmission
converters
magazine rings for Masurca missiles
WTB 21.5
seamen's mess nº 7
pedestal for launcher
washplace
store
store
store
WTB 12.8
hydraulics control
capstan machinery
hydraulics control
WTB 5.6
canvas store
deck store p&s
steering gear

75

Terrier RIM-2 missile, with a 'tandem' (in-line) booster, but was larger and heavier: the key characteristics are tabulated below:

	Terrier	Masurca
length (missile)	4.5m	5.41m
length (booster)	3.73	3.32m
total length	8.23m	8.52m
max span of fins	1.2m	1.5m
weight (missile)	540kg	840kg
weight (booster)	830kg	1,018kg
total weight	1,392kg	1,858kg
warhead	99kg	100kg

Like Terrier (but unlike the contemporary RN Seaslug, which had wrap-around booster rockets) Masurca was fired from a conventional twin-arm launcher. Each of the two horizontal revolving drums in the magazine carried 18 missiles ('blue' to port/'yellow' to starboard), of which 17 were combat rounds and the 18th a dummy round (*composite inerte de vérification*) essential for maintenance of the installation. The missile at the top of the ring was hoisted into a preparation chamber where the large booster fins were fitted and the control surfaces of the missile deployed. The missile was suspended from a loading rail, then rammed onto the missile launcher. A second magazine forward of the first held ten reserve missiles and ten boosters in two rows of five (see drawing). This made for a total of 44 combat missiles.

The booster burned for five seconds after launch, then was separated from the missile using a small explosive charge. At this point speed was 800m/s. The missile engine then took over; it was powered by a solid-fuel propellant and maintained a speed of Mach 2.5, rising to Mach 3.

The earliest operational variant of Masurca was the Mk 2 Mod 2 beam rider; it had a theoretical maximum effective range of 40,000 metres. A tracking radar cued by a powerful three-dimensional (3-D) target designation radar locked onto the target, and the missile was 'gathered' into the beam, its control surfaces being controlled from the ship using radio signals, until the warhead exploded in proximity to the target. There were two guidance channels each comprising a DRBR 51 tracking radar, a computer, and a test system to ensure that the missile was functioning correctly (*Ensemble de Test Automatique*

Suffren: Masurca Missile System

Inboard Profile

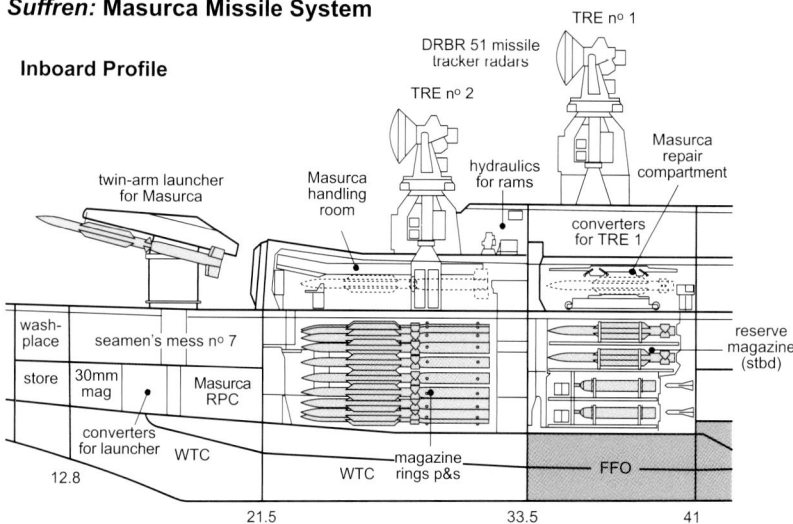

The main magazine housed two horizontal revolving drums each for 18 Masurca missiles; the drum to port was designated 'blue', the one to starboard 'yellow'. The missile at the top of the drum was hoisted into a preparation chamber where the booster fins were fitted and the missile control surfaces deployed; the missile was then rammed onto one of the twin arms of the launcher. Note the reserve magazine abaft the main magazine; it could hold a total of ten missiles in broken-down condition, the boosters being stowed separately. Directly above was a workshop for missile testing and repair.

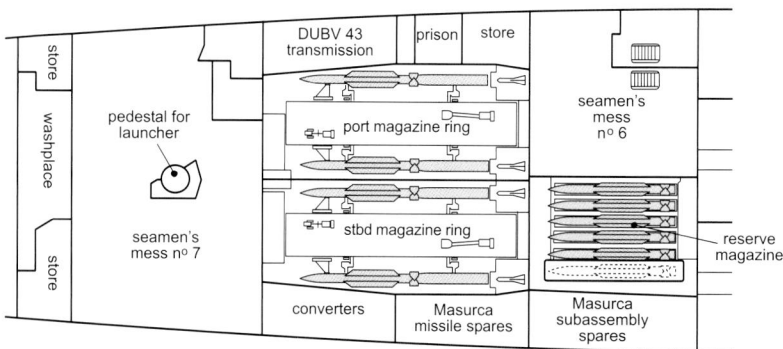

© John Jordan 2021

et de Détection Avarie Masurca, or ETADAM). Each of the two channels was capable of guiding one or two missiles to the target; a two-missile salvo could be fired every 30 secs. Three analogue computers were initially embarked (one for each of the missile guidance channels, the other for target designation).

The inherent disadvantage of the radar beam riding system, which was employed for all the area defence surface-to-air missiles that entered service during the 1950s and 1960s, is that the beam spreads as it travels outwards from the emitter; accuracy in the terminal stage of the missile's flight therefore becomes problematic at longer ranges. Another issue is that the guidance path of the missile is essentially a straight line to the target. This is useful for missiles with a great speed advantage over their target or where flight times are short, but for long-range engagements against high-performance targets the missile will need to 'lead' the target in order to arrive with enough energy to conduct its terminal manoeuvres.

The development that arguably hastened the end of beam riding was the advent of powerful ECM jammers, and anti-radiation missiles (ARM) that could be launched by the target aircraft down the beam of the tracking radar. The US Navy soon abandoned beam riding for its Terrier missile in favour of semi-active homing, in which the shipborne tracking radar simply follows and illuminates the target, and the missile homes on the reflected radar pulses. This results in a more economical flight path, and the width of the reflected illuminating beam diminishes as the missile approached the target, thereby improving accuracy in the terminal stage of flight (see graphic).

The French were slower to adapt due to the high costs associated with a relatively limited installation. There were successful trials of a new Mk 2 Mod 3 semi-active version of the Masurca missile aboard *Duquesne* in January 1970, and in February 1971 a range of 45.000 metres was achieved; the earlier Mk 2 Mod 2 missile was retired from service in 1975. After modernisations that took place between 1978 and 1985 the original analogue computers were superseded by digital models, resulting

in further improvements in performance: interceptions were carried out at ranges up to 55,000m and at heights of between 30 and 22,800 metres.

Target Designation, Tracking and Missile Guidance

The DRBI 23 L-band pulse-doppler 3-D radar, which was housed within a distinctive radome atop the bridge struc-

Suffren: DRBI 23 3D Radar

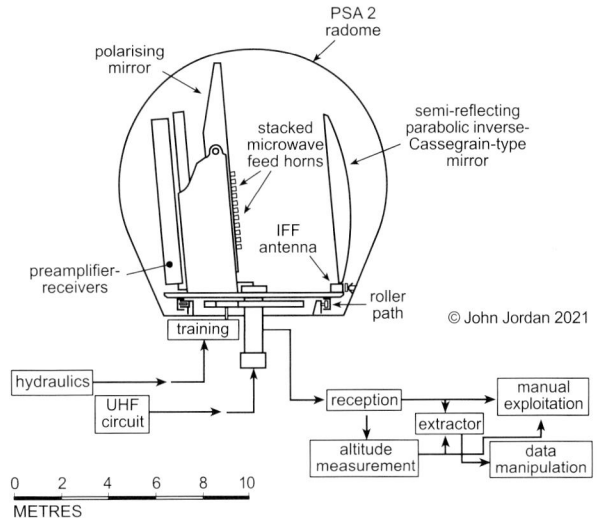

The antenna for the DRBI 23 radar was a large inverse Cassegrain reflector. An array of stacked microwave horns produced the stacked beam used for monopulse height finding. The horns illuminated a semi-reflective parabolic mirror. The mirror reflected onto a secondary reflector that twisted the polarisaton of the beam so that it could pass through the first reflector. This type of 'folded' optics made it possible to achieve considerable focusing of the radar beam in limited dimensions. Moreover, the elaborate feed did not block the main beam of the radar – an important consideration in any stacked beam or monopulse system.

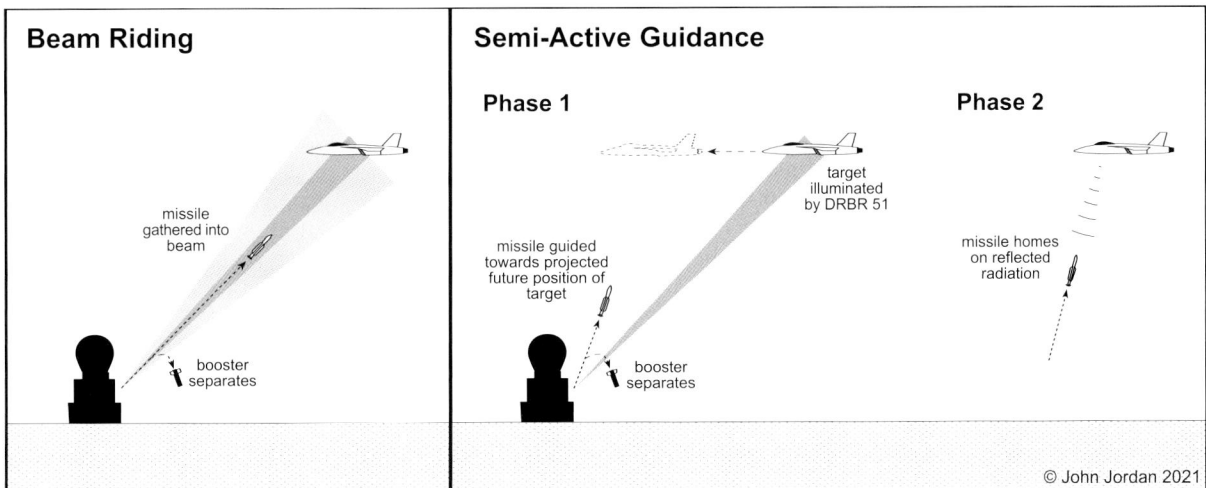

ture, was developed specifically to provide long-range target designation and tracking for Masurca and was fitted only in *Suffren* and *Duquesne*. The antenna comprised a large inverse Cassegrain reflector with a mirror behind and used stacked beams for monopulse height-finding (see drawing). It was developed from an experimental AC 12 3-D radar model from CSF which was followed by a 3-D C-band radar, also from CSF. The first elements were trialled at Fort Mengam in 1965 prior to installation in *Suffren* at the end of the same year. A smaller, less capable version, DRBV 13, was fitted in *Aconit* (see the author's article in *Warship 2022*, 191–97).

The L-band 23cm emitter was able to operate in the single frequency, dual frequency or frequency-agile modes to counter jamming. There were three types of receiver to cope with a variety of atmospheric conditions, and the radar could also be used in passive mode to detect hostile emissions. Exploitation was either manual or automatic using SENIT (see below). DRBI 23 was modernised and digitised 1988–91.

The ADC 6 3-D inverse Cassegrain antenna used 6MW of power. Total weight of the combined array was 28 tonnes; it was seated on a roller path with a diameter of 8 metres, hence the need to stabilise the ship. It rotated at 0–8rpm in both directions with the original hydraulic

training system; it was later powered by electric motors and rotation was fixed at 8rpm. A distinctive polyester radome 11 metres in diameter protected the array from smoke, wind, spray and rain.

The DRBI 23 array supplied height-finding data and target designation to two DRBR 51 guidance radars superimposed above the Masurca launcher. DRBR 51 was a C-band monopulse guidance radar equivalent to the US Navy's SPG-55. Each radar group comprised a pair of antennae: a large-diameter tracker (which could also measure the deviation of the missile from the line of sight) and a small-diameter command transmitter. A third antenna generated the initial gathering beam.

In the beam-riding mode DRBR 51 located and tracked the target, and measured the distance between the axis of the beam and the two missiles in the salvo, allowing the missiles to be gathered into the beam using command signals. The missiles were tracked at a wavelength of 5cm and command signals transmitted at 7cm; the tracking channels for the two missiles in a salvo were designated 'blue' and 'yellow' respectively. In the semi-active mode continuous wave (CW) power was injected into the main dish. DRBR 51 located, tracked and illuminated the target; the missile then homed either on the radar reflections or on the jamming emissions from the target (see graphic).

Duquesne at Queen Elizabeth II's Silver Jubilee Review at Spithead in June 1977. The 30mm cannon have recently been replaced by canisters for the MM 38 Exocet surface-to-surface missile. (John Jordan)

Medium and Close-range Air Defence Weapons

As a complement to Masurca, *Suffren* and *Duquesne* were fitted with two 100mm Mle 1953 mountings forward. This was the first French-designed automatic medium-calibre gun, and was fitted in the carriers *Clemenceau* and *Foch*, the experimental fleet escort *La Galissonnière*, and the *avisos-escorteurs* of the *Commandant Rivière* class. It had an analogue fire control system, and could be employed against aircraft, surface warships and the shore. Development began in 1953, with trials 1958 on the fast escort *Le Brestois* (E 50 type) from 1958.

The mounting, which weighed 22 tonnes, was built by Ruelle and had remote power control (RPC) for training and elevation. The mountings on *Suffren* were christened *Héro* and *Fantasque*, those on board her sister *Messine* and *Palerme*. The Mle 53 featured a 55-calibre gun with an elevation of -15°/+85° and a theoretical rate of fire of 60rpm, and fired a 13.4kg F1 HE shell with a 1.1kg bursting charge; the fixed round weighed 23.2kg with a 4.5kg propellant charge. Initial velocity was 855m/sec, and effective range 12,000m against surface ships and 6,000m against aircraft. A carousel with 32 ready-use rounds enabled the gun to open fire at one round per second; the carousel then needed to be replenished from the magazine via a single hoist (20rpm). Stowage

A view from the bow of *Duquesne* taken on 20 March 1998, with the single 100mm 64 CADAM guns guns and the radome for the DRBI 23 target indication radar prominent, flanked by two Syracuse SATCOM radomes. (Jean Moulin)

The distinctive radome for the DRBI 23 air surveillance & target designation radar is prominent in this photo, taken when *Duquesne* visited Portsmouth on 16 June 1997. DRBI 23 is flanked by radomes for the Syracuse SATCOM system, and there is a Malafon missile on the launcher amidships. (John Jordan)

The below-decks reload carousel for one of the two 100mm guns. (Jean Moulin collection)

was 836 rounds in no 1 magazine and 1,044 in no 2.

For gun fire control the two missile frigates initially embarked a stabilised director incorporating the DRBC 32A radar and a 4.2-metre optical rangefinder. Trialled in *La Galissonnière*, it was also fitted in the *avisos escorteurs* and the modernised fleet escorts of the T 47 type.

During a major 'half-life' refit *Suffren* (1988–89) and *Duquesne* (1990–91) had their original Mle 1953 mountings replaced by the 64 CADAM (*CADence AMéliorée*), which was similar in conception but capable of 80rpm with the same ammunition. At the same time the DRBC 32A fire control radar was superseded by a new lightweight FC director featuring a monopulse, frequency-agile radar (DRBC 33A), a TV camera with IR, and a laser rangefinder. When directing shellfire, the angular error between the shell and the target could be measured using scartometry. Computing was digital, with a dedicated console in the operations centre.

When first completed *Suffren* and *Duquesne* were also fitted with two single 30mm DP guns in an automatic gun mounting manufactured by Hispano-Suiza. The fully-enclosed gunhouse weighed 4 tonnes and was controlled remotely from an optical station. The 30mm gun fired a fixed round weighing 950g (projectile 420g) with an initial velocity of 1,000m/sec; belt fed, the gun was capable of 565rpm. The mountings would be removed between 1975 and 1980 to make space for launchers for the Exocet surface-to-surface missile (see below); they would be replaced by four 20mm Oerlikon cannon located at the 'corners' of the ship.

Anti-submarine Warfare

The French missile frigates, unlike their US counterparts of the *Farragut* class, were designed from the outset to carry out anti-submarine missions in support of the carrier task forces they were to escort, and were provided with a full complement of A/S weaponry and electronics.

During the immediate postwar period the primary A/S weapons were mortars and rockets with a moderate range that matched the high-frequency sonars of the period. The US and Royal Navies and the French *Marine Nationale* also invested heavily in long-range 'heavyweight' 533mm and 550mm homing torpedoes, sometimes with wire guidance to compensate for the 'dead time' between launch and arrival at the estimated position of the submarine, but these encountered a multitude of technical problems that would lead to their abandonment.

The late 1950s saw the advent of fast Soviet attack submarines powered by nuclear reactors that would ideally need to be engaged outside the range of the rockets and mortars, and which could in theory outrun (or out-manoeuvre) the long-range torpedoes. This led to two strands of development: torpedo-carrying missiles with a range of around 10,000 yards, with matching medium-/low-frequency sonars to provide underwater detection and target designation; and manned or drone helicopters cued by the new long-range sonars and armed with small homing torpedoes to attack the submarine. The US Navy initially opted for ASROC and DASH (Drone Anti-Submarine Helicopter) and the Royal Navy for the Australian Ikara missile and the Wasp helicopter (MAnned Torpedo-Carrying Helicopter, or MATCH), all of which entered operational service in the early/mid-1960s.

French developments followed the same lines of thinking. The last of the eighteen conventionally-armed fleet escorts of the T 47 and T 53 types, *La Galissonnière*,

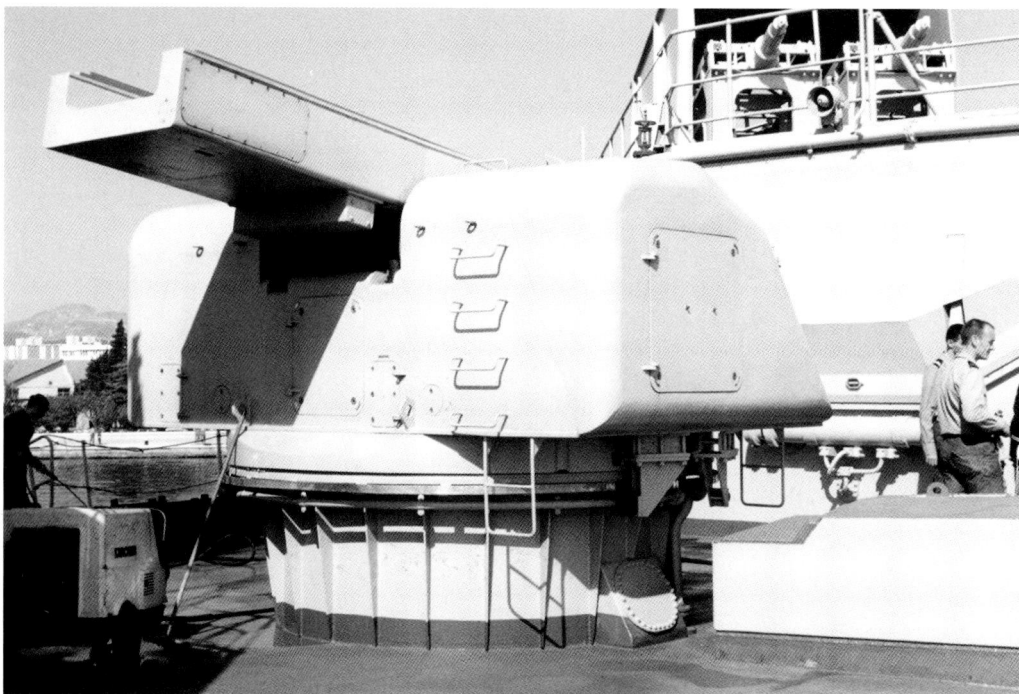

Close-up of the launcher for Malafon, which could be trained through 360 degrees but launched the missiles at a fixed angle of +15 degrees. (Jean Moulin)

Deployment of Malafon A/S Missile

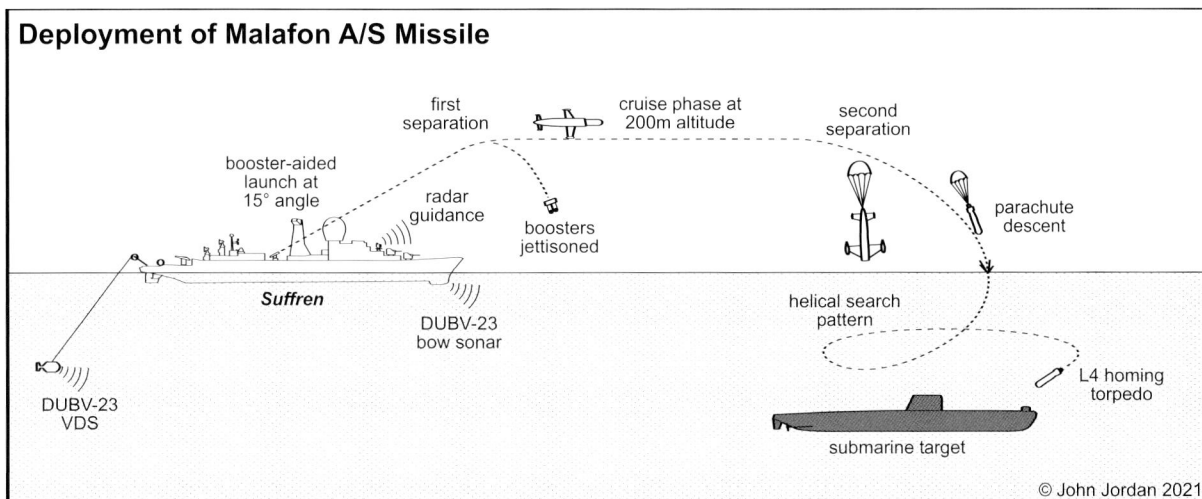

© John Jordan 2021

was rebuilt as a test ship for the new anti-submarine weapons and sonars, completing for trials in 1962. She was fitted with a flight deck and collapsible hangar for an Alouette light helicopter, a launcher for the prototype of the Malafon anti-submarine missile, a 305mm 4-barrelled A/S mortar of the type currently being installed in the latest 'fast escorts' (Type E 52B), 550mm trainable tubes for L3 long-range anti-submarine torpedoes, and complementary fixed and towed sonars (initially the American SQS-503 high-frequency sonar and the French experimental DRBV 40 Y, which used the same electronics as the fixed sonar). In this new configuration the ship was redesignated T 56.

As initially conceived, the FLE design was to have embarked a helicopter, but once the requirements of the Masurca launch system and handling arrangements became clear this idea was abandoned. The collapsible hangar of *La Galissonnière* proved complex and unwieldy. Doubts were also expressed about the all-weather capability of light helicopters and drones, meaning that for long periods of the year availability would be low, particularly in the hostile waters of the North Atlantic and the

A Malafon anti-submarine missile undergoing maintenance in the 'hangar'. (Jean Moulin collection)

Suffren: Plan of Malafon Hangar

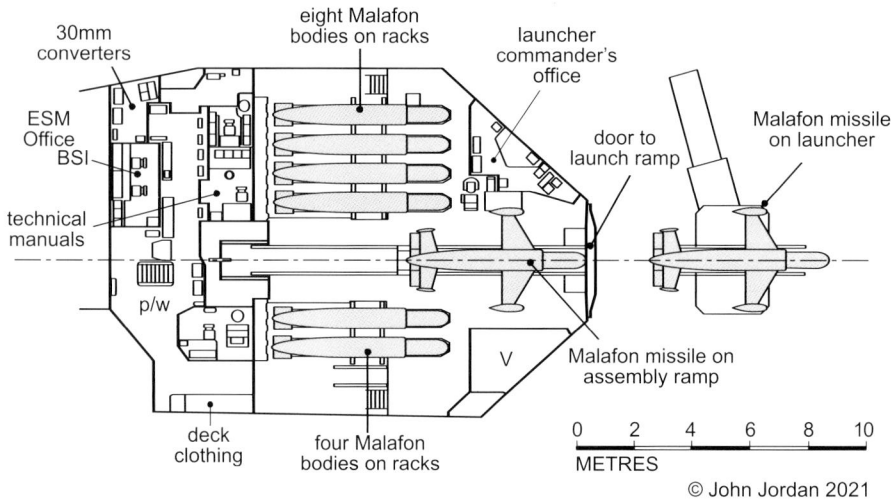

30mm converters

ESM Office BSI

technical manuals

eight Malafon bodies on racks

launcher commander's office

door to launch ramp

Malafon missile on launcher

p/w

V

Malafon missile on assembly ramp

deck clothing

four Malafon bodies on racks

The Malafon hangar could stow a total of 13 missiles, of which twelve were on racks and the 13th on an assembly ramp where the fins were fitted before the missile was run out onto the single launcher.

0 2 4 6 8 10
METRES
© John Jordan 2021

Norwegian Sea. It would be some years before more capable all-weather helicopters such as the Royal Navy's Lynx and the US SH-2 Seasprite would enter service. The anti-submarine weapon of choice throughout the 1960s was therefore the long-range A/S missile.

Malafon (MArine LAtécoère FONd) was a larger, heavier and more bulky missile than either of its foreign contemporaries, the Australian/British Ikara and the American ASROC (see table for comparison), but had greater range. Designed and built by the aircraft manufacturer Latécoère, it comprised an airframe wrapped around a short 533mm L4 homing torpedo that constituted its forward end, with twin detachable booster rockets slung beneath the tail-fins, and was housed in a 'hangar' located on the 1st Deck (see GA Plans), not in a box launcher (ASROC) or a below-decks magazine (Ikara). Twelve missile bodies were stowed on racks in two rows and a thirteenth on an angled 'assembly' ramp at the forward end of the hangar, where the fins were fitted before the missiles were run out. The single launcher, which was amidships and protected from the worst of the elements by the superstructures, was trained using hydraulic motors. A missile could be fired at a fixed angle of 15 degrees every 90 seconds.

Malafon was remotely controlled (for direction only) from the operations centre, using shipborne sensors; it was normally tracked by the DRBC 32A gunnery fire control radar. The boosters were ditched 4 seconds after launch; the missile then cruised at an altitude of 200m, releasing its torpedo 800m from the target (see graphic). The torpedo descended on a parachute, then followed a helical search pattern, using its acoustic homing head to locate the submarine. The L4 torpedo had a length of 3.13m and weighed 540kg including a 150kg warhead. It had a maximum range of 5,000m at 30 knots and was effective down to a depth of 300m.

Malafon was backed up by the L3 550mm heavyweight A/S torpedo, ten of which were stowed on four catapults and six cradles in a second hangar on the 1st

Suffren: Torpedo Room

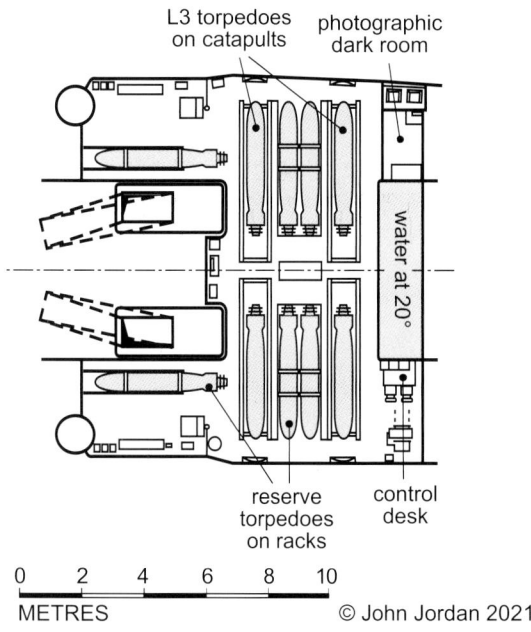

L3 torpedoes on catapults

photographic dark room

water at 20°

reserve torpedoes on racks

control desk

0 2 4 6 8 10
METRES
© John Jordan 2021

Anti-submarine Missiles

	Ikara	ASROC	Malafon
Type	cruise missile	ballistic rocket	cruise missile
Length	3.43m	4.5m	5.85m
Span	1.53m	0.84m	3.19m
Weight	600kg	430kg	1,330kg
Speed	260m/sec	340m/sec	230m/sec
Range	10,000m	9,000m	12,000m
Payload	Mk 44 324mm torpedo	Mk 44 324mm torpedo	L4 533mm torpedo

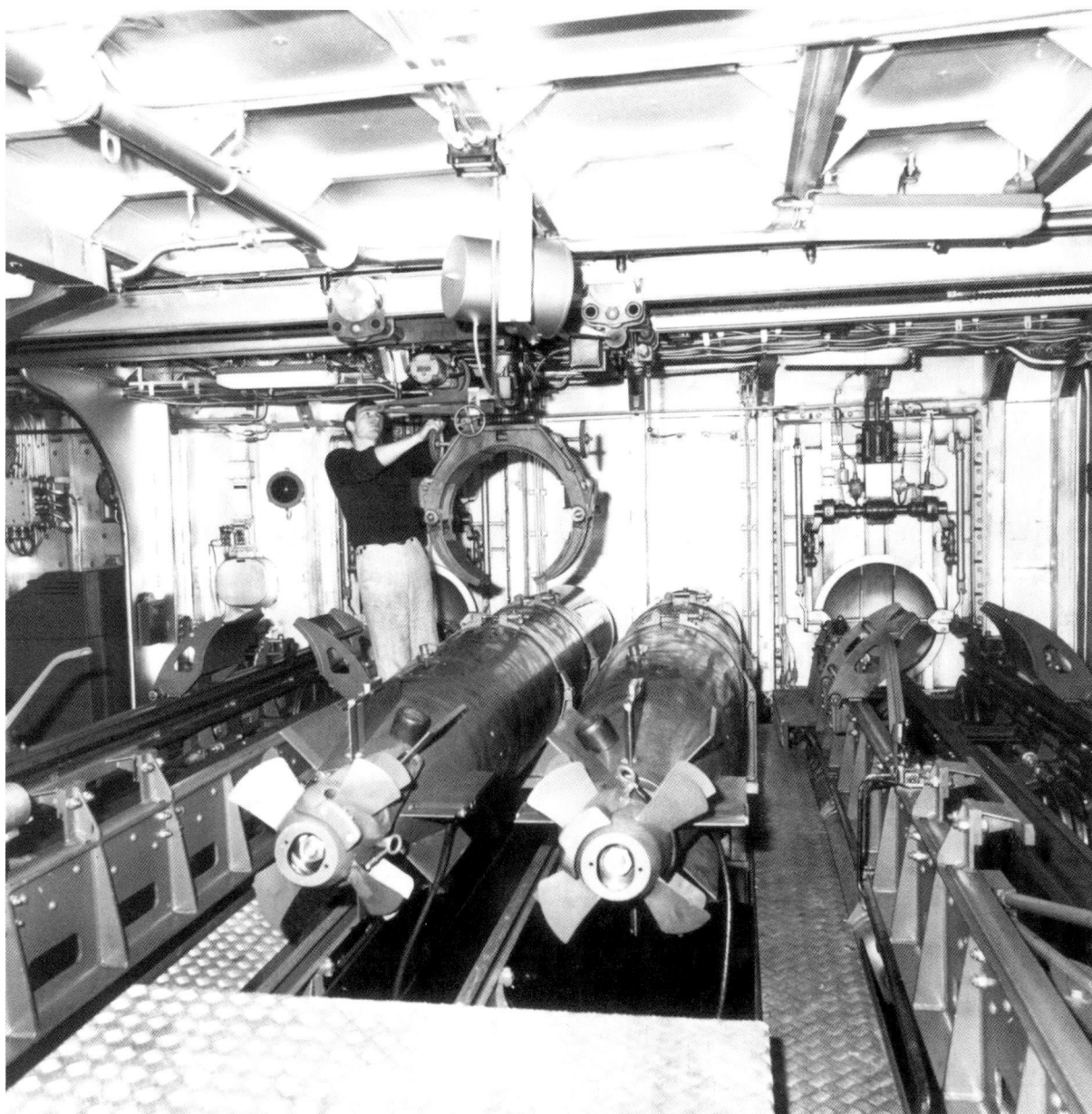

The torpedo room of *Suffren* in May 1968. The torpedoes in the photo are the 550mm L3 'heavyweight' anti-submarine model, which had a maximum range of 5,000 metres. On either side of them are two of the four catapults, which launched the torpedoes through circular apertures in the sides of the 'hangar' which were normally closed by hemispherical caps. (Jean Moulin collection)

Deck just forward of the boiler uptakes (see GA Plans). The torpedoes were launched by a simple pneumatic ejection system using compressed air through circular hatchways in the sides of the hangar. The catapults and four of the reload cradles were angled at 90 degrees to the ship's axis, and the torpedoes were handled by a system of overhead rails (see accompanying drawing).

The L3 was an electrically-propelled anti-submarine homing torpedo. A nickel-cadmium battery powered a 40kW motor that drove two contra-rotating propellers; speed was 25 knots and maximum range about 5,000 metres. The torpedo was launched in the direction of the last-known position of the target and ran at a depth of 30 metres, conducting a sinuous search after reaching its enable point. The active acoustic seeker from Alcatel had an acquisition range of 600 metres, and the torpedo could dive to a depth of 300 metres.

Introduced in 1960, the L3 torpedo was originally conceived for the 'short' 550mm torpedo tubes of the E 52 'fast escorts' and the *avisos-escorteurs*; it was 4.32m long and weighed 910kg (including a 200kg HBX-3 or TNT warhead). It was developed at Saint-Tropez and entered service in 1961. *Suffren* was the first ship to have the traditional trainable tubes replaced by catapults (Mle RD 59), which subsequently became standard on French escorts.

The L3 was superseded from 1972 by a new 533mm torpedo designated L5, designed to take advantage of the improved detection ranges of new sonars. The L5 Mle 4 torpedo had a length of 4.4m and weighed 935kg including a 150kg warhead; it remained therefore a comparatively 'heavyweight' model compared with the short-range 324mm A/S torpedoes fitted in contemporary RN and US Navy escorts. A silver-zinc battery was combined with a two-rotor electric motor to obtain a speed of 35 knots; maximum range was 7,000m. The L5 had an active acoustic homing head and could dive to a depth of 500m.

Sonars

The new stand-off weapons needed to be matched with a new generation of low-frequency (LF) sonars capable of submarine detection and target designation at long range. When the US Navy decided, at a relatively late stage, that their first missile frigates of the *Farragut* class would mount ASROC, the planned SQS-4 sonar (14kHz), which had a nominal range of only 5,000 yards, had to be replaced by the larger, more capable SQS-23 (4.5–5.5kHz). Because the frigates had a conventional hull it was considered prudent to house the new sonar in a hull dome, as a prominent bow bulb would have adversely affected shiphandling characteristics.

Suffren and *Duquesne*, on the other hand, were designed from the outset with a hull form suitable for mounting a large bow sonar and a matching variable depth sonar deployed from the stern. However, the French lagged behind the Americans in sonar technology, and the new sonars had to be developed from scratch during the 1960s.

The primary underwater sensors fitted in *Suffren* and *Duquesne* were the paired DUBV 23 and DUBV 43 sonars; these could be used independently or in combination (see below). The DUBV 23 bow sonar operated between 4.9kHz and 5.4kHz, which was at the high end of the low-frequency spectrum. The transducer weighed 7,500kg and had 48 staves with preformed beams. It was suspended for stabilisation within a bow sonar bulb 9.35 metres long and 1.85 metres wide. The French wanted to achieve reliable direct-path detection at ranges of 10,000–15,000 metres with a view to providing target data for the Malafon missile; long-range detection using the 'bottom bounce' or 'convergence' modes, which were key features of the contemporary US SQS-26, was of less interest. Accuracy was claimed to be 150m (±1%) in range and one degree in bearing. A single operator was needed for surveillance, but three other consoles were required to track the target. The DUBV 23 sonar was trialled successfully on *La Galissonnière* from 1965 and would be installed in all major ASW ships completed 1960s and 1970s.

Its partner, the DUBV 43 variable depth sonar, featured

Duquesne at Toulon on 13 July 1982. She retains the original TACAN aerial, but the 30mm cannon have beeen displaced by the launchers for Exocet SSMs and replaced by four standard 20mm guns in open pedestal mountings. Note the Syllex chaff dispensers abreast the DRBI 23 radome. (Pradignac & Leo)

Suffren: Command Spaces

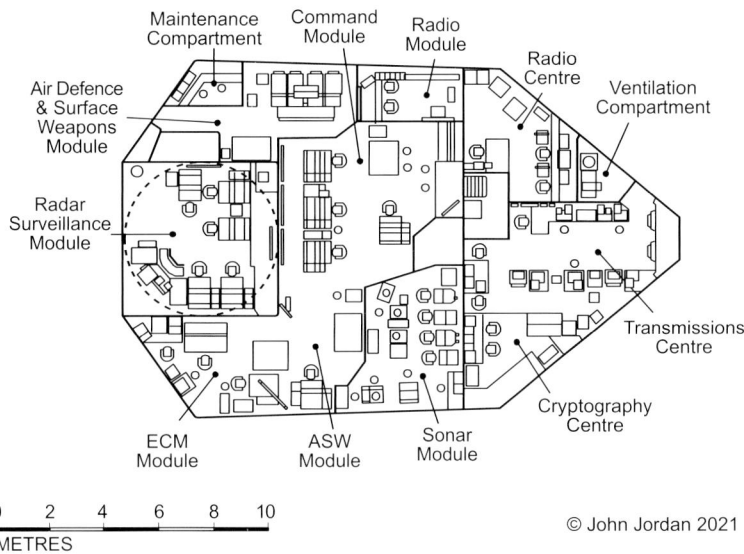

Maintenance Compartment
Command Module
Radio Module
Radio Centre
Ventilation Compartment
Air Defence & Surface Weapons Module
Radar Surveillance Module
Transmissions Centre
ECM Module
ASW Module
Sonar Module
Cryptography Centre

The drawing is based on the official plans, and shows the command spaces when the ship was first completed in the late 1960s. A central Command Module was surrounded by separate modules for Air Defence and Surface Weapons, Sonar and ASW, Radar Surveillance, Electronic Countermeasures (ECM) and Communications.

0 2 4 6 8 10
METRES

© John Jordan 2021

a large 'fish' housing a similar array to the DUBV 23, and the sonar and its handling mechanisms occupied a considerable part of quarterdeck. The transducer featured 24 staves divided into eight columns (192 elements) with preformed beams. The fish weighed 10 tonnes in the air and 7.75 tonnes submerged; dimensions were 5.5m (L) x 1.7m (W) x 3.7m (H). It was fully stabilised and could be towed to a depth of 200m at speeds between 4 and 24 knots using a cable 4cm thick and 250 metres long. The mechanical assembly, comprising a winch and a hydraulically-operated pivoting arm, was built by Ateliers & Chantiers du Havre.

The commitment of the *Marine Nationale* to the large variable depth sonar was prompted by its major areas of deployment; it was particularly useful for looking beneath the thermal layers prevalent in the Mediterranean. Development began in 1963, when a prototype VDS of French conception designated DUBV 40 Y was enbarked on the fleet escorts *La Galissonnière*, *D'Estrées* and *Vauquelin*. A pre-production model of the DUBV 43 was trialled on *D'Estrées* from 1968, and one of the first production models was fitted in *Suffren* during a maintenance period Oct 1970–Feb 1972. The DUBV 43 sonar would be modernised during the 1988–1991 refits.

The detection range of the DUBV 23 and 43 paired sonars was 8–10,000 metres in normal conditions, but 20,000 metres was possible. Performance was affected not only by thermal layers, but also by currents, salinity, depth of water and noise. There were four possible modes:

1 autonomous (both sonars operated independently)
2 23 synchronised: both sonars piloted by DRBV 23
3 43 synchronised: both sonars piloted by DRBV 43
4 43 attack: as no 3 but target data supplied by DRBV 43.

It was possible to conduct an attack while still performing an all-round surveillance scan.

Anti-ship Missiles

Staff requirements for the 'escort cruiser' of the mid-1950s specified Malaface anti-ship missiles in addition to the Masalca and Masurca surface-to-air missiles. Malaface (MArine LAtécoère SurFACE) was a subsonic missile that employed a conventional airframe and a liquid propellant and had four solid-fuel boosters. When the missile arrived within 500 metres of the target, it dived and released a 700kg streamlined bomb that had four stabilisation fins. A small TV camera in the nose allowed the aim to be adjusted.

Development began in 1951 with trials from 1954. These were a failure and were suspended for two years to allow some of the technical issues to be resolved. Further trials took place in 1957–58 but these highlighted problems with the guidance system, in particular the electronics and the electro-mechanical components. On 16 June 1958 Malaface-10 was successfully launched and struck a target 16,000 metres away. However, the missile was still considered unreliable, and despite a successful trial of Malaface-15 in March 1959 the Navy decided to terminate the programme in February 1960, shortly after the final specifications for *Suffren* and *Duquesne* were drawn up. On a tactical level it was now accepted that strikes on surface ships would be more effectively carried out by carrier aircraft, which could conduct their attacks over the horizon. In the FLE 60 design Malaface was effectively replaced by Malafon, which was likewise built by Latécoère but had solid-fuel propulsion and did not need active homing, being programmed simply to release its A/S torpedo at the end of its run.

The development of a French anti-ship missile was

revived following the Styx attack on the Israeli destroyer *Eilath* in 1967. There were three initial projects: Exocet (from Nord Aviation), Mercure 4S (Thomson CSF) and Mer-Mer 4 (Matra). Exocet had its origins in the CT 20 target missile, which was developed as the turbojet-powered Rb 08 for the Swedish Navy. Nord Aviation had considerable experience in short-range anti-ship missiles, being responsible for the air-launched AS 30 (1958–64), then the Kormoran ASM in collaboration with the German company MBB. The proposal for the MM 38 Exocet was ready as a paper project in 1967 and was accepted by the Navy on 16 June 1969; the homing head used an X-band monopulse radar with a Cassegrain antenna supplied by Electronique Marcel Dassault (EMD). On 1 January 1970 Nord Aviation merged wih Sud Aviation and SEREB to form Aérospatiale.

The first test firing of Exocet took place on the Ile du Levant on 8 July 1970, and the first at-sea launch (from the patrol boat *La Combattante*) a week later. Five more firings were followed by further trials on board *Ile d'Oléron*. The first firing using inertial guidance took place on 19 February 1971, and on 17 November of the same year a missile struck its target, the ex-German submarine support ship *Gustave Zédé*, at a range of 37km. The first operational firing took place on 26 January 1972.

MM 38 Exocet weighed 735kg (including a 165kg warhead) and was stowed in a 750kg lightly-pressurised container, so required no maintenance. Missile dimen-sions were 5.2m x 0.35m with a 1-metre span once the fins were deployed. It was launched at a fixed angle of +15 degrees. The solid-fuel booster burned for 2.5 secs and took the missile to Mach 0.93 at an altitude of 30–70m, and the 150kg solid-fuel cruise rocket then took over, burning for 110–120 secs. A radar altimeter kept the missile at 9–15 metres above the surface until the attack phase, when the seeker locked onto the target. The final phase was conducted at an altitude of 3–8 metres, depending on the sea state. Maximum range was 42,000 metres, corresponding to the radar horizon of a medium-size surface ship. During the Falklands conflict of 1982 it was found that the detonation of the warhead on impact was unreliable, but that the unex-pended solid propellant often resulted in serious fires, causing the loss of HMS *Sheffield* and serious damage to HMS *Glamorgan*.

Once the MM 38 entered service it was fitted in all major French surface warships. Two paired launchers were fitted in *Duquesne* 1975–76 (Brest) and in *Suffren* 1978–80 (Toulon). They were located atop the Malafon hangar and replaced the former 30mm automatic guns (see photos).

Surveillance and Electronic Support Measures

The DRBV 50 air/surface radar, the antenna for which was mounted on the central mack, was the standard

Duquesne departing Toulon on 4 April 2005, two years before she finally decommisiioned. During her half-life refit Syllex and the 20mm AA guns were replaced by two Sagaie chaff dispensers; the port-side Sagaie can be seen abaft the launchers for Exocet. (Pradignac & Leo)

A fine view of *Duquesne* in her final configuration. Note the saucer-shaped TACAN beacon which replaced the earlier model. (DR)

model of the period. A frequency-agile S-band radar with pulse compression, it could detect an aircraft at 110km and a large surface unit at 20–30nm. In addition to *Suffren* and *Duquesne*, it was fitted in the two carriers, the modernised fleet escorts and the cruiser *Colbert*. It was complemented by DRBN 32, a navigation radar from Decca, which was mounted atop the bridge. DRBV 50 would be replaced by DRBV 15 1984–89, and DRBN 32 by a Decca 1226.

The sensors for electronic support measures were distributed between the mack and a short lattice mast located atop the Malafon hangar. On completion the two ships were fitted with ARBR 10E ESM (atop mack) and ARBB 30/31 ECM jammers (lattice mast); there was also a Fanfare towed torpedo decoy. The outfit was completed in 1975 with ARBB 32, which created a false echo to deflect SSMs, and two Syllex chaff dispensers, derived from the British Knebworth Corvus, which were located on platforms abeam the radome for DRBI 23.

In 1986 ARBR 10 was replaced by ARBR 16 with an ARBX 10 frequency meter, and during their half-life refits 1988–91 the two ships received a complete ESM/ECM upgrade. ARBR 16 was superseded by ARBR 17 (DR4000), and the jammers by DRBB 33, the antennae being located on either side of the radome in place of Syllex. The decoy launchers were replaced by the French-developed Sagaie, which could be trained and

elevated: there were ten tubes each holding a 170mm-diameter, 45kg rocket capable of deploying infrared and electromagnetic chaff out to 3,000 metres. The two Sagaie launchers were located to port and starboard abaft the Exocet launch canisters. The Fanfare torpedo decoy was replaced by the US SLQ-25 Nixie.

Atop the lattice mainmast was a TACAN beacon that allowed aircraft to self-locate in elevation and azimuth in relation to the ship; range was 180km. The original model, which was housed in a radome, was replaced by the saucer-shaped NRBP 2A in 1987.

Suffren and *Duquesne* were among the first ships to have SYRACUSE (*SYstème de RAdio Communication Utilisant un SatellitE*), which was fitted from 1983. The associated satellites, from MATRA and Thomson CSF, were put into orbit by an Ariane rocket 1984–88. Each satellite had two receivers and twelve locators, and used super high frequency (SHF). The onboard installation comprised a shipboard shelter weighing 4 tonnes and two radomes each housing a 1.5m-diameter antenna. Syracuse was operational from 1985 to 1994.

From 1988–91 INMARSAT (*INternational MARitime SATellite*) was also installed. Four satellites provided telecoms facilities and there were 37 land-based stations. The *Marine Nationale* acquired 35 of a navalised version and two SPIN antennae were fitted in *Suffren* and *Duquesne* during their half-life refit.

Duquesne awaiting disposal on 22 June 2014. The gun mountings and all the sensitive items of equipment have been removed.
(Jean Moulin)

Data Exploitation

Suffren had the first SENIT (*Système d'Exploitation Navale des Informations Tactiques*) tactical data system installed on completion. Similar in conception to the British CDS and US NTDS, its key functions were:

– display of tactical situation
– threat evaluation
– assistance in decision making
– weapons deployment
– transmission of data to other units

SENIT was fully automated to prevent human error and avoid overload – a maximum of 15 targets could be handled manually. However, in practice SENIT supplied the fire control data, but the decision to fire the weapons was made by an officer.

The consoles and plotting tables were located in a large Operations Centre directly below the enclosed bridge (see drawings). Operations were supervised by the *Officier Chef de Quart*. Data transmission to other units was via NATO Link 11 or HF/UHF radio, while Link 14 would allow transmission to other units not fitted with SENIT. The modular system was divided into areas of warfare. There was a central 'decision-making' module for the flag and commanding officers with information displays which included the air picture, and surrounding modules for anti-aircraft, anti-submarine, anti-surface and electronic warfare, and for transmissions.

SENIT used three Univac 1206 computers (located directly below the operations room) capable of handling two million data elements and making 75,000 calculations per second. These were superseded by the Univac 1212 model during the major half-life refits of 1988–91, when SENIT 1 was upgraded to v2; it was now capable of handling 128 tracks.

Evaluation

Suffren decommissioned in 2001; her sister *Duquesne* soldiered on for a further six years. They were to have been replaced in the task force air defence mission by the two units of the *Forbin* class ('Horizon' programme), but due to delays these were not in service until 2011, leaving the gap to be filled by the two ships of the C 70 AA type, which were armed with the shorter-range US Tartar missile.

Suffren and *Duquesne* were costly to build and to maintain due to the small scale of the original carrier escort programme, which was the victim of multiple cancellations to compensate for the huge expense of developing an independent nuclear deterrent. The Masurca and Malafon missiles had a larger footprint than their foreign contemporaries and were arguably less technically advanced. Nevertheless, *Suffren* and *Duquesne* were instrumental in moving French weapon and sensor technology into the modern era, and they had a major impact on the French surface navy, heavily influencing the designs that followed during the 1970s and 1980s.

THE ESCORT DESTROYERS OF THE *MATSU* AND *TACHIBANA* CLASSES

The IJN destroyers of the *Matsu* and *Tachibana* classes were designed in late 1942 to replace the 'fleet' destroyers then being lost or disabled in large numbers. Simplification with a view to mass production was required. **Kathrin Milanovich** looks at the conception and execution of this major wartime programme.

Overview

By November 1942 the Imperial Japanese Navy (IJN) was in urgent need of destroyers to compensate for the ships lost during the attritional Solomons campaign. Lacking escort vessels, the IJN had been forced to use *Kō* (A) type ships as escorts for landing operations and convoys, and in the Solomons they also had to perform transport missions, resulting in a sudden rise in losses. Escort and transportation duties could be carried out by simpler and vessels better suited to these tasks, which had never featured in the staff requirements for the fleet destroyers with their high speed and heavy gun/torpedo armament. It was therefore decided to build a new type

of destroyer designed for rapid wartime construction with simplified plans, structures, fittings and an armament better suited to the intended missions. Nine designs of the new 'D' (*Tei*) type destroyer – 'A' was the standard fleet type, 'B' (*Otsu*) the anti-aircraft type and 'C' (*Hei*) type the high-speed type – were drawn up, and in February 1943 the eighth of these (F 55H) was approved. Noteworthy features of the new destroyer were:

– Much-reduced displacement in order to compensate for shortages of steel and personnel in the shipbuilding industry and promote rapid production; the number of curved plates was to be kept to a minimum.

The 'D'-type escort destroyer *Take*, probably off Yokosuka on 30 May 1944. Note the very different shape of the shields for the single 12.7cm HA gun forward and the twin 12.7cm aft. Note also the light tripod masts, the spacing of the funnels (reflecting the layout of the boiler and engine rooms), and the prominent shield for the 61cm torpedo tubes with the platform for two of the four Type 96 triple 25mm MGs directly behind it. (Kure Maritime Museum)

– Restricting the use of high-quality steel (HTS) to the upper deck and sheer strake, and the employment of mild steel (MS) for all other parts of the ship – Ducol ('D') steel was completely eliminated.

– Adoption of a simplified propulsion system comprising two Kampon-type watertube boilers and low-powered Kampon geared turbines similar to those that equipped the torpedo boats of the *Otori* class and which were well-suited to rapid fabrication.

– Arrangement of the machinery spaces in a 'unit' system (BR/ER/BR/ER) in response to war lessons, to increase survivability in the event of damage. (This was controversial, as it ran counter to the principle of ease of production.)

– Mounting of Type 89 12.7cm 40-cal high angle (HA) guns instead of the 50-calibre 12.7cm low angle (LA) gun, and reinforcement of the close-range AA machine guns in response to the increased threat of air attacks.

– Embarkation of two 10-metre landing craft (*shōhatsu*) for transport duties.

Following completion of the detailed design and preparation of the working drawings, the first ship was laid down in August 1943 and completed in April 1944; this exceeded the planned building time of five months by three and a half months.

In the meantime the war situation had deteriorated, and the replacement of destroyer losses could not be expected with such lengthy building times. Also, material shortages and the limited capacity of the shipyards, which were overloaded with repair work, were having an impact, and in March 1944 the shortening of the building period to three months was urgently required.

At the time the proposed simplifications embodied in the design of *Matsu* class were discussed it was argued that they went too far. However, even these drastic measures proved to be insufficient, and extreme simplification (compared to the standard of 1942) had already been implemented in the designs of new types of escort (*Kaibōkan*) and naval transport (*Yusōkan*). The latter had been built using prefabricated blocks, which required an extended application of the electric welding first practised in the construction of wartime standard freighters. The design of the Modified *Matsu* class broadly followed the same lines, and the ship was entirely prefabricated with all-welded block construction using exclusively MS steel. The employment of MS steel meant that plate thickness had to be increased, but the slight increase in the displacement that resulted could not be avoided, despite of the weight-saving implicit in a welded structure.

In this second series of vessels the propulsion system was revised and simplified by removing the separate cruise turbine. The armament remained the same, but radar capability was increased and the underwater detection sensors were of the newest types with much-improved performance. The first ship of the new design (F 55B) was *Tachibana*. She was laid down at Yokosuka NY in July 1944 and completed in January 1945. It proved possible to reduce the construction period to five

months in some cases, but even this greatly exceeded the planned building time of three months.

Once design F 55B was ready, it was decided that all ships were to be completed to this configuration. However, in April 1945 the construction of ships not expected to be completed in the first half of FY 1945 was halted, and ships authorised under the two building programs (42 and 32 ships respectively) but yet to be laid down were cancelled. In the end 32 ships were completed (18 *Matsu* class and 14 *Tachibana* class), and nine ships were incomplete on the building ways at the end of the war.

Following work-up they were assigned to destroyer divisions of the Combined Fleet and employed for escort and transport missions, in the execution of which ten ships were lost. Some ships survived heavily damaged, others were undamaged and used for the postwar repatriation of demobilised personnel.

Origins of the Design

The strategy of the IJN ever since 1905 had been based on the annihilation of the enemy main force in a decisive battle as at Tsushima. In the Washington and London Naval Armament Limitation Treaties the IJN's ratio of tonnage to the US Navy had been restricted to 60 per cent, and this ratio was considered insufficient to win that battle. The IJN therefore divided the decisive battle into several stages and adopted an attrition strategy in order to reduce enemy force strength during the US fleet's transit across the Pacific, the goal being to obtain parity or even a slight superiority before the decisive gunnery battle. Torpedo forces were to play a key role in these early stages. The initial phase involved repeated attacks by Japanese submarines shadowing the enemy fleet during its transit, reporting course and formation and wearing the enemy down. The second stage involved night attacks by flotillas of destroyers led by light cruisers likewise armed with torpedoes, supported by specially-converted light cruisers with a heavy torpedo armament. These attacks were to be mounted immediately prior to the gunnery engagement.

The design of the IJN's fleet destroyers was based on this concept and their crews were trained accordingly. After the conclusion of Washington Treaty in 1922, 24 destroyers of the 'special type', the *Fubuki* class, were built for fleet work. They mounted nine torpedo tubes, and this heavy torpedo armament was also a feature of the destroyers built under the limitations of the London Treaty of 1930 (*Hatsuharu*, *Shiratsuyu* and *Asashio* classes) and the post-treaty designs (*Kagerō* and *Yūgumo* classes and 'C'-type fast destroyer *Shimakaze*). Due to the inferior force ratio imposed by the treaties the IJN also adopted the principle of 'quality before quantity', which meant that individual ships were to be superior to their US Navy and Royal Navy counterparts. The destroyer category was no exception, and this principle was retained after the expiry of the naval arms limitation treaties, and adhered to even after the disastrous defeat in

The 'D'-type destroyer *Momi* shortly after completion. She has the antenna for the No 22 surface search radar, with its distinctive double-horn configuration, on a post at the after end of the bridge structure. She has the full complement of 25mm MG. (Shizuo Fukui collection)

the Battle of Midway. In the following months eight 'A'-type destroyers of the *Yūgumo* class and 23 'B'-type destroyers of *Akizuki* class were completed, the latter being the IJN's only destroyer type designed for the protection of carrier task forces in the new building programme.

However, individual superiority had its price and resulted in the construction of comparatively few ships, particularly when set against the huge wartime building programme of the US Navy. While it was an understandable principle in peacetime, it was not well-suited to the attrition warfare that developed around the Solomon Islands in late 1942. Heavy losses meant that the IJN now needed 'quantity before quality', particularly for the destroyer, which was the most versatile multi-purpose warship type. The focus upon torpedo attacks in the IJN's prewar strategy brought about the neglect of traditional destroyer duties such as escort (defence against submarines and aircraft), patrol and reconnaissance, and resulted in the mounting of outdated weaponry and crews that were insufficiently trained for these purposes. The IJN was aware of these limitations, but believed them to be acceptable for the short war for which it had planned, with the decisive gunnery battle taking place within months rather than years.

The success of the first phase of operations seemed to confirm this assumption, and only seven destroyers were lost up to mid-1942. These could easily be replaced with the current building programme, and even after the radical change of policy following the Battle of Midway the IJN was happy to continue the construction of the first-class fleet and anti-aircraft destroyers already projected. However, the situation changed with the landing of American forces on Guadalcanal in August 1942, and the fighting in the Solomon Islands developed into fierce attritional warfare. The supply of the Japanese forces on Guadalcanal proved to be extremely difficult, and the losses of transports increased. The IJN was forced to rely on destroyers and submarines for resupply, and fleet destroyers used as transports had to off-load their reserve torpedoes. The 'A'-type fleet destroyer was poorly suited to transport duties, but became the main player in the many small-scale engagements that transpired, which were of an entirely different nature to those envisaged for the decisive battle. Damage and sinkings suddenly increased, and from August 1942 until the end of January 1943 no fewer than seventeen fleet destroyers were lost and more than 40 damaged. Between May and December 1942 the Solomons theatre alone saw twelve destroyers lost and 35 damaged while acting as escorts of landing and supply forces, then as transports. These losses were unsustainable with the current programmes. If the attrition persisted the shortage of destroyers would limit the number and scale of the operations that could be undertaken, Rapid replacement and a dramatic increase in the number of ships to be laid down became urgent.

In recognition of this situation, in November 1942 the Navy General Staff (NGS) asked the Navy Technical Department (NTD) to design a new destroyer smaller than the fleet type to be used for landing, supply, and escort operations and to be well-suited to rapid mass production. In response the NTD drew up nine sketch designs numbered F 55A to F 55I, based on the following principles:

– reduction in size (with savings in material and construction time)

– adoption of construction methods suited to mass production (the complexities of past destroyer designs were to be avoided)
– a primary focus on defence against air attacks, but equal consideration accorded to anti-submarine capabilities
– war lessons to be absorbed, with particular attention to survivability in the damaged condition
– high speed less important.

Low-angle or dual-purpose main guns were to be discarded in favour of either two to four 12.7cm or 12cm (or long 8cm) HA guns and more than twelve 25mm MG as close-in AA weapons. The torpedo armament of the sketch designs ranged from two quadruple 61cm torpedo tubes to a single bank of sextuple 53cm tubes, six torpedoes being considered the minimum for an effective torpedo attack. This comparatively powerful torpedo armament indicates clearly that irrespective of the different character of the new ships, torpedoes continued to be regarded as the main armament of a destroyer. As for propulsion three different types of engine were projected, taking account of the limited production capacity of the Japanese machine industry:

– single-shaft propulsion using one V 7 type high-pressure and high-temperature (HPHT) turbine set rated at 26,000shp
– two-shaft propulsion using two HPHT turbine sets of 19,000shp as installed in the torpedo boats of the *Otori* class
– two-shaft propulsion using two newly-designed HPHT turbine sets.

Maximum speed would be from 28 knots to 32 knots, depending on the propulsion system selected, and range from 3,500nm to 6,000nm.

Single-shaft propulsion was rejected due to the need for survivability (particularly damage to the engine room). The design of a new type of turbine would require several months and might also interfere with urgently needed engines that were already in production. The second proposal was therefore preferred, despite the comparatively low speed of 27.8 knots and the limited range of 3,500nm at 18 knots. After much discussion design F 55H was finally approved.

Design and Building Plans, Budgets and Results

Following this decision the NGS, on 2 February 1943 (secret document No 37), requested the construction of 42 destroyers of the new simplified type as the second addition to the Modified Fifth Replenishment Program (*Maru gō keikaku*) from the Navy Ministry. These ships were to be completed from FY 1943 to the end of FY 1945 (31 March 1946) and were to have the following characteristics:

Displacement	1,250 tons standard
Speed	28 knots
Range	3,500nm at 18kts

Equipment:
– three 12.7cm HA guns (1 x II, 1 x I)
– twelve 25mm MG (4 x III)
– six 53cm TT (1 x VI) with six torpedoes
– 36 depth charges + two DCT
– sonar and hydrophone
– radar detector
– two 10-metre transports (*tokei unkasen*)

The preliminary design was rapidly completed with the characteristics detailed above as F 55, and the detailed design also proceded quickly in company with the working drawings and the ordering of materials. The first ship of the new class, *Matsu*, was laid down at Maizuru Navy Yard on 8 August 1943, launched on 3 February 1944 and completed on 28 April of the same year.

The laying down of *Matsu* anticipated the approval of the budget for 32 'D'-type destroyers of 1,260 tons standard with building costs of ¥9,326,000 per ship, a total of ¥298,432,000; these provisional expenses for FY 1944 were put before the 84th session of the Diet (opened 26 December 1943) and were approved on 15 February 1944. The budget for a further ten ships was submitted to the 86th session (opened 26 December 1944) and approved on 1 February 1945. The building costs were calculated as ¥9,614,000 per ship, total ¥96,140,000. These ships were included in the second addition to the building programme proposed by the NGS and given the provisional building numbers 5481 to 5522.

A request for a further 32 ships, also with a unit cost of ¥9,614,000, was submitted to the same 86th session of the Diet as part of the so-called Wartime Armament Programme (*Maru sen keikaku*) and approved as temporary military expenses for FY 1945. They were given the provisional building numbers 4801 to 4832 and all belonged to the modified *Matsu* (*Tachibana*) class.

Hull Form and General Arrangement

The hull form followed the traditional style adopted from the *Fubuki* class onwards, with a long forecastle. However, a simplification of the body plan and the structure was considered essential to promote more rapid construction. Flat plates and plates with curvature in only one direction were employed, resulting in a comparatively 'flat', straight-lined hull that avoided the complex and time-consuming fabrication of two-dimensional and three-dimensional curved plates used in earlier destroyers. The flare of the bow was minimal and the forefoot was cut up at an angle.

The employment of high-quality Ducol ('D') steel for the strength members was abandoned because it could not be produced in sufficient quantity and also because it was difficult to weld. High-tensile steel (HTS) was

Table 1: **Building Programme**

Programme	No of ships	Laid down	Launched	Completed	Not laid down
2nd Addition to Circle 5 Programme	42	31	28	26	11
Wartime Armament Programme	32	10	7	6	22
Total	74	41	35	32	33

Ships ordered under the Modified Circle 5 Programme

	Maizuru	No	Yokosuka	No	Fujinagata	No	Total
Laid down	*Matsu, Momo, Maki, Kaya, Tsubaki*	5	*Take, Kiri, Momi, Yaezakura, Yadake, Kuzu, Sakura, Kaki, Hinoki, Kaede, Keyaki, Tachibana, Tsuta, Hagi, Sumire, Kusunoki, Hatsuzakura*	17	*Ume, Kuwa, Sugi, Kashi, Nara, Yanagi, Kaba, Katsura, Wakazakura*	9	31
Launched	same as above	5	same except *Yadake, Kuzu*	15	same except *Wakazakura*	8	28
Completed	same as above	5	same except *Yaezakura, Yadake, Kuzu*	14	same except *Katsura, Wakazakura*	7	26

Wartime Armament Programme

	Maizuru	No	Yokosuka	No	Kawasaki Kōbe	No	Total
Laid down	*Nire, Shii, Enoki, Odake Hatsuume, Tochi, Hishi*	7	*Azusa, Sasaki*	2	*Nashi*	1	10
Launched	same except *Hishi*	6	–	–	same	1	7
Completed	same except *Tochi, Hishi*	5	–	–	same	1	6

adopted instead, and even then its application was limited to the upper deck and the sheer strake. Other parts of the hull were of ordinary shipbuilding (mild) steel, which could be produced in large quantities and was easily welded. However, the thickness of the plates had to be increased in order to maintain strength, so the weight of the hull increased to 32.9 per cent of the trial displacement – for *Kagerō* the figure was 29.2 per cent and for the *Akizuki* class 31.7 per cent. Fittings were also simplified, and every effort was made to fit standardised equipment that was easy to produce.

The bridge was located at the after end of the forecastle deck with a machine gun platform in front and a tripod foremast behind it. In contrast to the elegant, streamlined three-level bridge adopted for destroyers up to the *Yūgumo* class, it was a simple box-type structure incorporating only two decks, and reflected the attempt to reduce the use of curved plates to a minimum. Abaft the foremast there were two slim funnels with a quadruple torpedo mounting for 61cm torpedoes and a machine gun platform between. Abaft the second funnel there was a deckhouse with a searchlight platform and the tripod mainmast with the after MG platform attached. Abaft the after (twin) main gun mounting there were two Type 94 'Y-gun' type depth charge throwers and their loading platforms, with two DC rails at the stern. The forward (single) main gun mounting was located on the forecastle forward of the MG platform and the bridge. There were 6-metre whalers on either side of the forefunnel, and the two small transport boats were similarly disposed on either side of the second funnel. (For more detail see the plans.)

Machinery

The layout of the machinery spaces is shown in the accompanying drawing. The adoption of a 'unit' system was a particular characteristic of this class and constituted the IJN's first attempt to improve survivability in a ship of this size. Whereas in earlier destroyers the two engine rooms were abaft the two boiler rooms, in the new ships the boiler and engine rooms alternated and were located in individual compartments separated by transverse bulkheads. Each of the two boilers was combined with a single set of turbines to form an independent unit, and there were no cross-connections between the main steam lines of the two plants. This unit arrangement had been introduced by the French Navy and the US Navy for their lightly-built 'treaty' cruisers of the 1920s, and the layout had subsequently been adopted by the Royal Navy.

The adoption of the unit arrangement made it more likely that a ship suffering torpedo, bomb or shell damage amidships could retain its mobility, albeit at a reduced speed. Disadvantages included an increase in hull length and the different lengths and angle of inclination of the port and starboard shafts, which affected the turning circle and introduced greater complexity. In the *Matsu* class the port shaft was composed of no fewer than seven elements (thrust block, four sections of line shafting, a stern tube shaft and the propeller shaft), while the starboard shaft had only a single line shaft section. Concerns were expressed that the adoption of unit propulsion would hinder mass production, but survivability concerns eventually prevailed.

The turbines were offset to port (forward engine room)

On the docking drawing of "Take" the polygonal line is erased and the stem is straight (dashed line)

No. 22 radar

Searchlight controller

Depth charge release tracks

Type 94 depth charge throwers (Y type guns)

Profile and plan of the 'D'-Type destroyer escort *Take*. (*Gakken* No 43)

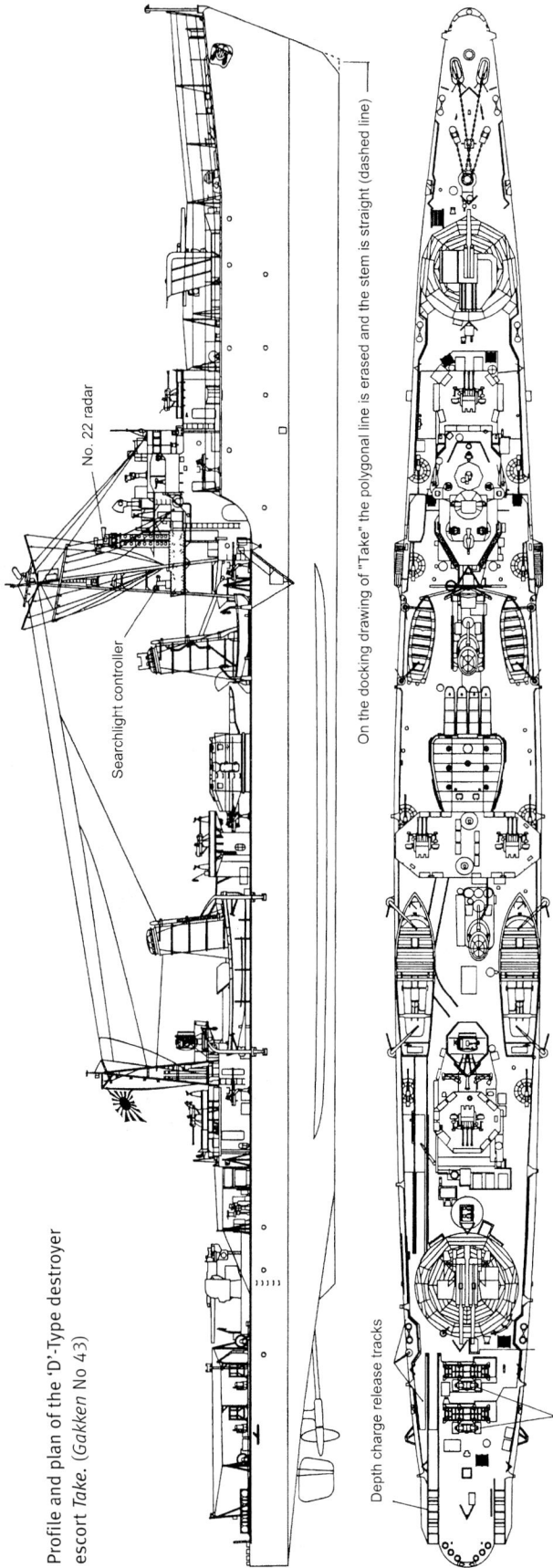

No. 22 radar

Searchlight controller

No. 13 radar

Single machine gun stand

Profile and plan of the Modified 'D'-Type destroyer escort *Tachibana*. (*Gakken* No 43)

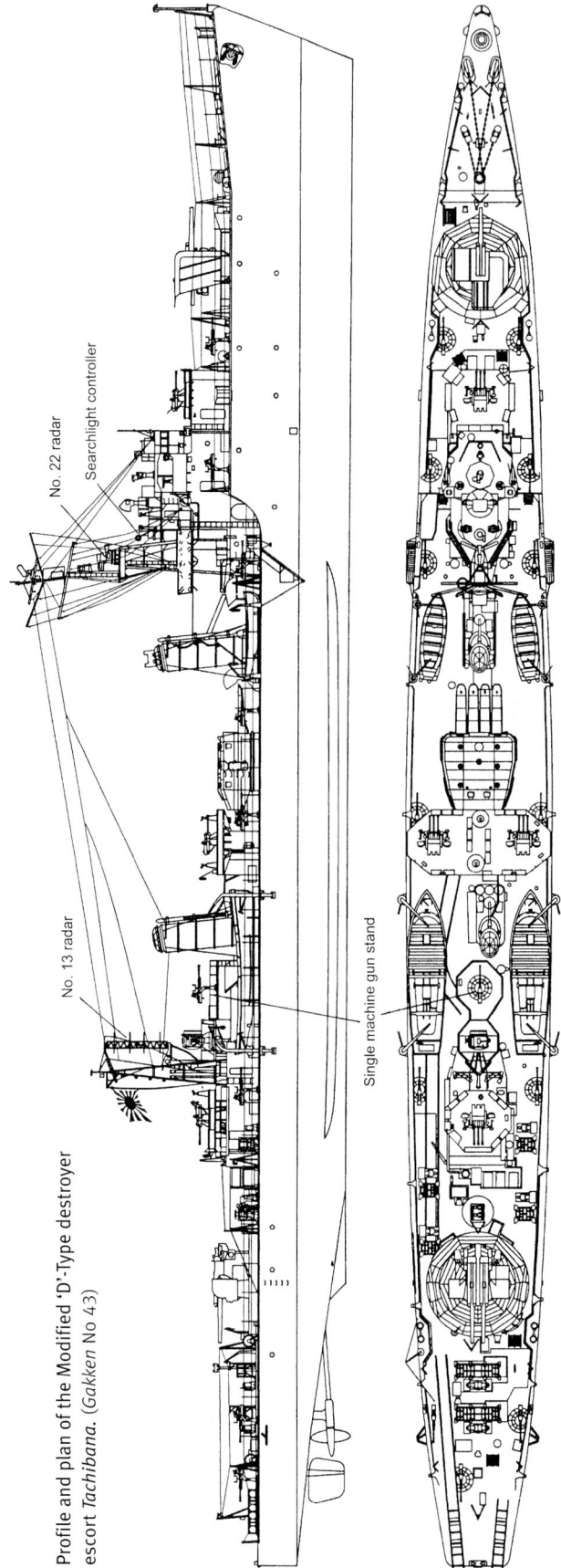

Bow

Midship

Knuckle

Midship

Bow

Bow view

Stern view

No upper deck camber

Bow view

Stern view

Upper deck camber

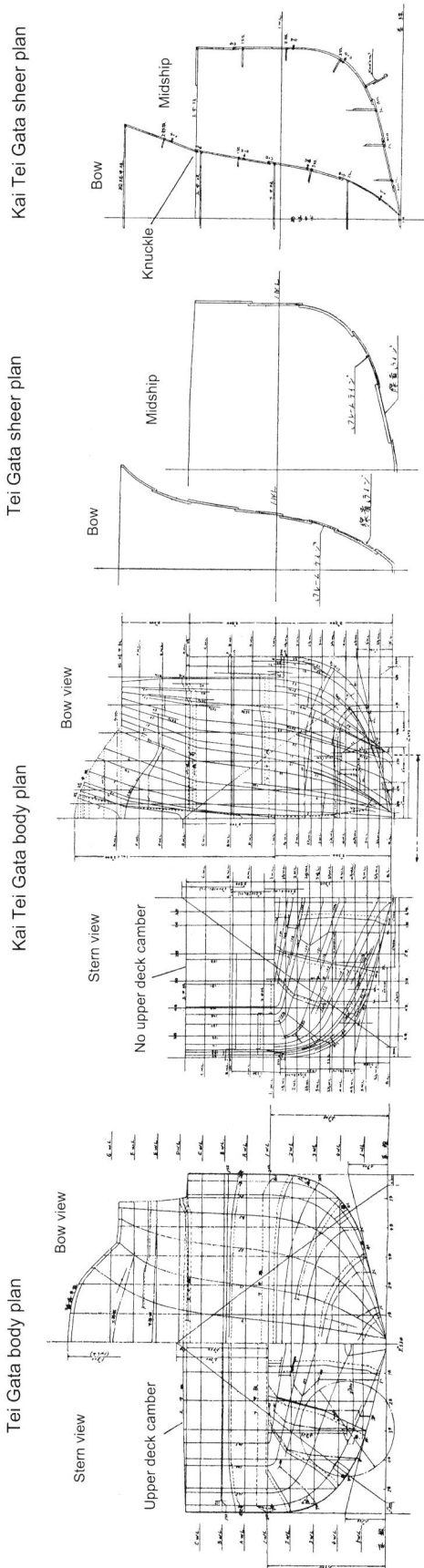

Comparative body and sheer plans of the 'D'-Type and Modified 'D'-Type destroyer escorts. The body plan shows the simplification of the lines intended to reduce construction times. Note the overlapping plates in the sheer plans. (*Gakken* No 43)

and to starboard (after engine room). Two diesel generators were mounted alongside the turbines to port in ER2 with a single unit to starboard in ER1. This division of the generators into two independent units, each with its own switchboard, likewise reflected the importance of damage control following the bitter lessons of the war.

The two main engines were Kampon-type single reduction turbine sets (No 3C, Model 5481) combining a high-, an intermediate-, and a low-pressure turbine. The HP turbine was attached to the front of the LP turbine, and the astern turbine was housed in the casing for the latter. A main condenser of the single-flow surface condensing type was suspended beneath the LP turbine. This type of machinery had been designed for the torpedo boats of the *Otori* class and was chosen because of the ease of production. However, the rotation (in revolutions per minute/rpm) of the propeller shafts differed from the torpedo boats, so that the main reduction gear had to be designed from scratch. As specified, the output of the turbines was 19,000shp with 400rpm, and they were to work with a steam pressure of 26kg/cm^2 and a steam temperature of 335°C with a vacuum of 700mm; steam consumption was 3.85kg of steam per shp/hr. The capacity of the astern turbine at 247rpm was rated at 2,000shp on a single shaft for a total of 4,000shp using both shafts (see Table 2).

The cruise turbine (No 3A, Model 5481) and cruise reduction gearing (No 1A, Model 5481) were connected to the port-side IP turbine in the forward engine room. There was no cruise turbine in the after engine room; the cruise turbine exhaust passed to the HP turbine in the after engine room, and from there followed the steam flow that normally entered the starboard engine via the ahead throttle. The rated capacity at cruise full power was 3,000shp for both shafts, but by using the main engines at the same time without disengaging the cruise turbine up to 280 shaft rpm (about 21.5 knots) the output could be doubled. However, the cruise turbine and its shafting was suppressed in the later units after it was found to be impractical for a ship of this size.

The two boilers were of the Kampon *ro gō* oil burning type (No 3B, Model 5481) with superheater and air preheater, and worked with a steam pressure of 32kg/cm^2 in the steam drum, 30kg/cm^2 at the superheater outlet and a steam temperature of 350°C. The pressure and temperature stated above in relation to the turbines were those measured at the turbine inlets, and the differences in pressure and temperature relates to the loss in the steam pipes from the boilers to the turbines.

The Kampon *ro gō* was the standard three-drum water-tube boiler with the steam drum above and the two water drums below in an 'A' configuration; they were connected by water tubes arranged on both sides. A characteristic feature was the comparatively small height of the boiler compared to its volume. The heating surface totalled 628m^2 with 416m^2 (saturated), 82m^2 (superheater) and 130m^2 (air preheater). The burners were the Ship Type 11 cone Model 5, and the fuel rate at specified full power was 7.3kg/m^2 in normal operation and

Matsu: Machinery

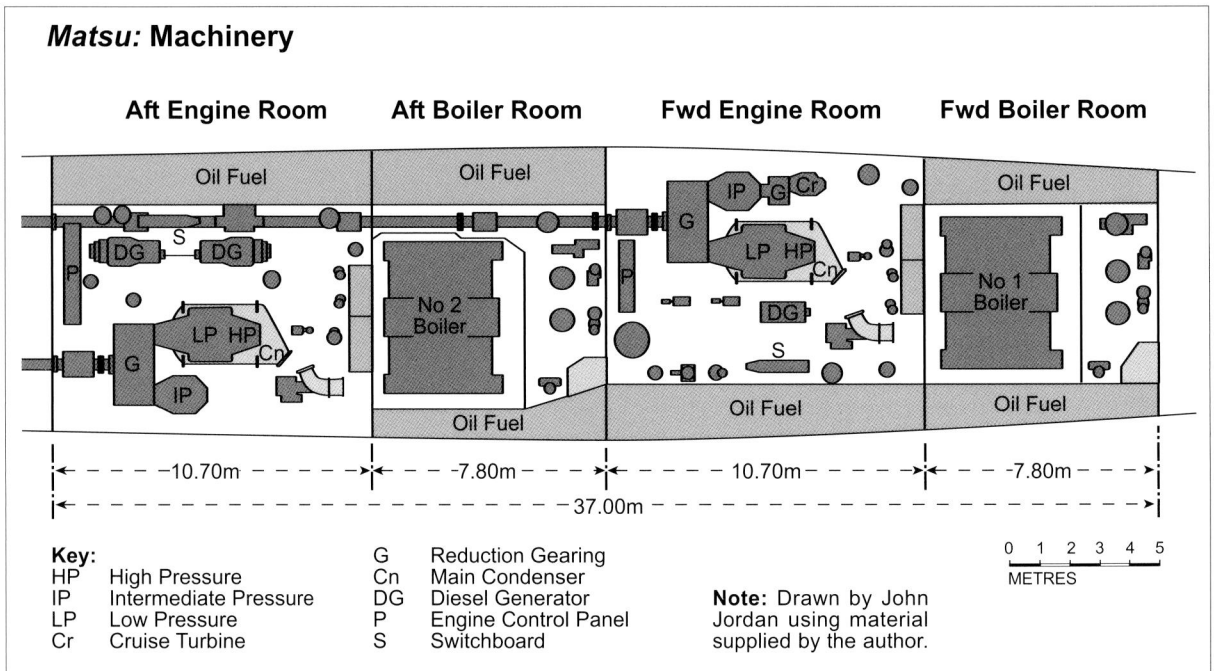

Aft Engine Room Aft Boiler Room Fwd Engine Room Fwd Boiler Room

Key:

HP	High Pressure
IP	Intermediate Pressure
LP	Low Pressure
Cr	Cruise Turbine

G	Reduction Gearing
Cn	Main Condenser
DG	Diesel Generator
P	Engine Control Panel
S	Switchboard

Note: Drawn by John Jordan using material supplied by the author.

9.1kg/m² at overload. The boiler rooms were of the 'open' type and each boiler was fitted with two forced draught blowers.

There was no cross-connection between the high-pressure sides of the main feed system of the two boiler rooms, nor between the HP sides of the fuel oil system. This was a characteristic feature of other Japanese destroyer types, but cross-connections in the main steam lines were omitted in the *Matsu* class for the reasons outlined above. There were numerous other particular modifications, such as the different float control tank in the condensate system. However, a key criterion for the design of the propulsion machinery was that fittings be simplified for 'speed of construction and ease of operation'.

Each of the two boiler rooms had a length of 7.8m; each of the engine rooms 10.7m. This made for a total length of 37.0m, which compared with 40.5m for the machinery spaces of the *Kagerō* class, which had three boilers and two sets of turbines with a rated output of 52,000shp. This shows the cost of the unit arrangement of the *Matsu* class in terms of overall length and volume.

The boiler in the forward boiler toom was located on the middle line but that in the after boiler room had to be offset to starboard to clear the port propeller shaft; the respective positions of the funnels reflected this arrangement, with the forefunnel on the middle line and the second funnel offset 0.75m to starboard. The separation of the two slim funnels was dictated by the unit arrangement of the machinery spaces. Fuel oil tanks were located on both sides of the boiler rooms, the starboard side of the forward engine room and the port side of the after engine room.

The propellers were of manganese bronze; they were three-bladed with a diameter of 2.65m and turned at 400rpm when the ship was running at full speed. The all-up weight of a single propeller was about 3.3 tonnes.

The steering mechanism was the usual electric oil-pressure plunger type capable of turning the rudder 70 degrees (*ie* from hard to port to full starboard) in 15 seconds. The electric oil pressure pump was placed on the starboard side, and on the opposite side there was a manual oil pressure pump with sufficient power to steer the ship in the event of failure of the electric pump. The capstan was of the reciprocating type and was capable of lifting a maximum weight of 9.8 tonnes at the rate of 9m/sec.

Repair facilities were based on the specifications for

Table 2: Engine Performance

Power	SHP	RPM	Steam pressure (bar)	Steam temp (°C)	Vacuum (mmHg)	Steam consumption (kg/shp/hr)
Specified full power	19,000	400	26.0	335	700 or more	3.85 or less
Overload full power	20,000	407	–	–	–	–
Cruising	3,000	238	–	–	–	–
Standard speed	3,200	231	26.0	310	720 or more	4.8 or less
Maximum cruising speed	6,000	280	26.0	310	720 or more	4.8 or less
Astern	4,000	247	21.0	320	665 or more	915 or less

2nd class destroyers. The principal equipment was a small lathe and a large drilling machine in each of the engine rooms, and a blacksmith's shop with a small portable forge located outside the machinery spaces.

The weight of the machinery, including water (32 tonnes) and oil (6 tonnes), was 363 tonnes, corresponding to 23.8 per cent of the trial displacement. The output/weight ratio was an impressive 52.3shp/tonne. The breakdown of machinery weight was as follows:

engines: 91 tonnes
propellers and shafting: 50 tonnes
auxiliary machinery: 34 tonnes
boilers: 73 tonnes
funnels and uptakes: 46 tonnes
pipework, valves and cocks: 43 tonnes
miscellaneous: 28 tonnes, plus water and oil as stated
   above.

Armament

Guns

Another characteristic feature of these ships was the mounting of three naval standard Type 89 12.7cm 40-cal HA guns. This marked a radical change from the LA and DP guns mounted in the conventional fleet destroyers of the 'A' type, and was a response to the lessons of war. There was a single gun on the forecastle in front of the bridge with a full shield to protect against the elements, and an unshielded twin mounting on the quarterdeck aft. Both mountings were controlled by the newly-adopted Type 4 Model 2 FC director, and there was a separate Type 97 high-angle rangefinder with a 2-metre base length atop the bridge.

For close-in defence against air attacks four triple and eight single Type 96 25mm machine guns were fitted, disposed as follows:

– triple mountings: one atop a deckhouse extending from the forward end of the bridge, two on a MG platform forward of the second funnel, and one atop the after deckhouse.
– single: two on the forecastle deck abreast the bridge, two on the upper deck abreast the foremast, two on the upper deck amidships and two abreast the after deckhouse.

In the later ships the number of single 25mm was boosted by five additional mountings: two sided mountings on

A 'D'-type destroyer of the *Matsu* class engages attacking US carrier aircraft at the the Battle of Cape Engaño. The IJN carriers of the Combined Fleet were escorted by the 31st Escort Group (flagship *Isuzu*), comprising four 'B'-type anti-aicraft destroyers of the *Akizuki* class (Destroyer Divisions 61 & 41) and four 'D'-Type destroyers (Destroyer Division 43): *Kuwa, Kiri, Sugi, Maki*. (US Navy, 80-G-284448)

the forecastle behind the main gun, two disposed *en echelon* on the quarterdeck, and a single mounting on a platform abaft the second funnel (see Profile & Plan of *Tachibana*). No fire control directors were fitted for the machine guns; the guns were aimed using the Le Prieur local sight on each of the mountings.

Two Type 99 light MG of 7.7mm calibre were carried; these could be mounted on the two 10-metre landing craft when required (the 7.7mm was the standard machine gun of the IJA).

Torpedoes

A Type 92 quadruple 61cm torpedo mounting was fitted between the funnels. In contrast to the fleet destroyers no reserve torpedoes were carried, so torpedo firings were

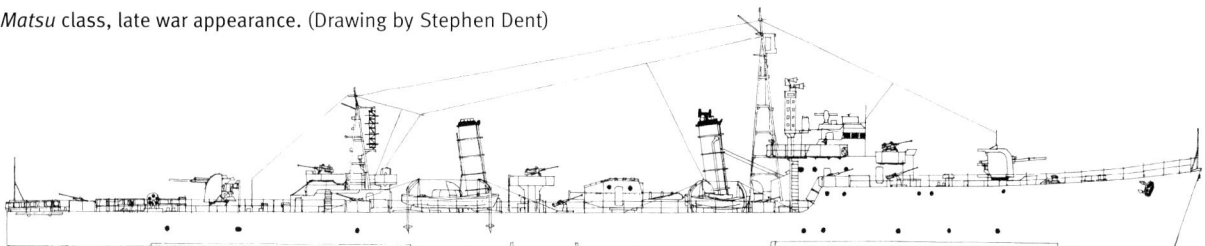

Matsu class, late war appearance. (Drawing by Stephen Dent)

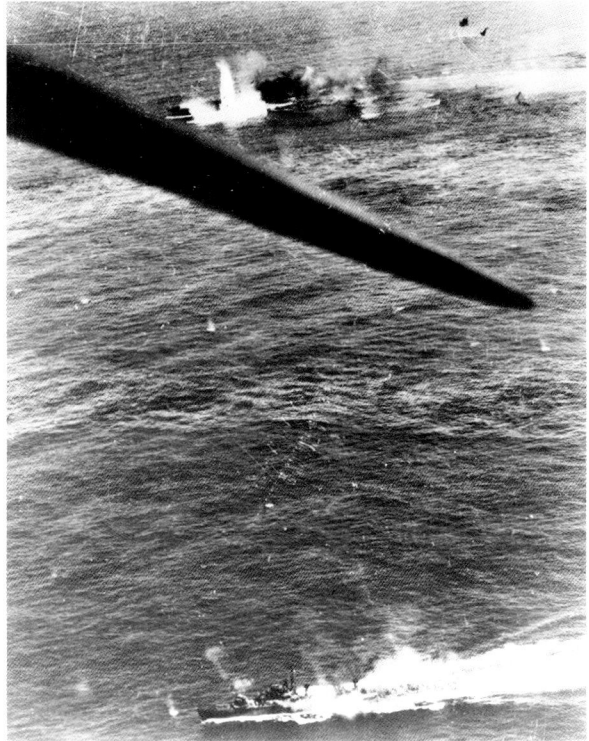

limited to the four Type 93 oxygen-propelled torpedoes carried in the tubes.

The initial intention had been to fit a 61cm quad mounting, as in the fleet destroyers, but there had been concerns about the small number of torpedoes that could be delivered in a single salvo, and it was proposed instead to fit six 53cm torpedo tubes to improve hit probability. However, as a result of the lessons learned in the night battles of the Solomons, the range and power of the 53cm torpedo were considered insufficient, and the final design reverted to the Type 92 quadruple 61cm model.

Anti-submarine equipment

Although the principal missions of this class were escort duties, the anti-submarine equipment was that of the standard Japanese destroyer, with two Type 94 DC throwers ('Y' gun) abaft the after main gun mounting and 36 depth charges on stern rails.

The early ships were equipped with a Type 93 sonar and Type 93 hydrophone, but these devices had limited capabilities for the detection of submarines and proved virtually ineffectual.

Radar

The radar equipment of the earlier ships comprised a No 22 surface search radar and a Type E-27 radar warning device. The horn antennae of the surface search radar No 22 were mounted atop a post at the after end of the bridge, while the antenna of the E-27 ESM gear, which had a similar configuration to the Greek letter theta [ɹ], was secured to the foremast.

The Modified *Matsu* Class

The 42 ships included in the Modified Fifth Replenishment Program were ordered from Maizuru NY (5), Yokosuka NY (24) and Fujinagata Shipyard (13) with the aim of reducing construction time to five months using 70,000 man-days. The first ship, *Matsu*, was laid down at Maizuru NY on 8 August 1943 and was completed on 26 April 1944. The second ship, *Take*, was built by Yokosuka NY from 15 October 1943 to 16 June 1944 and for the construction of the third ship, *Ume*, the Fujinagata shipyard needed from 1 December 1943 to 28 June 1944. Building times were nine, eight, and seven months respectively, and the man-days amounted to 85–90,000. Moreover, only five ships were laid down in 1943 (Maizuru 2, Yokosuka 1, Fujinagata 2), so the small number of destroyers of the *Matsu* class that would commission in the second half of 1944 would not be sufficient to replace losses.

The war situation steadily deteriorated, and the NGS became concerned that the Second and Third Operation Plans would be undermined, or even rendered impractical, by the shortage of destroyers. It now wanted construction times to be reduced to three months (March 1944). In response the Fourth Division of NTD began work on a further simplification of the *Matsu* design, and once this was approved the main personnel involved were moved immediately to the Design Section of Yokosuka NY to supervise the detailed design and the production of the working drawings, the two operations being conducted simultaneously.

The Modified 'D'-Type destroyer escort *Nire*. The photo was probably taken in January/February 1945, off Maizuru. Note the modified foremast and the position of the horn antennae for the No 22 radar. (Kure Maritime Museum)

The Modified 'D'-Type destroyer escort *Hatsuzakura* on 25 August 1945. Features to note: the depth charge racks on the stern, the large size of the 12.7cm HA twin gun mounting, the distinctive 'ladder' antenna of the No 13 air early warning radar on the mainmast, the slim funnels with the prominent galley pipe on the forward funnel, and the double-horn antennae on the foremast platform. (US Navy)

The new design, F 55B, was first implemented for the 19th ship of *Matsu* class, *Tachibana*, laid down at the Yokosuka NY on 8 July 1944 and completed on 20 January 1945. After that all construction was shifted to the new class: Maizuru NY from the 6th ship, *Nire*, and Fujinagata from the 7th, *Kaba*. The comparatively short period between the drawing up of the NGS requirement and the laying down of the first ship was achieved by completing the design work in parallel with construction. However, building times were not shortened as much as expected because of the general situation in the shipbuilding industry during the latter half of 1944; for the later ships completion would be shortened by one to two months, depending on the repair workload of the shipyards, the supply of materials, the available workforce and the disruption caused by air attacks

The designers prioritised simplification of the plans, structure and fittings and also the assembly of material. The anti-submarine and anti-aircraft capabilities would be improved by the adoption of newly-developed equipment. Generally speaking the design was changed in line with the principles introduced for mass production of coast defence vessels (*Ukuru*, No 1 and No 2 classes) and naval transports (No 1 and No 101 classes). The percentage of flat plating was increased by dispensing with the flare of the hull sides and the camber of the forecastle and upper deck; the stern was a flat transom, the bilge keels were of the flat plate rather than the built-up type, and there was a single instead of double bottom and a skeg-type cut-up.

The quality of the steel was further reduced with the adoption of mild steel throughout; this was better suited to electric welding and saved man-days. The downside of this was that the thickness of the plates had to be correspondingly increased: the plates used for the upper deck

were 18mm as compared to 12–14mm in the *Matsu* class, while the shell plating was 16mm thick instead of 12mm. Corrections to the permissible tensile strength also became necessary. All-welded block construction was adopted; blocks that were largely fitted out were welded together on the building ways to form the ship. It is reported that riveting was reduced to a few areas subject to strong vibrations, but this is disputed by some sources. The increase in the thickness of the steel meant a corresponding increase in weight that could not be entirely compensated by the adoption of electric welding, and the trial displacement of the modified type was 1,580 tonnes, an increase of 50 tonnes over the early ships.

The cruise turbine and its reduction gearing were eliminated. In part this was in the interest of simplification, but it was also recognised that separate cruising machinery was unnecessary in this type of ship; a further factor was the desperate situation of the engine production industry.

The gun armament is reported to have been identical to that of the *Matsu* class, although it incorporated the increase in the number of single 25mm MG from the outset. However, most data are based on the official orders; they do not always reflect the true condition of individual ships on completion, as the number of MG often depended on the efforts of the captain and the relationships established with the shipyard.

Underwater detection capabilities, on the other hand, were much improved by the installation of the new Type 3 hydrophone and Type 4 sonar. The performance of the latter was markedly superior to the Type 93, but in order to achieve it the 80 microphones, which were mounted in a circular, downward-facing array, had to be located in a compartment filled with fresh water. The fore part of the lower hull therefore had to be re-designed.

Close-up of the foremast of the Modified 'D'-Type destroyer escort *Shii*, showing the distinctive double horn of the No 22 air search radar. (*Gakken* No 43, courtesy of Hans Lengerer)

The sonar dome had a flat plate with a diameter of 3 metres at the bottom and was located beneath the deck-house for the triple 25mm MG. It was feared that this deformation of the lower hull would result in a reduction in speed. However, the improvement in ASW capabilities was an urgent requirement; considerable care was taken to minimise the drag effect of the sonar dome, and

following the completion of the first ship no significant decrease in speed was noted, and hull resistance proved less than anticipated.

The No 22 radar was complemented by a 'ladder' array for the No 13 air early warning radar, mounted on the forward side of the mainmast. At the same time the horn antenna of No 22 was relocated from the separate post on the after part of the bridge to the foremast; the structure of the latter was modified and the crow's nest suppressed. The different configuration of the masts and radar outfit is a key distinguishing feature between the ships of the two groups.

Production Statistic

Early in 1945 the war situation became desperate, and given the shortage of materials and skilled workmen, lack of fuel and the dramatic increase in air attacks on the homeland in company with the further advance of Allied troops, both the IJN and the IJA were forced to rely solely on the production of special attack weapons along-side the continued building of coast defence ships in much-reduced numbers. The construction of ships of the *Matsu* and *Tachibana* classes that had no chance of being completed in the first half of 1945 was halted on 17 April, and the ships yet to be laid down were cancelled.

In the event 26 ships of the Modified Fifth Replenishment Programme and six belonging to the War Armament Programme were completed, a total of 32. Of these 18 belonged to the *Matsu* class and 16 to the *Tachibana* class. At the end of the war a total of nine ships, five belonging to the former programme and four to the latter, remained incomplete on the ways at various stages of completion. Eleven ships of the Modified Fifth Replenishment Program and 22 ships of the War

The 'D'-type destroyer *Maki* enters Manila Bay on 29 October 1945 following the Japanese surrender. She had come to repatriate Japanese soldiers as part of the mobilisation operation. She retains her original foremast, but a 'ladder' antenna for the No 13 air early warning radar has been added to the mainmast. (SC 235251)

A profile and plan of the stern showing *Nashi* as a *Kaiten* carrier. Note the fixed tracks, which can also be seen in the lower photo of the salvaged *Nashi* in dock. (*Gakken* No 43)

Kai Tei Gata "Nashi" Kaiten carrier drawing (estimation)

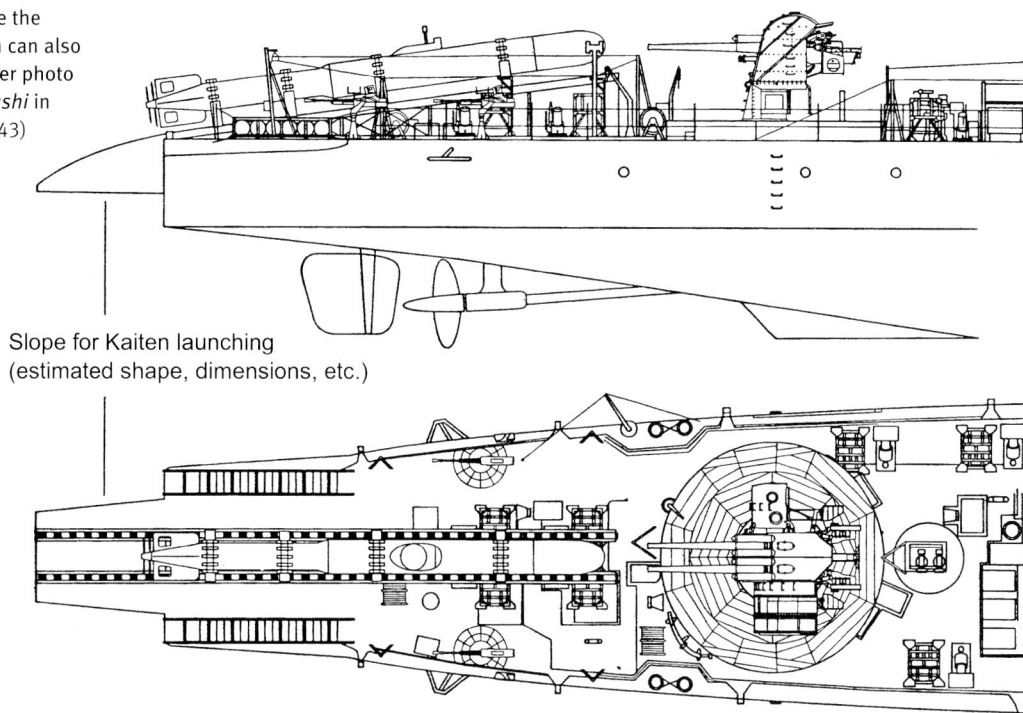

Slope for Kaiten launching
(estimated shape, dimensions, etc.)

Armament Program were cancelled before laying down.

The completed ships were assigned to the 11th Destroyer Squadron, and following work-up were assigned to divisions under the command of the Combined Fleet and utilised for front-line escort and transport duties.

Conversion into *Kaiten* Carriers

In the spring of 1945 it was planned to convert six ships as *Kaiten* carriers and, according to Fukui, *Take* and several other ships were fitted with rails on the upper deck aft and an overhanging stern ramp to carry and launch a single *Kaiten* human torpedo (see drawing). The length of the *Kaiten* Type 1 was 14.75 metres, so the DC throwers (but not the after 12.7cm HA gun mounting) had to be removed. Some sources claim that the depth charge rails were also landed, but they are still in place in the photo of *Nashi* following her salvage, albeit in shortened form.

Evaluation

The 'D'-type destroyers proved to be robust, seaworthy ships, but had only rudimentary anti-submarine and anti-aircraft capabilities; in particular, the HA guns were controlled by a simplified fire control system and no FC system was fitted for the machine guns. The failures with regard to building times were not the fault of the design, but should be attributed to Japan's insufficient ship-building capacity and to materiel and personnel short-ages. Arguably the Japanese ships yards persisted too long with conventional methods of construction. Despite these failings, the ships performed broadly as anticipated: many returned to base under their own power after sustaining considerable damage. The 'D'-type destroyer should therefore be considered a successful product of wartime mass construction.

The strategic duties assigned to the type – escort and transportation – could arguably have been better performed by the new-type coast defence ships and first class transports that followed them. However, in the context of the night- and day-time battles in the Solomon theatre that gave rise to the staff requirements for this type, it is understandable that the IJN tacticians would opt for a conventional destroyer armed with torpedoes.

On the other hand, the question remains as to why the IJN persisted in building this type once the Types 'C' and 'D' coast defence ships and first class transports were on the stocks, given that the rapid deterioration in the war situation had made their offensive deployment illusory, while the parlous situation in the shipbuilding and related industries urgently required the utmost simplifica-tion and standardisation not only of equipment but of the types of vessel. The diversity of ship classes and types with similar missions is often lamented by former IJN designers and constructors, and the negative influence upon productivity has long been recognised.

According to the late Hori Motoyoshi, the *Matsu* class were 'real escort ships which were far more adequate than the expensive destroyer for the decisive battle …'; however, the author does not entirely concur with this view given the serious defects, particularly in the anti-submarine role, when compared with *Ukuru* and the following classes of coast defence vessels.

Table 3: Building Data & Fates

Modified Circle 5 Programme

No	Name	Builder	Laid down	Launched	Completed	Notes
#5481	*Matsu*	Maizuru NY	08 Aug 1943	03 Feb 1944	26 Apr 1944	1944 escort and transport missions; 4 Aug 1944 while on transport mission to Chichijima engaged by US surface force (4 CLs & 7 DDs) NW of Mukojima (27°40N/141°48E); stricken 10 Oct 1944.
#5482	*Take* (II)	Yokosuka NY	15 Oct 1943	28 Mar 1944	16 Jun 1944	1944–45 escort & transport missions; 3 Dec 1944 torpedoed & sank USS *Cooper* (DD-695) Ormoc Bay and slightly damaged. Maizuru at end of war; stricken 25 Oct 1945. 1 Dec 1945 designated special transport & used for demobilisation. 28 June 47 allocated to United Kingdom as war prize. 16 Jul 1946 delivered at Singapore; scrapped Singapore 1948.
#5483	*Ume* (II)	Fujinagata	01 Dec 1943	24 Apr 1944	28 Jun 1944	1944–45 escort & transport missions; 31 Jan 1945 bombed & sunk by USAAF B-25s 20nm south of Eluanbi, Taiwan (22°30N/12°00E); stricken 10 Mar 1945.
#5484	*Momo*	Maizuru NY	05 Nov 1943	25 Mar 1944	10 Jun 1944	1944 escort and transport missions; 15 Dec 1944 while returning from transport mission to Leyte torpedoed & sunk by USS *Hawkbill* (SS-366) W of Luzon (16°40N/117°42E); stricken 10 Feb 1945.
#5485	*Kuwa* (II)	Fujinagata	20 Dec 1943	25 May 1944	25 Jul 1944	1944 escort & transport missions + Battle of Leyte Gulf; 3 Dec 1944 while on transport mission to Ormoc engaged by US surface force (3 DDs) and sunk by gunfire in Ormoc Bay (10°50N/124°35E); stricken 10 Feb 1945.
#5486	*Kiri* (II)	Yokosuka NY	01 Feb 1944	27 May 1944	14 Aug 1944	1944–45 escort & transport missions + Battle of Leyte Gulf. Kure at end of war; stricken 5 Oct 1945. 1 Dec 1945 designated special transport & used for demobilisation. 17 Jul 47 allocated to Soviet Union as war prize. 29 Jul 1947 delivered at Nakhodka and renamed *Vozrozdennyi*. Jun 49 target ship *TsL-25*; 3 Oct 57 renamed *PM-65*; stricken 20 Dec 69.
#5487	*Sugi* (II)	Fujinagata	25 Feb 1944	03 Jul 1944	25 Aug 1944	1944–45 escort & transport missions + Battle of Leyte Gulf. Kure at end of war; stricken 5 Oct 1945. 1 Dec 1945 designated special transport & used for demobilisation. 17 Jul 47 allocated to China as war prize. 31 Jul 1947 delivered Shanghai and renamed *Hui Yang*; never commissioned; stricken 11 Nov 54 and scrapped (parts used to refit ex-*Hatsuume*).
#5488	*Maki* (II)	Maizuru NY	19 Feb 1944	10 Jun 1944	10 Aug 1944	1944–45 escort & transport missions + Battle of Leyte Gulf; 9 Dec 1944 torpedoed & damaged by USS *Plaice* (SS-90) E of Danjo Gunto (31°57N/129°01E). Kure at end of war; stricken 5 Oct 1945. 1 Dec 1945 designated special transport & used for demobilisation. 17 Jul 47 allocated to United Kingdom as war prize. 14 Aug 1947 delivered at Singapore and scrapped.
#5489	*Momi* (II)	Yokosuka NY	01 Feb 1944	16 Jun 1944	03 Sep 1944	1944-45 escort & transport missions; while on transport mission in Philippines bombed & sunk by US carrier-based aircraft (CVEs of TG 77.4) 20nm SW of entrance to Manila Bay (14°00N/120°20E) 5 Jan 1945; stricken 10 Mar 1945.
#5490	*Kashi* (II)	Fujinagata	05 May 1944	13 Aug 1944	30 Sep 1944	1944–45 escort & transport missions + Battle of Leyte Gulf. Kure at end of war; stricken 5 Oct 1945. 1 Dec 1945 designated special transport & used for demobilisation. 17 Jul 47 allocated to United States as war prize. 20 Mar 1948 completed scrapping by Kasado Senkyo.
#5491	*Yaezakura*	Yokosuka NY	18 Dec 1944	17 Mar 1945	–	Cancelled 23 Jun 1945 when 60% complete; 18 Jul 1945 bombed & sunk by US carrier-based aircraft Yokosuka.
#5492	*Kaya* (II)	Maizuru NY	10 Apr1944	30 Jul 1944	30 Sep1944	1944–45 escort & transport missions. Himi (Yamaguchi prefecture) at end of war; stricken 5 Oct 1945. 1 Dec 1945 designated special transport & used for demobilisation. 28 Jun 47 allocated to Soviet Union as war prize. 5 Jul 1947 delivered at Nakhodka and renamed *Volevoy*; Jun 49 target ship *TsL-23*; 10 Jun 58 heating hulk *OT-61*; stricken 1 Aug 59.

#5493	*Nara* (II)	Fujinagata	10 Jun 1944	12 Oct 1944	26 Nov 1944	1944–45 escort & transport missions; 30 Jun 1945 struck a mine at the western entrance to Kanmon Strait (6nm WSW of Shimonoseki, 33°54N/130°49E) and bent her stern. Moored Moji at end of war; stricken 30 Nov 1945; scrapped May–Jul 1948.
#5494	*Yadake*	Yokosuka NY	02 Jan 1945	–	–	5 Feb 1945 named; cancelled 17 Apr 1945 when 60% complete. Early in May taken off slip & scrapped after end of war; hull used as breakwater at Kaminato Port, Hachijōjima.
#5495	*Kuzu*	Yokosuka NY	20 Mar 1945	–	–	Cancelled 17 Apr 1945; broken up & material used for midget submarine construction (also reported scrapped after end of war).
#5496	*Sakura* (II)	Yokosuka NY	02 Jun 1944	06 Sep 1944	25 Nov 1944	1944–45 escort & transport missions; 11 Jul 1945 struck a mine & sank 265° 5.6nm off lighthouse northern jetty Osaka; stricken 10 Aug 1945.
#5497	*Yanagi* (II)	Fujinagata	20 Jun 1944	25 Nov 1944	18 Jan 1945	1944–45 escort & transport missions; 14 Jul 1945 bombed & seriously damaged by US carrier-based planes (TF 38) Tsugaru Strait; towed to Ominato and there at end of war; stricken 20 Nov 1945. Scrapped Ominato Oct 1946–20 May 1947.
#5498	*Tsubaki* (II)	Maizuru NY	20 Aug 1944	30 Sep 1944	30 Nov 1944	1944–45 escort & transport missions; 14 Jul 1945 bombed & seriously damaged by US carrier-based planes (TF 38) off Okayama. Kure at end of war; stricken 30 Nov 1945. Scrapped by Harima Shipyard Kure Dock 1–28 Jul 1948.
#5499	*Kaki*	Yokosuka NY	05 Oct 1944	11 Dec 1944	05 May 1945	1945 work-up Inland Sea. Maizuru at end of war; stricken 5 Oct 1945. 1 Dec 1945 designated special transport & used for demobilisation. 28 Jun 47 allocated to United States as war prize. 4 Aug 1947 delivered at Yokosuka. 19 Aug 47 expended as target Yellow Sea (35°29N/123°35E).
#5500	*Kaba* (II)	Fujinagata	15 Oct 1944	27 Feb 1945	29 May 1945	1945 work-up Inland Sea. Kure at end of war; stricken 05 Oct 1945. 1 Dec 1945 designated special transport & used for demobilisation. 17 Jul 47 allocated to United States as war prize. 4 Aug 1947 delivered at Sasebo; scrapped Mitsui Shipbuilding 1 Mar–Oct 1948.
#5501	*Hayaume*	Yokosuka NY	–	–	–	Not laid down; cancelled.
#5502	*Hinoki* (II)	Yokosuka NY	04 Mar 1944	04 Jul 1944	30 Sep 1944	1944–45 escort & transport missions; 7 Jan 1945 while on transport mission in Philippines engaged by US surface force (4 DDs) and sunk by gunfire 50nm WSW of Manila Bay (14°30N/119°08E); stricken 10 Apr 1945.
#5503	*Katsura*	Fujinagata	30 Nov 1944	23 Jun 1945	–	Building halted 23 Jun 1945 when 60% complete; cancelled.
#5504	*Tobiume*	Yokosuka NY	–	–	–	Not laid down; cancelled.
#5505	*Kaede* (II)	Yokosuka NY	04 Mar 1944	25 Jul 1944	30 Oct 1944	1944–45 escort & transport missions. Kure at end of war; stricken 5 Oct 1945. 1 Dec 1945 designated special transport & used for demobilisation. 28 Jun 47 allocated to China as war prize. 6 Jul 1947 delivered at Shanghai and renamed *Heng Yang*; never commissioned; training hulk; scrapped 1960.
#5506	*Fuzi*	Yokosuka NY	–	–	–	Not laid down; cancelled.
#5507	*Wakazakura*	Fujinagata	01 Jan 1945	–	–	Building halted 11 May1945 when 7% complete; cancelled.
#5508	*Keyaki*	Yokosuka NY	22 Jun 1944	30 Sep 1944	15 Dec 1944	1944–45 work-up Inland Sea. Kōbe at end of war; stricken 5 Oct 1945. 1 Dec 1945 designated special transport & used for demobilisation. 28 Jun 47 allocated to United States as war prize. 5 Jul 1947 delivered at Yokosuka. 29 Oct 1947 sunk as target off Nojimazaki Lighthouse/Bōsō Peninsula (33°44N/140°01E).
#5509	*Yamazakura*	Fujinagata	–	–	–	[Initially named *Kusunoki*] Not laid down; cancelled.
#5510	*Ashi*	Yokosuka NY	–	–	–	[Initially named *Azusa*] Not laid down; cancelled.
#5511	*Tachibana* (II)	Yokosuka NY	08 Jul 1944	14 Oct 1944	20 Jan 1945	1945 ASW operations Tsugaru Strait. 14 Jul 1945 during patrol mission bombed & sunk by US carrier-based aircraft (TF 38) in Hakodate Harbour, Hokkaidō (41°48N/140°4E); stricken 10 Aug 1945.

Continued overleaf

No	Name	Builder	Laid down	Launched	Completed	Notes
#5512	*Shiitake*	Fujinagata	—	—	—	[Initially named *Fuzi*] Not laid down; cancelled.
#5513	*Yomogi*	Yokosuka NY	—	—	—	[Initially named *Ashi*] Not laid down; cancelled.
#5514	*Tsuta* (II)	Yokosuka NY	31 Jul 1944	02 Nov 1944	08 Feb 1945	1945 work-up Inland Sea. Sasebo at end of war; stricken 5 Oct 1945. 1 Dec 1945 designated special transport warship & used for demobilisation. 17 Jul 47 allocated to China as war prize. 31 Jul 1947 delivered at Shanghai and renamed *Hua Yang*. Never commissioned. Stranded Pescadores Islands en route Shanghai-Taiwan 1949 and laid-up there until stricken 1954.
#5515	*Aoi*	Yokosuka NY	—	—	—	[Initially named *Akane*] Not laid down; cancelled.
#5516	*Shiraume*	Fujinagata	—	—	—	[Initially named *Ashi*] Not laid down; cancelled.
#5517	*Hagi* (II)	Yokosuka NY	11 Sep 1944	29 Nov 1944	01 Mar 1945	1945 work-up Inland Sea. Maizuru at end of war; stricken 5 Oct 1945. 1 Dec 1945 designated special transport & used for demobilisation. 28 Jun 47 allocated to United Kingdom as war prize. 16 Jul 1947 delivered at Singapore & scrapped.
#5518	*Kiku*	Fujinagata	—	—	—	[Initially named *???*] Not laid down; cancelled.
#5519	*Kashiwa*	Yokosuka NY	—	—	—	[Initially named *Yomogi*] Not laid down; cancelled.
#5520	*Sumire* (II)	Yokosuka NY	21 Oct 1944	27 Dec 1944	26 Mar 1945	1945 work-up in the homeland. Maizuru at end of war; stricken 5 Oct 1945. 1 Dec 1945 designated special transport & used for demobilisation. 17 Jul 47 allocated to United Kingdom as war prize. 20 Aug 1947 delivered at Hong Kong; 1947 sunk as target off Hong Kong.
#5521	*Kusunoki*	Yokosuka NY	09 Nov 1944	16 Jan 1945	28 Apr 1945	1945 work-up Inland Sea. Maizuru at end of war; stricken 5 Oct 1945. 1 Dec 1945 designated special transport & used for demobilisation. 28 Jun 47 allocated to United Kingdom as war prize. 16 Jul 1947 delivered at Singapore & scrapped.
#5522	*Hatsuzakura*	Yokosuka NY	04 Dec 1944	10 Feb 1945	28 May 1945	1945 work-up Inland Sea. Yokosuka at end of war; stricken 15 Sept 1945. 1 Dec 1945 designated special transport & used for demobilisation. 17 Jul 47 allocated to Soviet Union as war prize. 29 Jul 1947 delivered at Nakhodka and renamed *Vetrenyi*; renamed *Vyrazitelny* 2 Oct 47; target ship *TsL-26* Jun 49; stricken 19 Feb 59.

1943/1944 Wartime Replenishment Programme

No	Name	Builder	Laid down	Launched	Completed	Notes
#4809	*Nire* (II)	Maizuru NY	14 Aug 1944	25 Nov 1944	31 Jan 1945	1945 work-up Inland Sea; 22 Jun 1945 bombed and damaged by USAAF B-29s Kure. Kure at end of war; stricken 5 Oct 1945 & used as floating pier. Scrapped at Harima Shipyard, Kure Dock Jan–20 Apr 1948.
#4810	*Nashi* (II)	Kawasaki	01 Sep 1944	17 Jan 1945	15 Mar 1945	1945 work-up Inland Sea; 28 Jul 1945 bombed & sunk by US carrier-based aircraft (TF 38) Mitajiri Saki, Kure (34°01N/132°30E); stricken 15 Sep 1945. Jul–31 Sep 1954 salvaged by Hokusei Senpaku Kōgyō (refloated 21 Sep); Sep 1955 repaired & rebuilt Kure. 31 May 1956 became JMSDF guard ship (later escort ship) *Wakaba*; stricken 31 Mar 1971 & scrapped.
#4811	*Shii*	Maizuru NY	18 Sep 1944	13 Jan 1945	23 Mar 1945	1945 work-up Inland Sea. Kure at end of war; stricken 5 Oct 1945. 01 Dec 1945 designated special transport & used for demobilisation transports. 28 Jun 47 allocated to Soviet Union as war prize. 5 Jul 1947 delivered at Nakhodka and renamed *Volny*; target ship *TsL-24* 17 Jun 49; renamed *OT-5* 11 Nov 59; stricken 8 Aug 60.
#4812	*Enoki* (II)	Maizuru NY	14 Oct 1944	27 Jan 1945	31 Mar 1945	1945 work-up Inland Sea; 26 Jun 1945 struck a mine off Obama Lighthouse, Fukui Prefecture, seriously damaged and settled on the bottom (35°28N/135°44E) and remained there until end of war; stricken 30 Sep 1945. Scrapped at Nanao Jun–Jul 1948.
#4813	*Azusa*	Yokosuka NY	29 Dec 1944	—	—	Building halted 7 Apr 1945; cancelled & scrapped 1948.
#4814	*Odake*	Maizuru NY	05 Nov 1944	10 Mar 1945	15 May 1945	1945 work-up Maizuru. Maizuru at end of war; stricken 5 Oct 1945. 1 Dec 1945 designated special transport & used for demobilisation. 28 Jun 47 allocated to United States as war prize. 4 Jul 1947 delivered at Tsingtao; expended as target off Tsingtao 19 Aug 1947 (35°29N/122°52E).

#4815	Hatsuume	Maizuru NY	08 Dec 1944	25 Apr 1945	18 Jun 1945	1945 work-up in the homeland. Maizuru at end of war; stricken 5 Oct 1945. 1 Dec 1945 designated special transport & used for demobilisation. 28 Jun 47 allocated to China as war prize. 6 Jul 1947 delivered at Shanghai and renamed *Xin Yang*; rearmed and commissioned Mar 48; refitted and rearmed 1954; stricken Dec 61.
#4816	Tochi	Maizuru NY	23 Jan 1945	28 May 1945	–	Building halted 28 May 1945 when 60% complete; cancelled. Scrapped after end of war but hull used as breakwater in Akita Harbour.
#4817	Hishi (II)	Maizuru NY	10 Feb 1945	–	–	Building halted 17 Apr 1945; cancelled, broken up, material used for midget sub construction (another source states 'scrapped after end of war').
#4820	Sakaki	Yokosuka NY	29 Dec 1944	–	–	Building halted 17 Apr 1945; cancelled, broken up, material used for midget sub construction (another source states 'scrapped after end of war').

Source: Endo Akira, 'Unknown Names of Never Completed Ships in Japanese Navy', *Sekai no Kansen*, 5/1968, 45.

Note:

The names of the ships with the provisional building numbers 4801 to 4808 were *Kōgiku, Hatsugiku, Akane, Shiragiku, Chigusa, Wakagusa, Natsugusa, Akigusa*; 4818 was to be *Susuki* and 4819 *Noguki*; for numbers 4821 till 4832 no names were allocated.

A postwar image of the 'D'-Type destroyer escort *Nashi* in dock. Note the straight stem and the single 12.7cm 40-cal HA gun. *Nashi* was salvaged from July to 31 September 1954 by *Hokusei Senpaku Kōgyō* (Shipping Industry) and refloated on 21 September. Following salvage she was repaired and reconstructed at Kure Zōshensho.

A later photo of *Nashi* in dock, being prepared for reconstruction. The superstructures have been removed. The *Kaiten* tracks at the stern are prominent; the mine rails remain in place. (Both photographs JMSDF, courtesy of Kitamura Kunio)

THE MAKING OF AN ARMED MERCHANT CRUISER: SMS *SEEADLER*

Dirk Nottelmann provides, as part of his wider study of cruisers of the Imperial German Navy, some fresh insights into the story of one of the most famous auxiliary warships in history: *Seeadler* – the only sail-powered commerce-raider. These focus on the ship's 'prehistory' as told by archival sources, rather than the oft-cited, and sometimes unreliable memoirs of her commanding officer, *Kapitänleutnant* Count Felix von Luckner,[1] and other later publications.[2]

At 09.25 on Christmas morning, Monday, 25 December 1916, the lookouts on HM Armed Merchant Cruiser *Patia* (Commander William G. Howard),[3] part of the 10th Cruiser Squadron's patrol line 'A', which was stretched out to the west of the Faroe Islands, spotted a sailing vessel in the overcast and gloomy distance to the northwest, at a position of about 63°0'N, 9°23'W;[4] the cruiser altered course to intercept. Without much haste, it set a converging course, while the crew were first called to Divine Service. This was in effect a breach of the standing orders for the squadron, which required ships to close immediately at full speed, flying a large red pennant, and to fire two blank rounds to raise attention before coming to a stop to the windward.[5]

At 10.45 the log was hauled in and the engines prepared to stop, which took a further five minutes. At 10.50 it was noted in *Patia*'s log:[6] 'Stopped. Boarded Norwegian ship "Hero" from Carlsund to Melbourne. Cargo lumber.' A boarding party was sent over, and two officers and one yeoman of signals climbed aboard[7] to find that the ship had suffered badly from a recent severe storm. The glass of several scuttles had been broken by the heavy seas and the interior of the cabins behind them, including the captain's, were still soaked, which

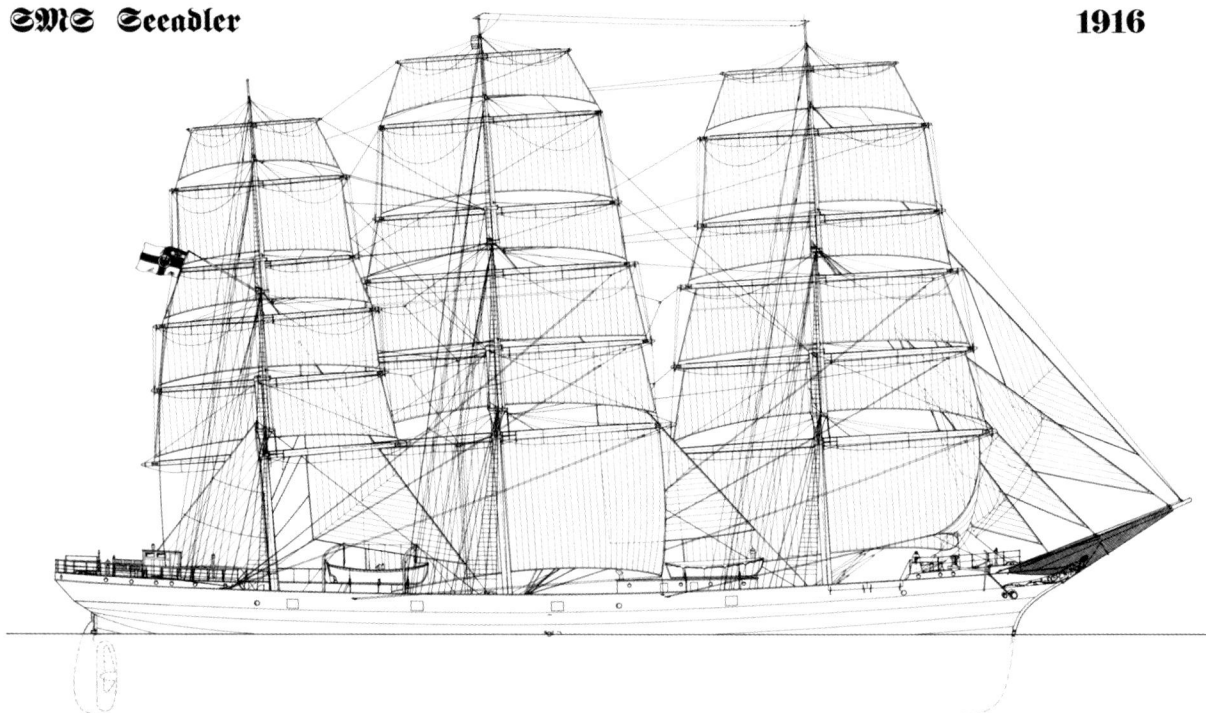

SMS Seeadler 1916

SMS *Seeadler* fully rigged. (Drawn by the author)

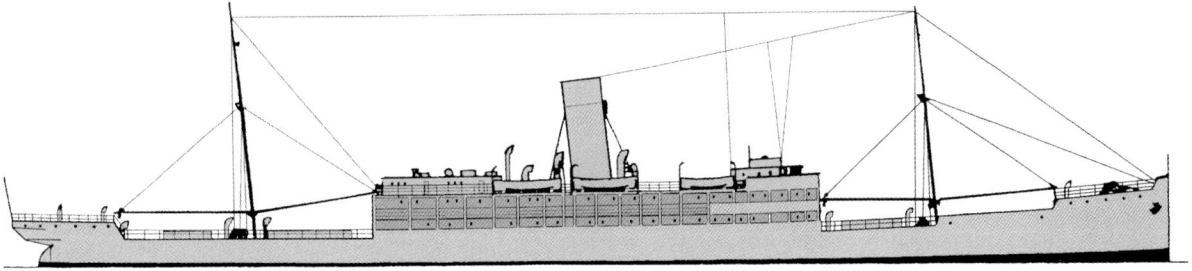

The auxiliary cruiser *Patia*. (Drawn by the author)

'unfortunately' had affected the trading certificates as well. After what appears to have been a somewhat superficial inspection, the result must have been communicated to the waiting AMC (not recorded in the log) because, at 11.20, the Senior Naval Officer (Acting Captain Frederic W Dean) aboard the AMC *Hilary* was informed by W/T that *Patia* had 'Boarded and detained Norwegian s.v. Hero, Carlsund to Melbourne, cargo lumber, request instructions, lat. 63°, long. 9°23'W, steering 260°, papers correct.'[8]

At the same time *Patia*'s boat crew was informed by semaphore: 'Ship to remain hove to pending instructions, about two hours, boat to return.' Accordingly, the investigation team re-boarded the boat alongside and rowed back to *Patia*, having ordered the crew of 'Hero' to wait for the prescribed time and to await further instructions. At 12.08 a reply came in from *Hilary*: 'Allow vessel detained to proceed.' This was not only communicated as a single order to proceed at 12.20, but was amended at 12.25 with the signal 'Pleasant voyage' by *Patia*. 'Hero' replied with a courteous 'Thank you!' while the AMC gathered speed and vanished in the haze in a south-westerly direction.

The memory of this seemingly-uneventful routine operation would be a source of some embarrassment to the Royal Navy when it learned, much later, the true circumstances and to which ship *Patia* had wished a pleasant voyage. We will return to some of the details below. However, the back-story to this episode had begun some 17 months earlier, coincidentally in an area not very far away from the position of these events.

Pass of Balmaha

The story of this full-rigged sailing vessel of later fame began relatively uncontroversially in 1888 on the slipways of the Robert Duncan & Co Shipyard, Port Glasgow, as Yard No 237. The hull was launched on 9 August under the name *Pass of Balmaha* for delivery to Gibson & Clark, Glasgow, later the same year. Built entirely from steel, the ship had, according to the Greenock report No 9563 of 29 December 1887, the following principal characteristics: length on the waterline 233ft 6in (71.2m), beam 38ft 8in (11.8m), draught 24ft 9in (7.5m), tonnage 1,571grt, 1,498nrt. The vessel received the official British Register No 95087 and the

Sail plan of *Pass of Balmaha*, taken from an original drawing supplied by courtesy of M Kastius. (Drawn by the author)

call-sign KTRP. For the subsequent 25 years, few people would recognise the existence of this moderately sized full-rigger, which had a multitude of sister-ships and half-sisters and never made it into the headlines,[9] as was true for 99.9 per cent of all vessels trading the seas.

Thus it may be said, albeit with a word of caution, that the major high points in the history of this vessel came following a series of changes of ownership after the turn of the century: first to the River Plate Shipping Co, Montreal, in February 1908; to the Pass of Balmaha Co Ltd, Montreal, in 1909; and, after a change to the United States Registry on 20 November 1914, the Harby Steamship Co Inc, New York, for a price of $30,000. This was a branch of the Harris-Irby Cotton Company, Boston, and her homeport would be Boston, MA. This last change of ownership makes sense in view of the circumstances outlined below.

From the moment that Great Britain declared war on Germany, a blockade of the German coast was declared; contrary to many expectations this would be a distant blockade. This is not the place to discuss the legal aspects of blockade, as no universal common understanding had been reached about this measure over the previous century, and the question rapidly degenerates into a purely academic one. The same may be said, in a way, about the approach of the US Government, which initially endeavoured to secure a balanced neutrality between the two main antagonists under the banner of an idealised 'freedom of the seas'. Our focus will be limited to one particular cargo in which the US Government had great economic interest, as Germany had been one of its main trading partners before the war: cotton. The threat to this lucrative trade became a reality when Britain

unilaterally declared cotton to be contraband in October 1914. The US Government protested, with the result that cotton was removed from the list before returning gradually from March 1915 when the 1914–15 cotton harvest, said to be one of the largest in history,[10] was over. Until then a fortune could be made by transferring ships to the US flag and sending them to Germany loaded with cotton.

Events even proceeded to the point that *Pass of Balmaha* and a company companion, *Vincent*, were matched against each other for a transatlantic 'cotton race' to the German port of Bremen. Substantial bets were placed when both vessels left New York on 25 January, but *Pass of Balmaha* let down her backers, being beaten by *Vincent* by several days; the latter arrived outside the German mine barrage on 3 March. At least one source claims that *Pass of Balmaha* had been intercepted by British patrols and sent into Kirkwall en route, but this is not confirmed by other sources.[11] The only definite information to emerge thus far regarding her whereabouts stems from the war diary of the North Sea picket flotilla: '16 March 1915, Wind NW 5–6 Beaufort; (...) W/T 2057 from CiC. American full-rigger has grounded south of Westerland [Island of Sylt]. Refloating must be supported by all means, result to be transmitted.'[12] But it was not that straightforward. When three tugs, sent from Cuxhaven the next day, failed to pull her off at the first attempt, it was decided to send some lighters to remove the cargo first. By 29 March, 950 tons of cotton had been transferred to lighters, and the following day *Pass of Balmaha* was refloated and subsequently towed to Geestemünde/Bremerhaven to unload the remaining 550 tons. She was then briefly

Pass of Balmaha during salvage operations off Sylt. (Heidrich collection)

docked at the local Seebeck Shipyard for a bottom survey before setting sail back to the USA some time in May to embark another load of cotton.[13] This would become a story of its own from the outset.

In the process of determining the true owners and destination of this second load of cotton before the prize court at Hamburg, after *Pass of Balmaha* had been seized by Germany, a letter from Dr Heinrich Albert[14] arrived at the end of August revealing surprising information. Albert reported to have been approached by the owners of Harris-Irby Cotton Co not long before, and they had:

… expressed the wish of shipping cotton to Germany. After having discussed each possible means of shipping, they frankly declared themselves willing to take any risk; if necessary, they would even declare the cargo as being bound for Arkhangelsk, Russia …. I see now that these gentlemen have transformed their said plan into reality. It is only natural that, for the conduct of the correspondence of the whole affair, the cargo's *bona fide* destination will always be Arkhangelsk; otherwise the company would possibly be exposed to police raids and the payment of penalties. In the event of a satisfactory outcome, the company may be favourably disposed to further enterprises of this nature.[15]

It must be admitted that the German Admiralty had some doubts regarding the seriousness of this letter; they saw it more as an attempt by the cargo's owners, Harris-Irby, to regain their property. We will return to this below.

What is certain is that the *Pass of Balmaha* had been loaded with 4,615 bales of cotton, worth 2.5m Marks in German currency, equivalent to about US $625,000,[16] before leaving New York with the stated destination of Archangel some time in June 1915. Her skipper, who would play a leading role in the forthcoming events, was again the US citizen William J Scott. However, on 21 July 1915, *Pass of Balmaha* would eventually arouse the suspicion of the lookouts of the AMC HMS *Victorian*, at about 59°57'N, 12°42'W, despite her flying the Stars and Stripes. Just one hour before, the cruiser had put an armed guard aboard an intercepted steamer when the American sailing vessel came in her way, to be stopped for a search at 16.45.[17] What happened during this inspection has never been fully revealed. It can only be assumed that the boarding team found some irregularities or suspicious terms in the shipper's papers (*cf* the above-mentioned letter), besides entries in the logbook which showed that the ship had been to Germany on its previous voyage. Suspicions raised by the boarding team may have been supported by the unusual southerly position for a vessel heading straight from New York to Archangel. However, the investigation would eventually last for more than two hours before it was decided to send another armed guard, under Sub-Lieutenant CE Harris RNR, on board the *Pass of Balmaha* to sail her either to the port of Lerwick in the Shetlands or to Kirkwall in the Orkneys for a more thorough inspection.

The role of such armed guards on board neutral shipping must be dealt with briefly here, as their existence would, once it had become recognised, cause a stir among some German authorities, in the context of the struggle between the advocates of unrestricted U-boat warfare and those promoting a war along prize rules. *Pass of Balmaha* had been the third in a succession of vessels where armed guards had been placed, the other two having been Norwegian ships. Each of the crews, interrogated about the procedure of boarding, agreed that the leader of the guards immediately took command of the vessel, having raised the prize pennant and the White Ensign in place of the vessel's own national flag,[18] and in one case had assured the captain that he would try to ram any U-boat attempting to stop the vessel. This information, gathered by the Germans orally, would add to and explain some of the suspicions that the *Marinekorps Flandern* had already developed on the basis of intercepted W/T traffic in the Channel. In several cases during the recent weeks, they had listened in to transmissions stating: 'SS *Xxxx* now proceeding with armed party on board' or similar. In consequence, they saw these guards as an attempt to ambush U-boats by using neutral shipping, especially as in one case a machine gun had been part of the equipment of the boarding party. However, for the time being these concerns were minimised by the German Admiralty, who did not see sufficient proof for a breach of international law, as German conduct would have been quite similar.

As Captain Scott himself related when he reached Cuxhaven, shortly after his arrival in Germany: 'The cruiser ordered us, guided by a crew of 1 officer and 4 men of his company, to steer to Kirkwall, because this would be safer. The boarding party came aboard, the officer armed with cutlass and pistol, the men armed with rifles and pistols'.[19] Noteworthy is the omission of any mention of being instructed to exchange the Stars and Stripes for the White Ensign, as alleged by one source;[20] this would be crucial two days later. For the next two days, during which the ship had proceeded some 160nm towards Kirkwall, nothing noteworthy happened on board, although on the evening of 23 July a German submarine was seen to cross the path of the *Pass of Balmaha* without taking notice of her immediately, instead first chasing and sinking the French steamer *Danaë* and, on the following morning, a series of British fishing-vessels: *Cassio*, *Roslin* and *Strathmore*.[21] The subsequent events are related by Captain Scott:

The next morning, about 3 a.m., we saw a steamer blow up at about 5nm distance. Afterwards the U-boat approached us again. When the boat was about 2nm away, the British climbed into the bosun's store below, taking with them their pistols. They left their rifles on deck. I had the hatch closed and lashed. Through a small door within the hatch, I called the lieutenant, insisting that everybody hand over his arms. He complied and each pistol as well as his cutlass were given to me. Furthermore, I insisted that he and his men remain below deck, as my personal life was

This is how the famous illustrator Claus Bergen saw the situation on the morning of 24 July 1915. (Author's collection)

sacred to me. I would gladly supply the party with everything necessary, mainly provisions of course. Afterwards, I had the pistols thrown overboard, followed by the rifles, which had been hidden within one of the lifeboats under cover of canvas. (...) All of this happened before the U-boat came close; as afterwards, there would have been insufficient time. The crew begged me continuously, 'Captain, captain, do not hand us over'. They insisted on hiding in the store, and I complied, just to have them out of sight. Had the British not agreed to surrender their arms, I would have left them to starve or, in the event of an attempted break-out, I would have stopped it by force. Next, we had to send our papers by boat to the U-boat. When it returned, it had the petty officer aboard.[22]

The submarine in question, *U 36*, under the command of *Kapitänleutnant* Ernst Graeff, had left Heligoland on 17 July for a patrol, under orders to focus his attacks against naval ships with the secondary mission of conducting warfare against commerce to the west of the Hebrides. Graeff came close to his main objective only once when, on the morning of 22 July, operating west of the Shetlands, he fired two torpedoes at the approaching AMC *Columbella*, both of which would miss. Afterwards he switched target, mainly to fishing vessels, sinking eight of them plus three cargo vessels, up to the morning of 24 July. For what occurred afterwards, first on board *U 36*, later on board *Pass of Balmaha*, we have to rely on the report of *Steuermannsmaat* Lamm, the only first-hand account, made after arrival and certified as correct by Captain Scott:

Our boat was about 120nm[23] to the west of the Hebrides when we caught the American sailing vessel [full-rigger] Pass of Balmaha. It was impossible to spare a bigger prize crew because the boat was to continue its cruise. Our CO therefore ordered me to move over to bring the vessel to the Lister Deep. I was selected due to my rank as a petty officer and because I understand English, especially navigational English. I did not take any arms with me, supposing I would fare better without them. All this happened on

24 July at 4.30. When I came aboard, I took command and ordered the vessel to steer in a northerly direction, because I wanted to circumnavigate the Shetlands at a minimum distance of 120nm. I told the officer of the watch, and later the captain as well, that any resistance would be futile, German U-boats were stationed at distances of about 50nm, and each had been informed of the proceedings. Consequently, the course to Lister Deep had to be adhered to: any deviation might cause the vessel to be torpedoed. This story was spread by the Officer of the Watch and, additionally, by an older German, who by coincidence were on board, so that after a short while the whole crew had understood. In the beginning, we sailed mainly in calm winds, which shifted and rose, later setting a more southerly course with moderate winds from aft, until we stood by my own reckoning, in the evening of 31 July, about 50nm off the Horns Reef.

Before continuing with Lamm's report, the above-mentioned – admittedly thin – part of the story needs to be double-checked briefly. If Lamm had proceeded in line with his own intentions, rounding the Shetlands to the north and never getting closer than 120nm, the distance he had sailed to Horns Reef would be about 850nm in 156 hours, resulting in a (reasonable) average speed of about 5.5 knots. Unfortunately, it seems that the logbook of *Pass of Balmaha* has not survived, so we cannot be absolutely certain. Unbeknownst to him, Lamm had already passed the patrol lines of the 10th Cruiser Squadron and was essentially safe, except from coincidental encounters. Some time during the forenoon of 28 July, he must have passed the Shetland-Bergen narrows which, however, was not then of the same importance as in subsequent years.

What of the alleged submarines with which – according to some accounts, including the British Naval Staff monographs[24] – Lamm had come into contact? On the days in question, three boats were in the vicinity of *Pass of Balmaha*'s supposed track. *U 6* was operating halfway between the Firth of Forth and the Skagerrak, before returning to Heligoland on 30 July, too far south to be of any significance. *U 41* had had to break-off her operation after having been damaged by gunfire from the armed trawler *Pearl*, and could dive only with difficulty.

The encounter with *U 28* would surely have aroused some comment, given the latter's distinctive (and unique) paint scheme. (Author's collection)

On 28 July she was still to the west of the Shetlands, before returning to Heligoland on the same date as *U 6*. She clearly then had other concerns than taking an interest in an old sailing ship clearly bound for Germany. The only boat that could probably have met *Pass of Balmaha* was *U 28*, as she is reported to have circled the Shetlands westbound on 26 July 'at a wide distance.' Besides the fact that neither Lamm nor Scott mention any direct contact with another submarine, their stories invite further questions (see below).

Lamm would get into an argument with Captain Scott on 30 June, since the latter, fearing for the safety of his ship in a freshening northwesterly wind, an approaching lee-shore downwind, and darkness coming up, brought the vessel onto a more westerly course, albeit without setting more sail. Lamm suspected an attempt to sail the ship to the UK, or at least in the direction of the Channel. In his words:

My intention was to proceed immediately to Horns Reef, but the captain refused, fearing that the Horns Reef light would not be lit. We agreed to sail back and forth until daybreak. I took the opportunity to get a little more sleep. For the duration of our now one-week-long voyage, I had had few opportunities [for sleeping], remaining on the bridge almost permanently and getting little rest. For these few moments I had given standing order to call me as soon as something happened or a ship was sighted. This order was obeyed to the letter. The Americans suspected U-boats all around and, naturally, I was not inclined to challenge their belief that I was observing them continuously.

About 5.30 next morning, 31 July, I went to the bridge

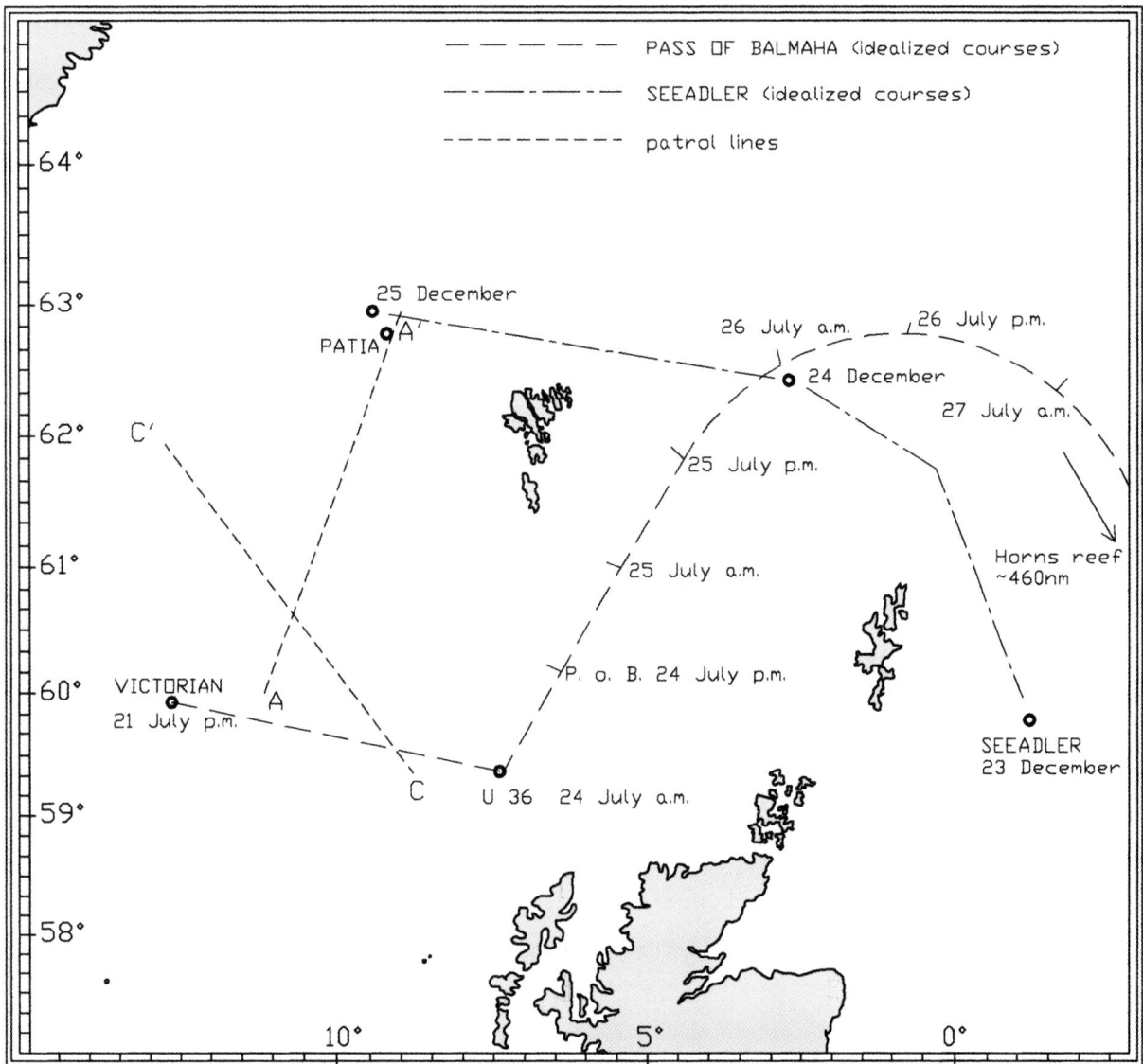

Chart of the locations north of the British Isles showing the respective tracks followed by *Pass of Balmaha* in 1915 and *Seeadler* in 1916. The tracks are approximations, as are the positions of *Pass of Balmaha* after the release by *U 36*, while the positions of *Seeadler* are taken from her log. (Drawn by the author)

and saw that the ship was steering a more westerly course, about 120nm to the west of the nearest shore.[25] The captain, questioned about the reason for this, did not wish to alter course, arguing that the wind was unfavourable for closing the shore. I disagreed, suspecting he would steer for the Galloper lightship or the Dutch coast. The captain ordered me to go below, which I refused, ordering the mate to tack instead, set the upper topsail and the topgallant, and to set course for Horns Reef light again. We approached the lightship during the evening of Saturday 31 July but, with darkness approaching, we decided to remain in our current position, sailing back and forth again, until setting course for Lister Deep on the island of Sylt, which we approached without further problems. Having dropped anchor, I had the recognition signal set in combination with flag to summon a pilot. When nothing happened between 11am and 3pm I added the distress 'NC' signal, after which a pilot boat came and undertook the necessary procedures.

We were reported as being a prize vessel, and the next day two tugs, 'Retter' and 'Enak', would tow us to Cuxhaven.

The subsequent proceedings were recorded in the war diary of the boom defence supply vessel 'D' (formerly the salvage tug *Enak*):

2 August 1915
06.50: Stood by the full-rigger, which was lying at anchor off Westerland. Sent over three guards, had the naval ensign raised, began to tow at 07.30 in the direction of Cuxhaven.
10.00: The salvage tug *Retter* joined to assist.
12.30: Passed the Eider Light Vessel.
15.50: Passed war-L.V. 'A'.
17.30: Entered Cuxhaven.
19.00: American full-rigger moored and secured in the *Amerikahafen*.

After ensuring that the vessel was safely moored at Cuxhaven, Lamm would finally be confronted with the unwelcome secret of his unknown shipmates:

While still underway, the mate and the carpenter had made some indications that I must immediately inform them in the event something should happen to me. Despite my trying to gain more precise information, they gave only a vague reply, declining to reveal anything incriminating while the captain was on board. Until our arrival in Germany, they would not reveal any more. After arrival, I called the carpenter, insisting that he tell me the truth now. In the presence of the commander of the Elbe boom defence, he finally revealed that there were one British officer and four sailors from H.M.S. 'Victorian' hidden within the bosun's store. These were brought up onto the open deck immediately, having been imprisoned there for about 11 [sic!] days.

He closed his report with the statement:

The crew behaved very well in my presence. I had the feeling that they preferred being brought to Germany instead of Arkhangelsk.[26]

Captain Scott would add little to Lamm's story. He mainly reported his view on the brief misunderstanding at Horns Reef, stating that it would have been easy for him to flee to Norway if he had so wished, but had preferred to sail to Germany. Confronted with the question of why he had not revealed the presence of the armed guards, he offered some additional detail:

I have tried to avoid any confrontation by all means, and still believe to have done the right thing in concealing the existence of the British before a single German and locking them up to nullify their superior number. I would have done the same had the situation been reversed.

Regarding the question of guarding the entrance, I should say that during night time the hatch was locked, keeping the British inside. At daylight, we instead kept the hatch under constant observation; I inspected it personally roughly every hour. The British were only allowed to use the heads, the door of which is immediately adjacent to the hatch of the bosun's store.

They were provisioned by us, having forgotten, in their haste, to take their own rations below.[27]

Interlude

We will not touch on the many subsequent questions as to the ownership of the ship and its cargo. Ultimately, on 18 December, the court at Hamburg would decide that the ship fulfilled the criteria for being seized as a prize because of her change of flag and ownership from British to US after the declaration of a state of war, according to paragraph 12 of the prize regulations. On the other hand, the cargo was viewed as still being US property, so Harris-Irby received full compensation for the cotton, which had already been sold on the German market.[28]

From the moment that it became known that a full load of cotton had reached Cuxhaven, the cotton traders of Bremen made requests to the Navy that the cargo be placed in the care of their specialist warehouses, either at Bremerhaven or Bremen. The Navy had few options other than to take up these offers. At 04.55 on 11 August, the tug *Schulau* took *Pass of Balmaha* in tow for the short transfer to Bremerhaven, which she entered at 13.55. The unloading of the cargo would take until 18 August, and that same morning the ship was towed back to Cuxhaven. It is interesting to compare the individual war diaries of the relevant institutions, which even disagree on the name of the tug that had her under tow. The stopover at Cuxhaven would be brief, just sufficient for changing tugs, this time to the former salvage tug *Simson*, as it had been decided to lay up the vessel in the port of Hamburg. Accordingly, on 19 August, close to midnight, *Pass of Balmaha* was moored there in the sailing ship basin (*Segelschiffshafen*).

Of some interest may be the fate of the crew, the

THE MAKING OF AN ARMED MERCHANT CRUISER: SMS *SEEADLER*

composition of which is, unfortunately, not fully revealed by the files consulted. Immediate action had to be taken regarding crew members whose country was at war with Germany: two Finnish (then Russian) and two Japanese citizens. According to paragraph 104 of the prize regulations they were to be escorted over the border provided they agreed not to take up arms against Germany for the duration of the war. The situation was complicated by the fact that the two Japanese refused to be discharged because Captain Scott had only paid them until 2 August: they insisted on being paid until 21 August. The British armed guards, on the other hand, as belligerents, were delivered to the prisoner-of-war camp at Sennestadt. As a side-note, before he left for the United States on 31 December, Captain Scott complained of not having received any reward for his goodwill in sailing the ship to Germany. As his plea was supported by the authorities involved, it may well have been successful – but this, again, lies beyond the content of the consulted files.

A Revolutionary Idea

It is generally agreed among modern researchers that the idea of converting a sailing ship into an AMC did not originate with Count von Luckner. The idea is most commonly attributed to *Leutnant der Reserve* Alfred Kling (born 1882), future first officer of *Seeadler*. While it is true that Kling came up with that idea, he had a (now forgotten) predecessor, *Oberleutnant der Reserve* Hinrich Hermann Hashagen (born 1885), then commanding a torpedo boat in the Weser harbour flotilla. On 9 December 1915 Hashagen had submitted to his superiors a memorandum entitled 'Proposal for outfitting a sailing ship as an AMC':

The war on commerce has shown thus far that the cruisers employed frequently experienced difficulties in obtaining sufficient coal: in particular, S.M.S. Dresden was immobilised for several weeks down at Cape Horn. Additionally, the cruise of [Prinz] Eitel-Friedrich has demonstrated that high speed is not necessarily an advantage. Despite her maximum 15 knots, she was able to hold the high seas for seven months.

My plan, founded on these observations, is to fit out an ordinary sailing ship of 1,500-2,000 tonnes with three concealed 10.5cm guns, four torpedo tubes and mines. The principal tasks of this ship would be, firstly, laying mines at different points along the English west coast before, secondly, relocating to the area of the northeasterly tradewinds to undertake a war on commerce.

Due to the speed of 12 knots, which a well-trimmed sailing ship will easily attain in the tradewinds, it would be easy to prey even on steamers sailing to the leeward, and to sink them by gunfire. (...)

The ship will need to sail around the northern tip of England, having been towed through the northen part of the mine barrage. It could, after having reached the Skaw,

Seeadler under full sail. This image must have been 'staged', using one of the captured ships as the photographer's stand-point. Additional evidence for this is that both motor boats are missing from their stowage position aft. Amidships, the name 'SEEADLER' is clearly visible, while the neutrality markings have been covered up. (Author's collection)

easily pass for a Norwegian sailing ship, given that there exists a lively lumber trade between Norway and England. If the ship were to be accompanied by U-boats to the coast of England, it might as well serve as valuable trap for enemy cruisers.[29]

Hashagen did not fail to mention his ten years' experience on sailing vessels, having served the last three years as navigating officer on the North German Lloyd sail training ship *Herzogin Cäcilie*. This background possibly helped to secure the forwarding of his proposal to the Admiralty. As there is a gap in the records, one can only assume, given that it was favourably received and initial measures ordered, that it then fell victim to creeping bureaucracy, as the next mention of the idea would be a response on 15 March by the Hamburg senior port captain *Kapitänleutnant* dR Hermann Langkopf to an inquiry – clearly made earlier – as to what kind of square-rigged sailing vessels were laid up, loaded or in ballast in that port. Only two days earlier, the Admiralty had received – initially unofficially – another memorandum regarding 'fitting out a sailing ship as a privateer-cruiser', endorsed by the commanding officer of the II. Naval Aircraft Division, *Korvettenkapitän* Friedrich Brehmer. This memorandum had actually been the brainchild of his subordinate, Alfred Kling, then commanding the II. Naval Aircraft Division's tender *D4*, but had been discussed and shaped in cooperation with *Kapitänleutnant* Friedrich Wolf, first lieutenant of the AMC *Möwe*, and *Kapitän zur See* Hermann Bauer, then Leader of U-boats. This may be the reason why it emerged as a more elaborate proposal than that of Hashagen. Only the key points will be noted below:

A medium-sized, solid sailing vessel of the type reported to dock in numbers at the port of Hamburg, possessing favourable sailing qualities, is to be fitted out as a privateer-cruiser in the following manner:

Armament: 3-4 long-range 15cm guns, favourably positioned, as well as one 8.8cm QL gun; two torpedo tubes mounted on deck. The mounting of both types of armament will be much simpler than on steam ships, most probably entirely in centreline mounts.
Engine: Diesel engine providing speed for 8–10 knots for use in calms and to provide extra manoeuvrability in combat.
Rigging: Brand new rigging (preferably of the 3-mast barque type), to be disguised as worn by painting and patching. The topgallants must be prepared for immediate striking. Plenty of spare sails, spars, and rope.
W/T: Provision made for quickly taking in the antennae.
Accommodation: provision made for 200 prisoners, galley, heads, sick-bay, etc, on the lower deck.
Ballast water: Provided for fresh water supply.
Boats: A sturdy motor boat provided with a machine gun or, better, a 5cm QL gun.
Provisions: For the crew plus 200 prisoners; sufficient ammunition, torpedoes and bunkerage for 18 months. (In an otherwise empty vessel, this will be easily accommodated.)
Complement: 75 men maximum, possibly only 50; hand-picked (volunteers with experience on sailing ships in sufficient numbers). Four officers (including one experienced W/T officer) besides the CO, one MO. Well-proven gunners and torpedomen. (…)

Arrangements for the outward-bound voyage to be as follows:

The vessel will be provided with neutral papers (either Swedish or Norwegian) and a name that closely resembles that of a neutral sailing ship in the lumber trade departing at the same time. (It can safely be assumed that those vessels will be reported to England.) Guns and torpedo tubes will be covered by a high, well-prepared deckload of lumber (usually protruding about 1m above bulwarks.) This deckload is to be prepared in such a way as to permit the guns and the torpedo tubes to fire during the passage of the blockade lines. Afterwards, it will be removed entirely. Each suspicious fitting will be removed and stowed below, together with the surplus crew, in inaccessible rooms. Only the captain and the nucleus crew in their appropriate disguise will remain on deck.

Prepared in the above manner, the ship will sail along the Norwegian coast, having been towed initially by a torpedo boat, with the diesel engine being engaged at night time. It will round Iceland to the north, then sail south between Iceland and Greenland. (…) In the event that the ship is stopped in the blockade line and personally searched, even while only carrying an unsuspicious deckload of lumber, there is a good chance of a successful deception. If not, arms must decide! …[30]

The Admiralty seems to have found the proposal sufficiently convincing to prompt a personal discussion with Kling on 13 May regarding further details and, subsequently, to install him as supervisor for the proposed conversion of a selected vessel. Indeed, the Admiralty had already approached port captain Langkopf with a request that he might pick the most suitable vessel from his existing list. Under various pretences, Langkopf went aboard several vessels, inspecting their logbooks and making transcriptions of documents and plans which he would forward to the Admiralty on 6 May. *Pass of Balmaha* must have come out best – although there is no formal record of exactly why she was chosen other than general comments regarding the suitability of her loading capacity and speed. One prevailing argument for her being chosen may have been the truly 'average profile' of the vessel, despite the known German preference for barque-rigged vessels. In any event, on 18 May Kling received an offer from the famous Tecklenborg shipyard at Bremerhaven-Geestemünde, signed by the equally famous naval architect Georg W Claussen,[31] for the conversion of *Pass of Balmaha* along the proposed lines. The price for the conversion was estimated by

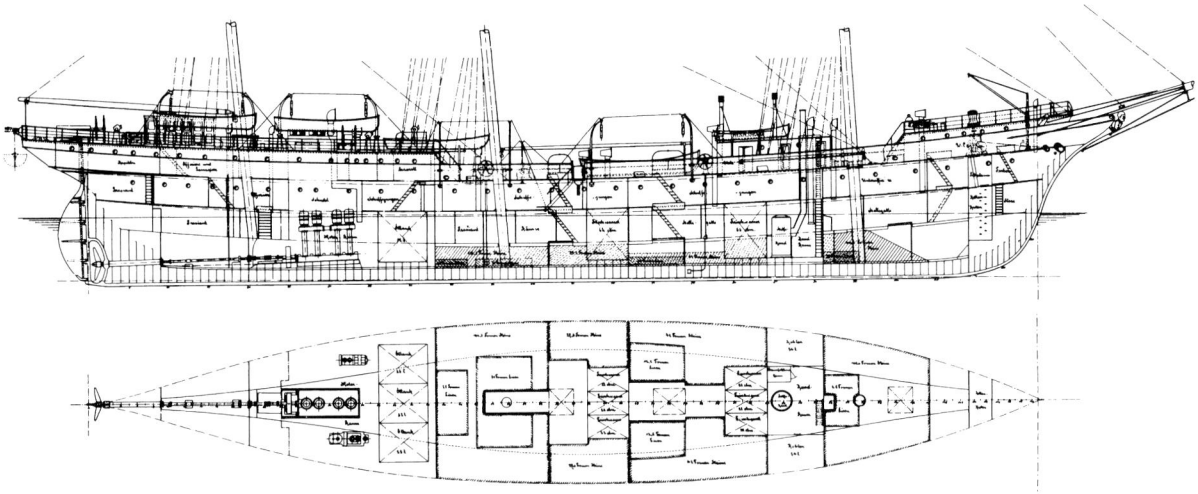

Longitudinal section of *Grossherzog Friedrich August*. Note the engine, employed for the first time in a German sail training vessel, and of a similar type to the one mounted in *Seeadler*. It was rejected for the raider because it was considered to be insufficiently powerful. (*Schiffbau* 1913/14, courtesy of the author)

Claussen at 550,000–600,000 Marks. To save precious time – a reduction from six months to four – it was initially proposed to install the 550bhp four-stroke diesel fitted in the new sail training vessel *Grossherzog Friedrich August*,[32] built by Tecklenborg in 1914. In this ship it provided a speed of 9 knots but, according to Claussen, would propel *Pass of Balmaha* at only 7.5–8 knots because of her fuller hull. This proposal was therefore turned down. The basic work was specified as follows:

1 Construction and installation of a *c*400bhp diesel engine, the exhaust of which would be led through the mizzen.

2 Replacement of the standard stern-post by one of cast steel designed to mount a propeller and shaft.

3 Installation of tanks for about 400 tonnes of diesel fuel, subdivided to allow the replacement of oil consumed by seawater.

4 Installation of several tanks for fresh water of about 135m^3 capacity, plus a device for melting ice and the necessary pipework for transfer of the water into the tanks.

5 Provision for accommodating about 200 men in hammocks on the lower deck, together with ten cabins each fitted with three bunks.

6 Installation of wash bowls, WCs, etc, for about 250 men.

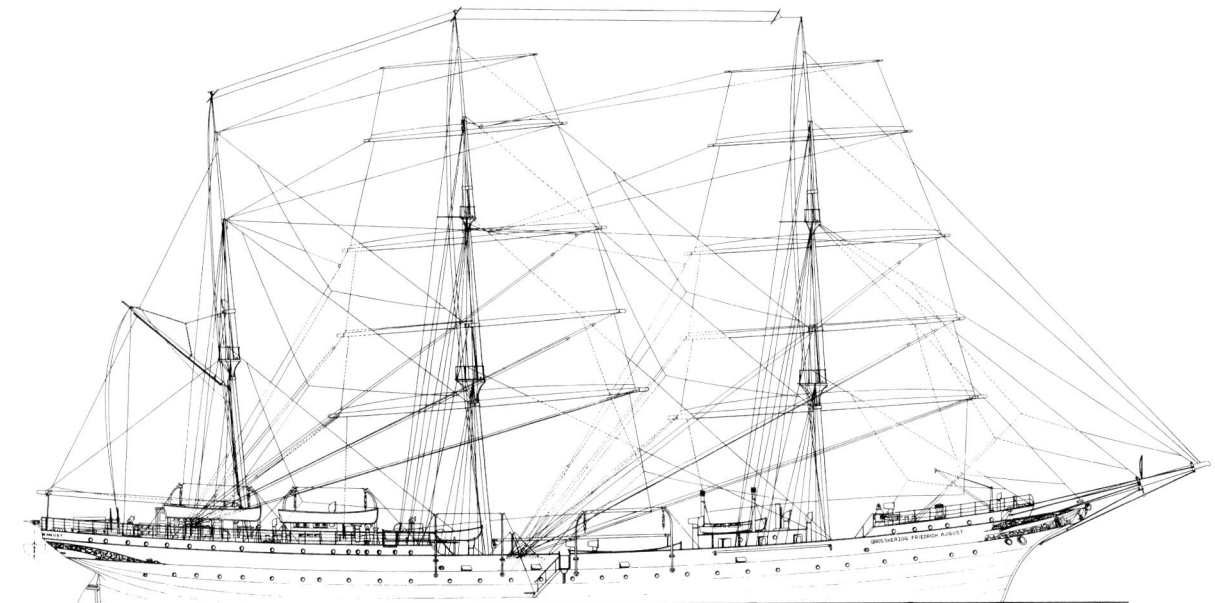

Sail plan of *Grossherzog Friedrich August*. It would have been almost impossible to disguise this handsome training vessel as an ordinary merchantman, as had been initially suggested. (*Schiffbau* 1913/14, courtesy of the author)

One of the most deceptive images in naval history! In most publications, it is described as 'Seeadler, fully equipped and ready to start her epic voyage' (or similar wording), and was even used as standard for a succession of ship models. In fact, Seeadler never looked like this! The image is of Pass of Balmaha, fully loaded and looking well-kept and pristine. (Author's collection)

7 Mounting of a steam boiler and radiators for heating on the lower deck.
8 Installation of electric lighting where appropriate.
9 Installation of stores for dry and wet provisions at appropriate locations.
10 Erection of two gun mounts aft of the forecastle, plus installation of a magazine equipped with a flood valve and an ammunition trunk.[33]

In mid-July, Pass of Balmaha would be transferred from Hamburg to Geestemünde, into the capable hands of the Tecklenborg shipyard. From now on, she would be known as the picket steamer Walter. Consequently, all further proceedings regarding the ship bore the designation 'most secret' or 'to be passed from hand to hand', as no detail whatsoever could be allowed to leak out. This may in part explain why the archival situation regarding the conversion is full of gaps – although the loss of most of the archives of the Tecklenborg yard may also be responsible.

Walter

Shortly after her arrival, a major decision was made regarding the propulsion system. It should be noted that the official history of the naval war, published by the German Admiralty, would state in 1937 that:

The sailing ship 'Seeadler' may be counted, to a certain extent, as a general cargo ship, because the sailing rig served principally as a decoy for a motor ship. It was intended to enable the ship either to proceed long distances or remain in a certain area under sail alone, without consuming fuel. Each attack, however, would be carried out with support of the engine.[34]

Thus, the choice of a suitable engine had been viewed as critical. On 26 August the Reichs-Marine-Amt (RMA – Reich Navy Office) intervened:

Because the motor of the training vessel 'Grossherzog Friedrich August' is quite small and, additionally, its removal would interrupt the planned continuation of her training cruises, a motor found and confiscated at Ghent has been examined and found suitable once certain modifications have been made. It is of the same type that had been constructed by the Tecklenborg concern for the motor ship 'Rolandseck', and was built by the Belgian firm of Carrels, partners of Tecklenborg. It is of the 6-cylinder type and, in its present form, too large and heavy. Accordingly, two cylinders will be removed, leaving a remaining output of c900bhp.[35]

This 'find' must be considered fortunate, as few viable alternatives would have been available in time. The firm of Carrel Frères in Ghent, Belgium, had been approached in 1913 by the British Admiralty for the delivery of two

Manufacturers' photo of the 6-cylinder diesel engine for the M/V *Rolandseck* of the Hansa Line. For installation on *Seeadler* of a similar engine, the central pair of cylinders had been removed. (Author's collection)

1,500bhp shipboard motors operating on the two-stroke principle. Before delivery, they were exhibited at the Ghent Fair and run there for about half a year. Contrary to the impression given in the file, the starboard engine was purchased by the German Navy for delivery to the Tecklenborg shipyard, where cylinders 3 and 4 would be removed. Two significant features of this engine had been its large flywheel and the solid, egg-shaped cylinder casings – which can still be observed, submerged, on the reef of Mopeha Island.[36]

The conversion had been in full swing since the beginning of August, mainly by adjusting the initial proposals by Claussen to actual circumstances. Under pressure of time, it was decided by the RMA to replace the sternpost made of cast steel with one constructed from 30mm steel plating. Similarly, central heating was dropped in favour of individual stoves, allowing the space planned for the boiler to be used as an additional provisions store. It was still intended to lead the engine's exhaust into the mizzen, but this was dependent on the actual design of the motor, which had not yet been finalised; in the event, this proposal was dropped, as Kling feared that discharge of the exhaust gases so high up would be visible from afar.

An exhaust to the sides below main deck level was briefly considered, but ultimately they were led forward and upward, through the central deckhouse, into the empty funnel for the planned heating boiler. Two diesel-driven 5kW generator sets, mainly to serve the W/T installation, were ordered from the Grade company, Magdeburg, for delivery within eight weeks. The size of the W/T office was the same as in torpedo boats, with the aerials spread between the main and mizzen masts, ready for quick retraction. On a special request from Kling, in addition to boats propelled by oars, two motor boats were provided. Their dimensions were set at a length of 8–10 metres, 2 tonnes maximum weight, engine power at about 6hp; they were to be double-ended.

In parallel with these technical activities, it had to be decided who would be in command of the future AMC. While he was qualified to supervise the conversion and would serve in the future as a very capable first officer, doubts were raised from the outset as to Kling's qualifications to be appointed as commanding officer – especially as he belonged to the Naval Reserve. The decision would be not be an easy one for the Naval Cabinet, which was responsible for personnel. Who, among the whole officer's corps, would be capable of commanding a sailing vessel, given that sail training had been abandoned ten years earlier? The choice of Luckner was perhaps inevitable, as he had no realistic competitor among the Navy's career officers.

The history and personality of Luckner has been well documented elsewhere.[37] However, one rarely-considered aspect should be mentioned here: going through the list of commanding officers of German AMCs, there are few, if any, with a clean service record.[38] Consequently, it is not unfair to view them as being somehow 'expendable' on their special missions, with Luckner no exception. However, the choice was made, and on 23 September Kling was finally informed of his new superior.[39]

Another important detail came into focus at the beginning of September: the fake cargo. In 1916 a full deck-

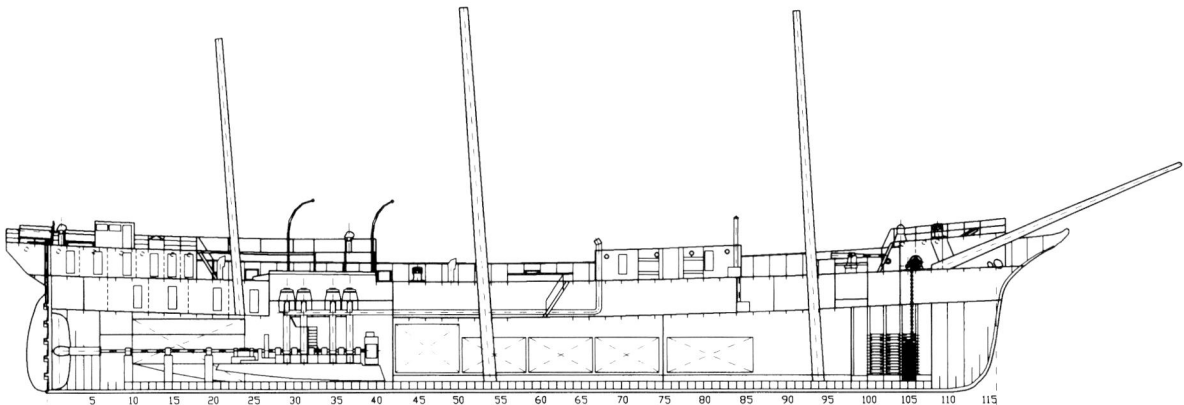

A GA profile of *Seeadler*, reconstructed from all available sources. Unfortunately, no precise information regarding the layout of the lower deck, which housed the accommodation for part of the ship's crew and the crews of the ships sunk, could be found. (Drawn by the author)

This poor-quality image is one of the few known photos that depict *Seeadler*'s deck-load. Of the disguised crew, only a few members are identifiable with any certainty: Lieutenant Pries (first from left, front row seated), Lieutenant Kling (second from left, front row seated), and Commander Luckner (second from right, front row seated). (Author's collection)

load of timber would be a not-inconsiderable quantity in a country fighting for survival and steadily becoming war-weary. Thus, the timber was purchased in Sweden and shipped to Hamburg on board the steamer *Greif*, where it would be temporarily loaded onto barges while the constructors made up their minds as to how best to distribute the timber on board *Pass of Balmaha*. A special part of the cargo would be the wooden raw material for the alteration to her rig which, at first glance, seems

strange: having endeavoured to find a vessel with an 'unsuspicious look,' the German Navy now changed a standard British rig into a distinctly German one, with double topgallant sails and royals. On the other hand, it now appears that this may have been to make possible another deception, in that the new rig was designed to be modified at sea at will. Besides the order for three new topgallant spars and three royal yards with sails, there was also an order for a set of split topgallant sails, as well as another set of new single topgallant sails. These would make little sense if the planned rig was not convertible. And she would leave Germany rigged in the latter way, with single topgallants.

At the end of October, the ship's guns were available to be delivered on board.[40] Contrary to early plans, there would be no mounting on the centreline; instead, the two mounts would be placed aft of the forecastle on each side, providing better firing arcs than a mounting on the centreline hatches owing to the potential wooding by the rigging. Initially, however, they were to be stowed below deck. Regarding the magazine, the full capacity had originally been set at 100 rounds for both guns, but this was first increased to 500 rounds before the Admiralty proposed a further augmentation to 1,000 rounds, if possible. The latter was confirmed by the shipyard on 5 November.[41] The main magazine was located below the waterline directly beneath the guns, and was equipped with forced ventilation and a 300mm flood valve with drenchers. For scuttling the ship, a second 300mm flood valve was mounted in the engine room, while several scuttling charges were prepared to be distributed in the holds if required. Aft, there was a

This remains the best image of the type of gun mounted on *Seeadler*: the 10.5cm/40 C/97 – here employed as a waist gun on the small cruiser *Bremen*. (Author's collection)

smaller magazine for machine guns, rifles and scuttling charges, integrated into one of the provisions stores.

The values for trim and stability, evaluated at the end of October, are most interesting:

Condition I

1	Ship without ballast	1,055t
2	Steelwork (tanks, decks, bulkheads, etc)	300t
3	Engine	268t
4	Bunker	455t
5	Potable water (fwd)	200t
6	Potable water (aft)	130t
7	Guns and ammunition	25t
8	Inventory	20t
9	Provisions (fwd)	90t
10	Crew	25t
11	Ballast (beneath tanks)	150t
12	Ballast (beneath engine)	85t
13	Coal	60t
14	Sand	16t
15	Lube oil	15t
16	Petroleum	2t
Total		**2,896t**

In this condition the ship would have a mean draught of 5.4m, with an aft trim of 0.39m, resulting in a draught forward of 5.21m and aft of 5.6m.

Condition II (with additional 200t deck-cargo)

1	[Ship as Condition I]	2,896t
2	Deckload of *c*300m³ of lumber	200t
Total		**3,106t**

In this condition the ship would have a mean draught of 5.7m, with an aft trim of 0.77m, resulting in a draught forward of 5.3m and aft of 6.1m. To compensate for the aft trim, and to settle the draught closer to the maximum load-line, about 200 tonnes of iron ballast would additionally be stowed in the tween deck, giving a maximum mean draught of 5.9m (5.68m after dropping the deck-load) and an aft trim of 0.5m (0.1m), resulting in a draught forward of 5.73m (5.64m) and aft of 6.23m (5.73m).

Seeadler/Hero

On 15 November it was ordered that the ex-*Pass of Balmaha* be added to the list of German warships under the name *Seeadler*; this was confirmed on 21 November.[42] The same day, Luckner was officially appointed captain of the ship. On 18 November the sailing order had been issued. It contained few surprises, as most situations simply could not be foreseen. The departure date could be adjusted to coincide with heavy weather, with the intention of reaching the latitude of the Skagerrak as soon as possible to enhance the credibility of the Norwegian flag being flown as a disguise. On the way north, special preparations would be made for German picket flotillas not to interfere with the possibly

Seeadler praying for a 'catch' while drifting idle on the ocean. It is not recorded whether the photo was taken in the Atlantic or the Pacific, although the state of the paintwork suggests the latter. (Author's collection)

suspicious vessel. Should any hostile ships be encountered, Luckner was explicitly ordered not to attempt any evasive manoeuvres, instead relying on deception and risking a search. In the event of failure of the deception, the ship was to be scuttled, but not before the German ensign and the command pennant had been hoisted and the officers had dressed in proper uniforms.

A problem arose when, even before the ship's commissioning, which was planned for the first days of December, the carefully chosen decoy identity for *Seeadler*, the Norwegian full-rigger *Maletta*, put to sea.[43] Since early November she had been moored in the Danish port of Aalborg, while being prepared to sail to Buenos Aires with a load of cement. A few days later, there was a change of plan in favour of a voyage in ballast to the USA, but she was then further delayed, waiting for a further 570 tonnes of ballast – although neither cargo would have been a good fit with *Seeadler*'s deck-load of timber. However, on 30 November, her departure from Aalborg was reported and verified by the German embassy – not without further notice that Danish authorities had played their own game of deception: the local newspapers had been fed with information that *Maletta* had sailed to the port of Randers to load cement. Ironically, *Maletta* would be stopped and searched, possibly twice, on her way: on 8 December by a German submarine,[44] and two days later by HMS *Avenger*.

The next choice of disguise, under the alleged name of *Carmoe*, raises some questions. The British naval staff dismissed this information when they learned it from Luckner's recollections after the war, as no vessel of this (or similar) name been taken into Kirkwall for inspection in December.[45] However, a ship under the similar name of *Karmø* had been extant,[46] even if not sent to Kirkwall. According to *Seeadler*'s war diary, the name *Kalliope* was chosen next, before it was found out that this 'alias' ship would possibly soon be crossing the blockade lines eastbound, and the risk of possible detection was deemed excessive.[47] However, there seems to have been a broad range of possible further choices, the next best being another Norwegian full-rigger named

Kalliope, alias *Seeadler*, riding at anchor in the North Sea before taking on her deck-load of lumber. Her reduced rig made sense in more than one way: besides presenting a more common profile, lumber as deck cargo is notorious for having a high centre of gravity, thereby adversely affecting the righting moment – an important consideration for a sailing vessel preparing to navigate the northern reaches of the North Atlantic in winter. Note the canvas rigged as a wind screen around the poop deck. (Author's collection)

Hero,[48] homeport Kristiania.[49] Of 1,709 tonnes and built in 1873, she was ultimately chosen because she was still located in Norway, being offered for sale. Consequently, from 14 December *Seeadler* would be temporarily named *Hero*.[50]

On 24 November the first crew members had arrived on board, mainly from the technical department, to enable them to familiarise themselves with the engine as soon as possible. Two days later, the ship departed the yard for a first 6-hour trial run. The following day a second run had to be abandoned because of unspecified engine problems, which would not be resolved until the 30th. During this enforced break, the remaining ballast iron was taken aboard by the crew. On 1 December, the ultimate trial run under diesel power was undertaken, lasting nearly eight hours. Rather than returning to the yard, the ship then anchored in Blexen Roads, upriver of Bremerhaven, the trials being declared as successfully completed.[51] Finally, on Saturday 2 December 1916 the war diary of SMS *Seeadler* opens with the subsequent entry:

> 9.30 a.m. on Blexen roads S.M.S. "Seeadler", three-masted, full-rigged vessel, was commissioned. It carries an auxiliary-motor, installed by Tecklenborg, Geestemünde. Complement: 64 including commanding officer.[52]

The next morning saw her leaving the Weser estuary for one of the numerous streams within the remote mud-flats area along the west coast of Schleswig-Holstein. The lighter with the deck-load of timber was already moored in the Norderaue, and the crew immediately set to work loading it on 6 December, a task that would be completed on 11 December. Meanwhile the name *Kalliope* was applied on both sides of the bow and on the stern, together with the large identifier 'NORGE' on both sides amidships, accompanied by the customary red, white and blue stripes. Interestingly, the ship had become heavier than calculated, with a measured draught of 6.3m forward and 6.8m aft. On 15 December *Seeadler*, now under her new disguise as *Hero*, was ready for departure, awaiting orders. Two days later the torpedo boat *S 128* would come alongside with the written order that the ship could depart at her Captain's discretion. On the morning of 21 December the forecast seemed to be favourable for an unseen breakthrough, and the ship weighed anchor. The entries in the war diary report a strong gale from SE and later SW for the subsequent three days until, on the morning of 25 December, the wind would calm down. At 09.10 a large steamer was reported on the port side aft. This brings the story full circle.

Endnotes:

[1] Felix Alexander Nicolaus Georg Count von Luckner (9 June 1881 – 13 April 1966): SMS *Panther*, OoW October 1914;

SMS *Kronprinz*, OoW August 1916; SMS *Seeadler*, CO November 1917.

2 In particular, L Dinklage, *Das Geheimnis der Pass of Balmaha*, Enßlin & Laiblin (Reutlingen, 1940).

3 A variety of names have been offered over the years as to which AMC had undertaken the subsequent inspection fuelled by Luckner's version, which claimed that he had been intercepted by *Highland Scot*. This particular name had allegedly been mentioned by one of the ship's boat's crew and was noted in *Seeadler*'s war diary. Other authors would claim that *Highland Scot* had only been a decoy name for the *Avenger*. Neither of these names is correct.

4 *Seeadler*'s position had been quite similar, at 62°54'N, 8°47'W.

5 TD Lilley; *Operations of the 10th Cruiser Squadron: A Challenge for the Royal Navy and its Reserves*, (PhD thesis, University of Greenwich, 2012), 140.

6 TNA ADM 53/54565 and *Naval Staff Monograph XVIII – Home Waters*, pt VIII, (Naval Staff, 1933), 15–16.

7 *Seeadler*'s logbook reports the boarding of two officers only. However, this procedure would be another deviation from the rules, as the 'instructions for boarding officers' called for at least one officer, an armed guard and a search party. See Lilley, *op cit*, 140.

8 Interestingly, this contact was not mentioned in *Hilary*'s log.

9 There seems to be some mention of her in local newspapers over the years, highlighting a particular crossing made in a remarkable time and so forth which is not regarded as relevant here.

10 GH Williams; *The United States Merchant Marine in World War I*, McFarland (Jefferson NC, 2017), 46–47.

11 *Ibid*. The author states that the ship was seized on 9 March and sailed to Kirkwall afterwards, before being released on 19 March and subsequently sailed to Germany. This is problematic. Besides the fact that on 19 March *Pass of Balmaha* was already stuck fast in the surf of Sylt, the period of ten days for seizing the vessel, sailing it into Kirkwall and having it thoroughly investigated seems to be overly brief. This view is supported by Lilley, *op cit*, 42, where the average time of detention in the port of Kirkwall in 1915 is stated to have been nine days.

12 NARA Roll T1022-243 PG-62765-NID, 573ff.

13 B Langensiepen and D Nottelmann, 'Roll the cotton down', *Marine-Nachrichtenblatt* 19 (2015), 58–59.

14 Dr Heinrich F Albert had been agent for the *Zentrale-Einkaufs-Gesellschaft* (Central Purchasing Agency), tasked with purchasing and shipping American goods at the expense of the German government, as well as acting as commercial attaché for the German embassy.

15 BAMA RM5/2809.

16 Compare this to the cost of the vessel stated above!

17 ADM 53/67377.

18 This procedure seems to have been the reason for protests in more than one case, as standing orders would be amended (CIO 333/1916), so that, according to XVIII (a) of 'Instructions for Boarding Officers and Prize Officers in Wartime', the White Ensign should only be raised above the vessel's national flag; this was revised again by order of

This model of *Seeadler*, located in the German Naval Academy, is said to have once belonged to the von Luckner family. It must certainly have been constructed by a modeller with a good knowledge of the ship, as some of the details depicted are of the original vessel, and have since been lost to the general public. (Marineschule Mürwik)

Admiral Tupper in May 1917, when he decided that boarded vessels should fly only its neutral colours. See Lilley, *op cit*, 150.

19 BAMA RM5/2809.

20 Williams, *The United States Merchant Marine*, 46–47. It is not known how the author aquired this information.

21 O Lörscher, *U 36 Chronology* (unpublished manuscript).

22 BAMA RM5/2809. [It should be noted that the protocol of Captain Scott's interrogation meant that his account was translated from English into German. It has subsequently beeen re-translated from the German into English, which makes it very likely that some nuances may have been lost.]

23 120nm must be regarded a significant overestimate – see map.

24 *Naval Staff Monograph* VII – *10th Cruiser Squadron*, Naval Staff (1922), 51.

25 This figure would be seriously questioned by his interrogators as being much too far out for the prevailing situation. A position 75nm off Horns Reef would be seen as realistic, being much more in line with the reckoning by the crew.

26 BAMA RM5/2809.

27 *Ibid*.

28 Robert M Harris had personally travelled to Germany and stayed for three months to argue his rights regarding the cargo – in which he would be ultimately successful.

29 NARA Roll T1022-655 PG-75107-NID, 227ff.

30 *Ibid*, 288ff.

31 Georg W Claussen had been the designer of the 5-masted barque *Potosi* and the 5-masted full-rigger *Preussen*, pride of the German merchant fleet.

32 *Grossherzog Friedrich August* was the third vessel of the *Deutscher-Schulschiff-Verein* (German Sail Training Ship Association), for the education of boys for the merchant fleet, and is still extant as the Norwegian sail training vessel *Staatsrad Lehmkul*. She had briefly been considered for conversion instead of *Pass of Balmaha*, but this proposal was soon abandoned as it would have been much more difficult to disguise her as a freight-carrying sailing vessel.

33 *Ibid*, 302ff.

34 E von Mantey, *Der Krieg zur See – Kreuzerkrieg,* Bd 3, Mittler (Berlin, 1937), 326.

35 It must be attributed to the special design of this early diesel engine, originally consisting of six cylinders mounted in pairs, that Luckner sometimes referred to the ship having two engines. The remaining two blocks of two cylinders could give the impression – to the inexperienced eye at least – of two individual engines coupled together.

36 It is to the credit to the most reliable model of *Seeadler*, in the collection of the German Naval Academy, that even the propeller is shown to rotate in the correct direction for a former starboard-side engine. Likewise, the number of propeller-blades is correctly shown, as the mounting of an even number of propeller blades was prohibited when using a motor with an even number of cylinders.

37 According to Luckner's ultimate surviving qualification report from 1915, his general conduct had been: 'Average. Insufficient thoroughness.' (K Franken, '*Qualifikations-berichte der Seeoffiziere*', Norderstedt, 2022).

38 N von Frankenstein, '*Seeteufel' Felix Graf Luckner*, DSV (Hamburg, 1997), 34.

39 This date alone contradicts claims made by Luckner about his involvement in the planning of *Pass of Balmaha*'s conversion.

40 Contrary to the information given in most publications, the guns were not of the L/45 type, as these would not have found sufficient space aft of the forecastle. In reality, they were of the 10.5cm L/40 C/97 type, introduced on the small cruisers of the *Gazelle* class before the turn of the century, most of which had already been retired from frontline duty, making their guns available.

41 Even if lacking definite numbers, from the weight distribution figures recorded below it is clear that it would have been impossible to stow 1,000 rounds. While the available space in the magazines would have allowed a stowage of about 800 rounds, the outfit cannot have significantly exceeded 600 rounds, given the recorded weights.

42 The naming was only made possible by the fact that the former small cruiser/gunboat *Seeadler* – though still on the navy list – had been in use exclusively as a non-active floating mine depot at Wilhelmshaven. In this capacity she would eventually be lost to an explosion on 19 April 1917.

43 *Maletta* (ex-*Clan Robertson*) had been built from steel as a full-rigger in 1884 at the same yard as *Pass of Balmaha*, being a little larger and registered at 1,702rt (Norsk Maritimt Museum – Malmstein register).

44 This dubious information originates in a story told by her crew to the boarding team of HMS *Avenger*, and is quoted in *Naval Staff Monograph* XVIII, 14. It cannot be confirmed by German sources.

45 *Ibid*.

46 *Karmø* (ex-*Circe*) had been built in 1885 as a full-rigger in Glasgow and registered at 1,619rt (Norsk Maritimt Museum – Malmstein register).

47 *Kalliope* seems to have been another ideal choice, having similarly been built from steel as a full-rigger in 1888 in a Glasgow shipyard and registered at 1,684rt (Norsk Maritimt Museum – Malmstein register).

48 *Hero* (ex-*Anemone*), a full-rigger again built from steel by Robert Duncan in 1873, was registered at 1669rt (Norsk Maritimt Museum – Malmstein register).

49 The alleged homeport of Arendal does not correspond with the information from the Norsk Maritimt Museum – Malmstein register.

50 The complicated search for a suitable name contradicts another of Luckner's oft-cited hair-raising stories, of stealing the *Maletta*'s log book in the port of Copenhagen.

51 BAMA RM99/413.

52 HD Schenk, *Graf Luckners 'Seeadler'– Das Kriegstagebuch einer berühmten Kaperfahrt*, Die Hanse (Hamburg, 1999).

THE BATTLESHIP *BOUVET*, MARTYR OF THE DARDANELLES

To conclude his series on the French battleships of the *Flotte d'échantillons*, **Philippe Caresse** focuses on *Bouvet*. Despite her reputation as the most successful ship of the series, *Bouvet* would strike a mine during the Allied attempt to force the Dardanelles and sink in less than a minute.

Bouvet was to be the last of five 1st class battleships with similar general characteristics but of independent design subsequently known as the 'Fleet of Samples' (*Flotte d'échantillons*). Like her half-sister *Masséna* (see the author's article in *Warship 2023*) she would have her 305mm and 274.4mm main guns upgraded from Modèle 1887 to the new Mle 1893, which was of more advanced construction, and the four 65mm QF guns of the original three ships would be superseded by eight guns of the new 100mm calibre.

The distinguished constructor Charles Huin (1836–1912), who had been responsible for the first ship, *Charles Martel* (see *Warship 2020*), was selected to design *Bouvet*, which would emerge from the building yards with a very different silhouette and internal layout to *Masséna*, built in parallel to a design by Louis de Bussy. Apart from the number, calibre and model of the main guns, the only major feature the two ships had in common was their three-shaft machinery, approved on 19 March 1892, which would set the pattern for subsequent French battleship construction prior to 1909.

The order to build the ship at Lorient Naval Dockyard

Bouvet on the slipway at Lorient. Note the wooden backing for the full-length belt, which would be fitted after launch. Decoration using shrubbery was very much in vogue for new ships and buildings at the time. (DR)

Bouvet: **Profile & Plan**

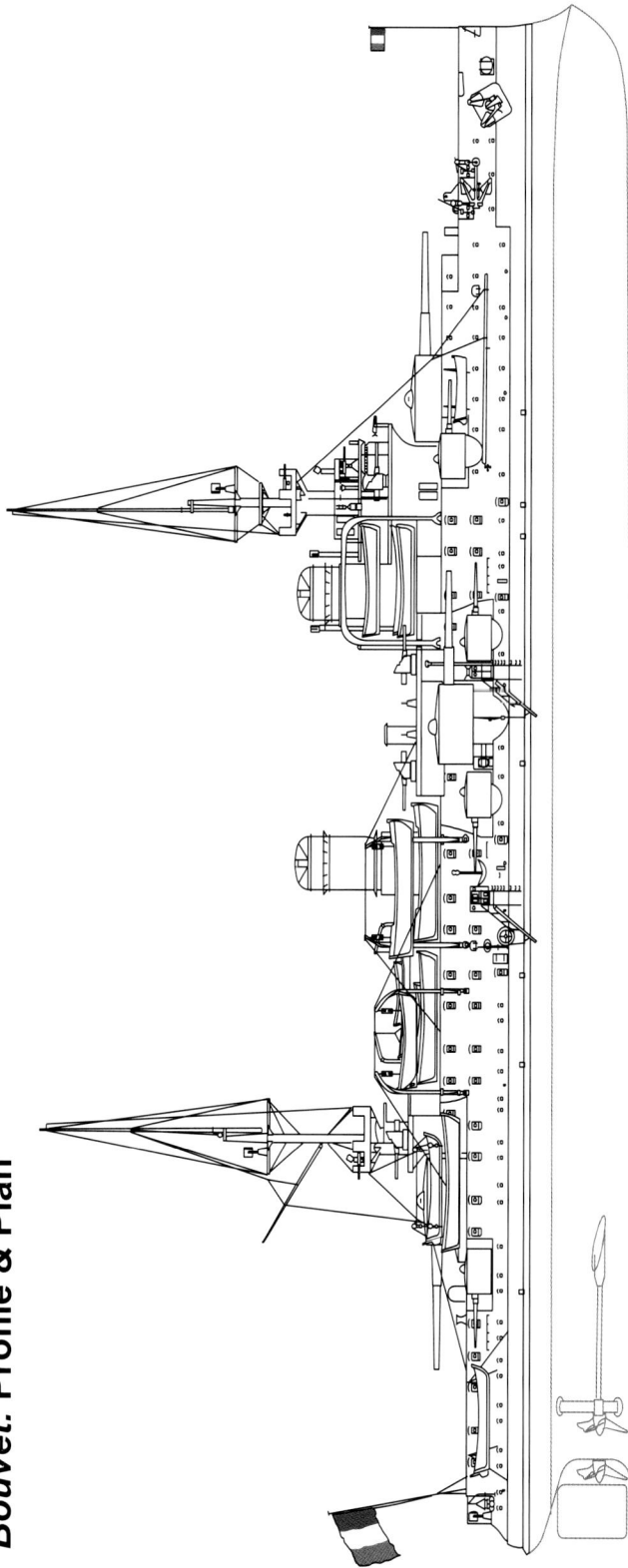

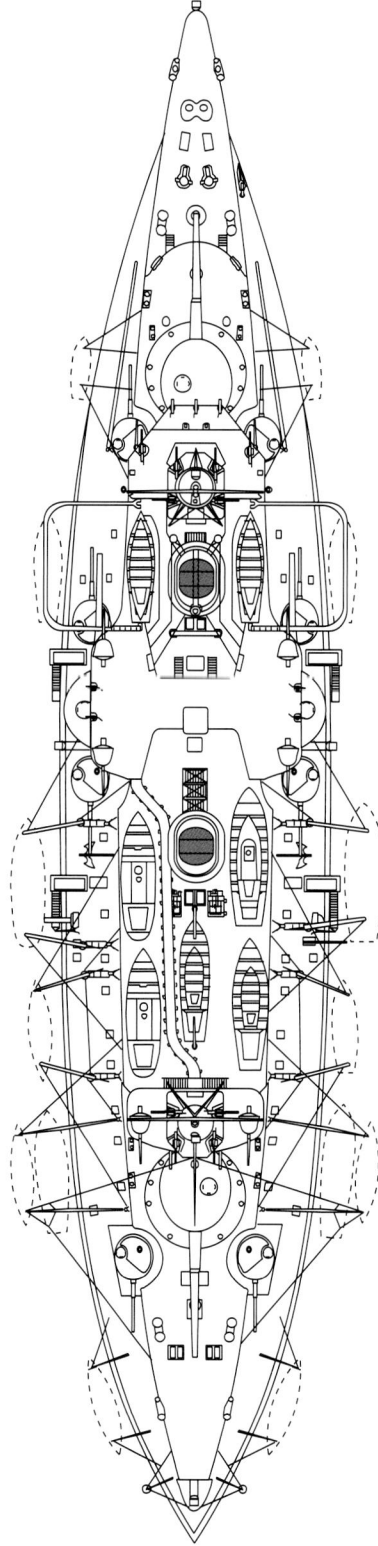

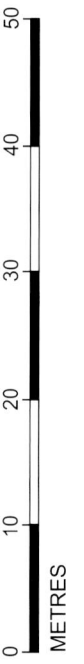

Note: Adapted from plans dated Lorient 25 January 1898.

© John Jordan 2021

0   10   20   30   40   50

METRES

Bouvet is launched on 27 April 1896. (DR)

was signed on 8 April 1892 and the keel was duly laid on 16 January 1893. The number of working days was estimated at 1,700,000 and the cost of construction of the hull at 5,644,000 francs (equivalent at the time to £226,000 sterling).

Certain characteristics would be modified during the studies for the battleship in order to save weight. The thickness of the belt would be reduced from 450mm to 400mm, the armour on the barbettes of the main turrets would be thinned by 10mm to 310mm, the conning tower would have smaller dimensions and the handling arrangements for the anchors would duplicate those of the earlier *Brennus* (see *Warship 2019*). Weight-saving measures extended even to the washing facilities for the crew: the number of washbasins was reduced from the original figure of 46 (for 130 men) to the 24 (for 127) on *Brennus*.

On 9 January 1893, the *Direction du Matériel* proposed electrical power for the turrets, as in *Jauréguiberry*, but this was rejected three days later by the *Direction de l'Artillerie*, which insisted on the retention of the original hydraulic machinery for training.

General Characteristics

The construction of *Bouvet* was supervised by naval engineers Dupré, Minel, Lejeune and Gelly. The weight of the hull, which was exclusively of steel, was 3,699 tonnes.

The armoured belt, which was of a new toughened homogeneous steel developed by Schneider, was 400mm

Table 1: Hull & General Characteristics

Length oa	122.60m
Length wl	121.01m
Length pp	117.90m
Beam (wl)	21.40m
Depth of keel	7.80
Draught	7.69m fwd, 8.40m aft
Freeboard at bow	6.00m
Normal displacement	11,982 tonnes
Full load displacement	12,052 tonnes
GM	1.09m (normal displacement, designed)
Surface of underside	3,322m²
Complement	
as private ship	21 officers, 10 midshipmen, 591 seamen
as flagship	31 officers, 10 midshipmen, 651 seamen
Protection	
Main belt	400mm (200mm bottom edge); 2m high
Upper belt	80mm; 1.2m high (2.55m at bow)
Armoured deck	70mm (extra-mild steel)
Splinter deck	30mm mild steel
Main turrets	370mm sides, 70mm roof, 60mm floor, 310mm barbette
138.6 turrets	100mm sides, 20mm roof, 60mm floor, 100mm barbette

amidships, reducing to 300mm forward and 260mm aft, the plates being tapered below the waterline to 200mm amidships and 120mm fore and aft. The belt was 2 metres deep amidships, reducing to 1.64 metres at its after end, and was mounted on a teak backing with a maximum thickness of 200mm; the plates were ordered from Marrel Frères. The light upper belt (*cuirasse mince*) was of 80mm special steel and was secured directly to the shell plating: there was a single strake 1.2m high over the cofferdam amidships, and a double strake 2.55m high at the bow; the plates were ordered from the Ateliers et Forges de la Loire.

The armoured deck was of 70mm 'super soft' steel secured to a double layer of 10mm mild steel; the plates were ordered from the Compagnie Générale Transatlantique (later Penhoët) of Saint-Nazaire. Beneath it was a splinter deck of 30mm 'boiler quality' steel. The armoured coamings around the hatchways and funnel uptakes that penetrated the main deck were of 300mm

special steel at the base, tapering to 200mm, and were 0.5m high.

The conning tower, from which the ship was fought under action conditions, had plates of special steel 320mm thick on the face and 300mm on the outer walls, and weighed 56.4 tonnes. The communications tube was protected by hoops of armour 200mm thick. The compass platform was located atop the conning tower.

The total weight of protection, including the turrets and barbettes (see below), was 4,241 tonnes.

Armament

Bouvet was armed with two 305mm 45-calibre Mle 1893 guns and two 274.4mm 45-calibre Mle 1893 guns, all

Table 2: **Armament**

Two 305mm 45-cal Mle 1893 BL guns in two single turrets
Two 274.4mm 45-cal Mle 1893 BL guns in two single turrets
Eight 138.6mm 45-cal 1891 QF guns in eight single turrets
Eight 100mm 45-cal Mle 1891 QF guns in open single mountings
Twelve 47mm 40-cal Mle 1885 QF guns in open single mountings
Five 37mm 20-cal QF guns in open single mountings
Three 37mm revolver cannon
Two a/w, two sub 450mm torpedo tubes (10 Mle 1892 torpedoes)

Calibre	Shell weight	Muzzle velocity	Firing cycle
305mm	292kg CI	815m/s	1rpm
	340kg APC	780m/s	1rpm
274.4mm	216kg CI	815m/s	1rpm
	255kg APC	780m/s	1rpm
138.6mm	30kg CI	770m/s	3rpm
	35kg APC	730m/s	3rpm
100mm	14kg CI	740m/s	6rpm
	16kg APC	710m/s	6rpm
47mm	1.5kg	650m/s	9–15rpm
37mm	0.5kg	435m/s	20–25rpm

CI cast iron
APC armour-piercing, capped

Calibre	Angle of elevation	Range
305mm	+14°/-5°	12,500m
274.4mm	+14°/-5°	11,600m
138.6mm	+20°/-6° (Nos 1&4)	9,700m
	+11°30/-6° (Nos 2&3)	
100mm	+20°/-10°	9,500m
47mm	+20°/-20°	4,000m
37mm QF	+24°/-40°	2,000m
37mm revolver	+15/-35°	

Bouvet: 274mm Magazine & Handing Room Layout: Starboard Side

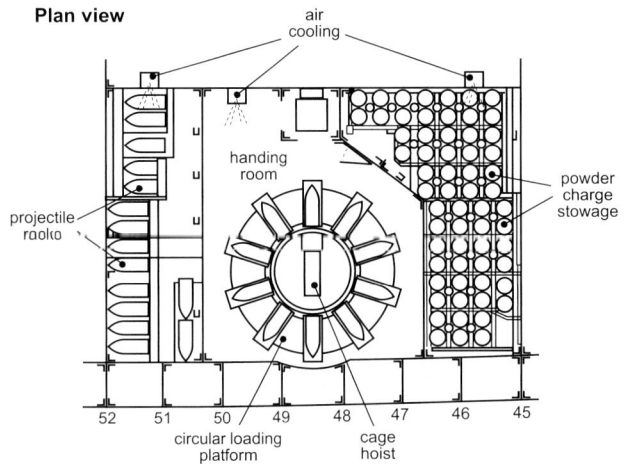

Note: Adapted from plans dated Lorient 28 January 1898.

© John Jordan 2015

Bouvet: Midship Half-Section Frame 48 *bis*

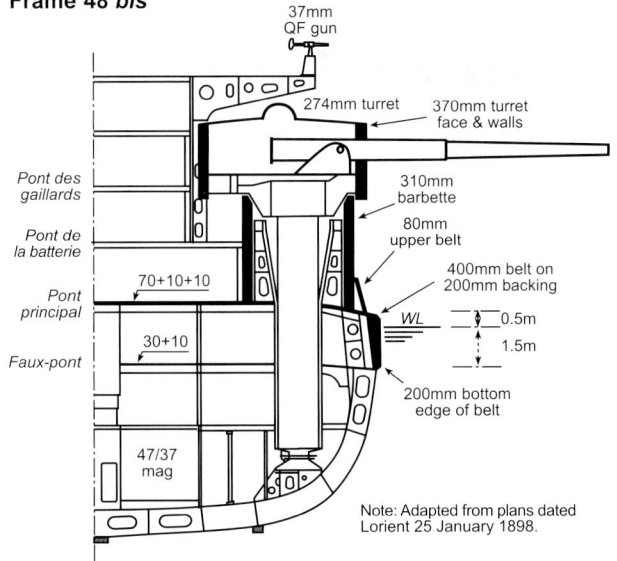

Note: Adapted from plans dated Lorient 25 January 1898.

© John Jordan 2015

A 305mm gun is embarked in the forward turret of *Bouvet*. (DR)

The after 305mm turret of *Bouvet*, with skylights and a small ventilation cowl in the foreground. Above the turret, on the after super-structures and on the lower platform of the mainmast, are 47mm ATB guns flanked by two of the eight 100mm QF guns. (DR)

mounted in single enclosed turrets designed by the Société des Batignolles. The 305mm turrets were mounted fore and aft, the 274.4mm turrets in the wings in the classic French 'lozenge' arrangement.

The main turrets were protected by 370mm plates and had 70mm roofs and sighting hoods, and there was 310mm protection for the barbettes. The armour plates

for the 305mm turrets were supplied by Schneider; those for the 274.4mm turrets by Saint-Chamond.

New optical sights for the turrets were fitted in May 1902, and Marbec breech flushing systems using compressed air were installed in February 1906. Two Barr & Stroud 1.37-metre (4ft 6in) rangefinders were embarked in 1908.

Bouvet moored in the Mourillon basin, on the eastern side of of Toulon Naval Dockyard. Behind her can be seen the covered building ways for submarines and torpedo boats. (DR)

The eight 138.6mm 45-calibre Mle 1991 guns were in single turrets, and were disposed in pairs on either side of the ship: two forward, two aft, and four amidships, fore and aft of the 274.4mm wing turrets. The turrets were protected by 100mm plates on the face and sides and had a 20mm roof, the armour plates being supplied by Châtillon & Commentry. As with the main guns, the turret mechanisms were supplied by the Société des Batignolles. New optical sights for the turrets were fitted in 1907.

The total weight of protection for the artillery was 4,215 tonnes.

Bouvet's armament was completed by two trainable above-water 450mm torpedo tubes and two fixed submerged tubes. The above-water tubes were mounted on the main deck abaft the midship 138.6mm turrets. The height of command was 2.0 metres, and training arcs were between 55 and 131 degrees. The two submerged tubes were located in an athwartships torpedo room on the lower platform deck forward of the pivot for the forward 305mm turret; they were perpendicular to the ship's axis and were 3.69 metres beneath the waterline. The four sighting positions for the torpedo tubes were equipped with sights specifically designed for major vessels.

The Mle 1892 torpedo had a length of 5.05 metres and had an all-up weight 501kg that included a 75mm warhead. When completed *Bouvet* carried a total of ten combat and two exercise torpedoes. The above-water tubes were disembarked in June 1906.

Machinery

Steam for the the propulsion machinery in the first four ships of the series was generated by 24 Lagrafel & d'Allest large watertube boilers. However, for *Bouvet* the Navy opted for the new Belleville boiler, which was built by Delaunay Belleville & Compagnie and rated at 17kg/cm^2; each had a single furnace. There were no fewer than 32 steam generators, distributed equally between two large spaces fore and aft. Each of these spaces was subdivided by a centreline bulkhead into two boiler rooms, each housing eight Belleville boilers disposed back to back (see GA plans). The two identical funnels had a cross-section area of 16m^2 and their tops were 18 metres above the boiler gratings. The first lighting of the boilers took place on 24 July 1897.

The boilers supplied steam for three vertical triple-expansion (VTE) engines built by Indret. The three-

The port-side wing turrets of *Bouvet*, with the single 274.4mm flanked by two 138,6mm guns. The maximum elevation of the latter guns was reduced from 20 degrees to 11°.30 to enable the 274.4mm gun to fire above them without obstruction. (DR)

cylinder (HP>IP>LP) engines were in three independent engine rooms located side by side, and drove three shafts. The three-bladed propellers were of 'special bronze'; those on the wing shafts had a diameter of 4.50m, that of the centre shaft 4.40m.

The engines were designed to deliver 14,000CV, but this was exceeded on trials (15,462CV). *Bouvet* achieved a maximum speed of 17.2 knots with natural draught and 18.2 knots with forced draught; 17.8 knots were sustained on a 4-hour trial. At 8.08 knots the turning circle was 1,425 metres. At a speed of 16–18 knots, with 15 degrees of helm, the angle of heel was 2 degrees.

Coal bunkerage was 620 tonnes at normal load, giving a theoretical endurance of 2,660nm at 10 knots, and 800 tonnes at full load for 3,420nm at 10 knots.

Pumps available in the event of flooding included four Thirion pumps rated at 1,000 tonnes per hour, five Thirions rated at 30t/h, a 30-tonne hand pump powered by four pistons, plus two small Le Testu pumps to empty the main drain. An early report on provision warned:

> The design of the ship did not include a means of righting the ship in the event of a major ingress of water, particularly if the side compartments beneath the 274.4mm turrets were opened up by a collision or by the explosion of a torpedo warhead. In the latter circumstances, despite connections between the two side compartments, there is a serious risk that the ship would capsize …

The twelve ventilators for the boiler rooms had a unit output of 55,000m³. The two ventilators for the engine rooms were from Rateau. Electricity for the ship was supplied by four Sautter & Harlé 32kW dynamos feeding a 400A/80V circuit; the dynamos were housed in a large centreline space on the 1st platform deck,

Table 3: **Machinery**

Boilers	32 Delaunay-Belleville
Engines	three 4-cylinder VTE built Indret
Propellers	three 3-bladed 4.40m/4.50m
Rudder	non-balanced, 24.93m²
Designed power	14,000CV (15,462CV trials)
Maximum speed	17.20 knots (18.18 knots trials)
Coal	620 tonnes normal
	800 tonnes full load
Endurance	2,770nm at 8.1kts (13 days)
	2,560nm at 12.1kts (7.33 days)
Equipment	
Searchlights	six 60cm Mangin 75A projectors
Boats	two 10-metre steam pinnaces
	one 7.6-metre White steam launch
	one 11-metre pulling pinnace
	one 10.5-metre pulling cutter
	one 10-metre admiral's gig
	two 9-metre pulling cutters
	four 8.5-metre whalers
	two 5-metre dinghies
	two 3.5-metre punts
	two 5.6-metre Berthon canvas boats
Anchors	two Marrel 13.10-tonne bower anchors
	one 6.60-tonne sheet anchor

Bouvet at her moorings in Villefranche roads. (Author's collection)

Bouvet: Propulsion Machinery

Hold

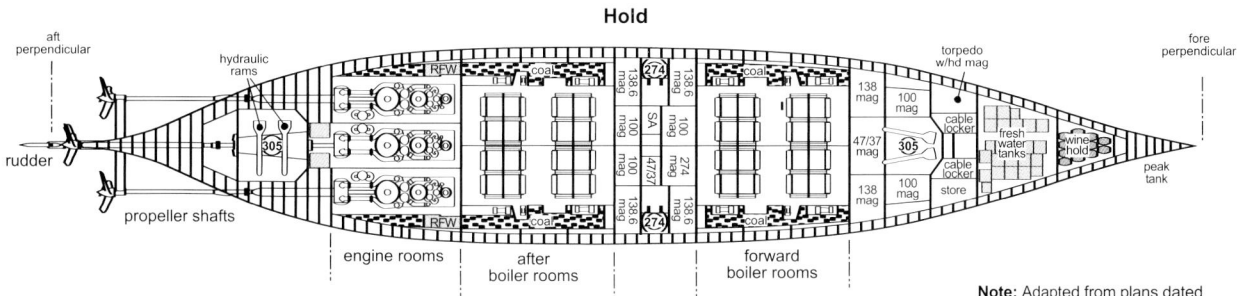

Note: Adapted from plans dated Lorient 25 January 1898.

© John Jordan 2015

Bouvet moored off the coasts of Provence. (Author's collection)

A fine image of *Bouvet* taken before 1908. (Author's collection)

Bouvet, seen from *Saint-Louis* in 1900. (Author's collection)

Bouvet in 1901, in company with *Gaulois* and *Jauréguiberry*, seen here to the right and left respectively. (DR)

Bouvet and *Charlemagne* in April 1906, photographed on their arrival at Naples, when they brought humintarian assistance following the eruption of Vesuvius. (Author's collection)

between the ammunition trunks for the 274.4mm turrets.

Equipment

There were six Mangin 60cm searchlight projectors: two atop the military masts (*ligne haute*) and four on the lower level (*ligne basse*). The two upper searchlights were installed on a fixed base and were operated remotely using electric controls. The four lower searchlights were mounted on trolleys with electric controls; when not in use they were stowed within the hull and superstructures to protect them from the elements. One was located on the battery deck in the bow, another at the stern, and the remaining two on either side amidships, immediately abaft the 274.4mm turrets. All were operated remotely using an electro-mechanical linkage.

The 11-metre pulling pinnace could be armed with a 37mm QF cannon. The 10-metre pulling cutters could be equipped with a 30cm searchlight projector and a 37mm cannon, and could also embark spar torpedoes. The larger pinnaces and cutters were handled by two large gantries abeam the bridge and by three pairs of davits amidships; there were pairs of smaller, simpler folding davits fore and aft for the whalers and 5-metre dinghies (see drawing). The larger davits were powered by two Samson two-cylinder steam winches each with a 2,500kg capacity.

The chain cables for the bower anchors had a diameter of 60mm and a length of twenty-six 30-metre shackles. The main capstan was a two-cylinder steam-powered model from Bossière rated at 36,000kg. Besides the two main anchors there were two 3.7-tonne stock anchors, which were fitted with four shackles of 30mm cable.

Provisions could be stowed sufficient to sustain a crew of 666 men for 45 days at sea. Fifteen cubic metres of wood were embarked to fuel the galleys, stowed in the wine hold.

In a report dated 1899, the CO of the ship, *Capitaine de vaisseau* Noël stated:

> … the cabins for the officers and CPOs were barely habitable when the scuttles were closed. Those for the senior officers in the after part of the ship, which are located between the after guns, are narrow and poorly laid out. I consider them to be unsuited to the accommodation of officers of this rank.

The Early Career of *Bouvet*

Bouvet was launched at Lorient Naval Dockyard on 27 April 1896 in the presence of senior civilian and military personnel of the 3rd Maritime District. Her

left: View aft from the forecastle of *Bouvet*, with the forward 305mm turret, the conning tower and the navigation bridge in the centre. Note the 47mm Hotchkiss ATB guns on the forward superstructure decks and on the lower platform of the military mast. (DR)

machinery would subsequently be installed from 15 October. The first static trials took place on 24 November 1897. *Bouvet* was commissioned on 5 March 1898; her Executive Officer was a certain CF Guépratte, with whom we shall become further acquainted at the Dardanelles.

It was originally envisaged that *Bouvet* would run her sea trials from Brest; however, due to overcrowding in the latter port she was despatched to Toulon on 9 March, dropping anchor there on the 20th. Trials continued throughout April and May. Gunnery trials continued until September due to problems with hydraulic pressure.

Following her entry into service, which took place officially on 27 July, *Bouvet* was docked, then sailed for Crete on 12 September to reinforce the French contingent on the island. The troops were disembarked in Souda Bay on the 16th and set about restoring order after the rebellion of the islanders. *Bouvet* returned to Toulon on the 21st and entered full commission (*armement définitif*) on 1st October, being allocated to the 1st Division of the Mediterranean Squadron alongside *Masséna,* flying the flag of Vice Admiral Fournier, and *Brennus*.

During the Fashoda crisis, Admiral Fournier led his squadron to nearby Les Salins d'Hyères from 26 to 28 October, but with the improvement in the international situation the exercises continued to their conclusion.

Following a docking in February/March 1899, *Bouvet* sailed with the squadron for Sardinia on 7 April. There were gunnery exercises from 17 July, followed by port visits to Barcelona and Port Mahon (Menorca). On leaving command of *Bouvet* Captain Noël wrote in his report:

> *Bouvet* is an excellent sea-boat in every respect. She responds well to the helm, keeps a steady course, and manoeuvres well. She behaves well in all weathers; in rough weather, with a head sea, she rides the swell with no significant pitching. She has a gentle, steady roll. However, in rough weather, with a beam sea, the waves break around the centre turrets, throwing up large quantities of spray. From a military point of view, *Bouvet* has considerable offensive power. The turret-mounted guns are easily trained and are admirably protected. *Bouvet* has manual loading for her main guns, which for the 305mm guns allows a firing cycle of one round in just under a minute, while for the 274mm guns one round every 50 seconds has been achieved.
>
> The conning tower is well-protected and relatively spacious but remains cramped.
>
> The compartmentation is comprehensive, but the question of righting the ship following a significant ingress of water has not been sufficiently studied, and should be a matter of concern.
>
> To sum up: with her high speed and excellent sea-keeping qualities, her powerful, well-disposed guns, her three robust engines, which are well mounted and reliable, *Bouvet* represents a formidable combat unit. This is a very successful battleship design, susceptible to improvement only with regard to a few details.

It should be noted, however, that stability concerns had already been raised during *Bouvet*'s trials. These concerns, which were never properly addressed, would ultimately result in her dramatic loss on 18 March 1915.

With the entry into service of the new battleships *Charlemagne* and *Gaulois*, *Bouvet* was assigned on 1st January 1900 to the 2nd Division with her half-sisters *Charles Martel* and *Jauréguiberry*. On 20 June, in preparation for the annual grand manœuvres, Admiral Gervais, C-in-C of the Mediterranean Squadron, hoisted his flag in *Bouvet*. The fleet sailed on the 22nd, passed through the Strait of Gibraltar on the night of 28/29, and effected a rendez-vous with the Northern Squadron the following day. After two weeks of manoeuvres, the combined squadrons visited Brest, and on 13 July they were in Cherbourg, where they welcomed President Loubet. The Mediterranean Squadron returned to Toulon on 14 August, and on 11 October a government delegation embarked on *Bouvet* for an inspection voyage to Tunisia. Port visits were made to Ajaccio, Porto Vecchio, Bizerte, Bastia and Calvi. *Bouvet* returned to Toulon on 22 October.

In early 1901 there were exercises along the coast of Provence, together with port visits to Corsica and Villefranche. From 27 June Admiral Gervais again embarked in *Bouvet* to command the grand manœuvres that took the fleet to Oran. President of the Council Waldeck-Rousseau and Minister of Marine Lanessan then embarked on the flagship. During the exercises that followed, *Bouvet* was 'torpedoed' by the submarine *Gustave Zédé*. The Squadron anchored at Les Salins d'Hyères, then simulated an attack on Ajaccio; it finally returned to Toulon on 1st August. The 2nd Division then sailed for the Atlantic, calling in at La Pallice on the 29th, Cherbourg on the 31st, and Dunkirk on 15 September. The ships returned to Toulon on 1st October. On 10 December *Bouvet* visited Ajaccio, followed by Golfe Juan on the 17th.

From February to April 1902 there were visits to Ajaccio, La Goulette, Algiers and Hyères. In May, *Bouvet* joined with other ships of the Squadron (*Jauréguiberry*, *Saint-Louis*, *Gaulois*, *Charlemagne*, *Iéna*, *Brennus*, *Masséna*, *Hoche* and *Marceau*) on a cruise of the coasts of Provence. The 7 July saw the return of Admiral Gervais on board to command the summer grand manoeuvres. Following a docking in September and October, *Bouvet* was placed in care and maintenance with a reduced crew.

On 29 January 1903, during exercises, *Gaulois* collided with *Bouvet*, striking to port close to the after accommodation ladder. This resulted in a minor ingress of water, and the hull plating and frames were pushed in over a length of two metres. Following this incident, Miniser of Marine Thomson ordered the two commanding officers to be relieved of their posts.

On 3 April the Squadron sailed for North Africa to escort President Loubet, then visited Cartagena from 23 to 27 June, where the Spanish King Alfonso XIII was received on board *Bouvet*. In September the battleship was in Corsica, with successive visits to Ile Rousse, Ajaccio, Porto Vecchio and Bastia. In October she was in the Balearics, after which she was docked from 1st to 16 January 1904. On 24 April President Loubet embarked for a visit to Genoa; *Bouvet* then continued on to Tyre, Beirut, Smyrna, Mytilene, Salonika and Phalerum, before returning to Toulon on 3 July. The battleship visited Saint Tropez 26–27 July, returning to Corsica in November.

During 1905 there were port visits to Ajaccio, La Goulette, Bizerte, Bône, Philippeville, Bougie, Algiers and Mers el-Kebir. In October, after the customary grand manœuvres, the 2nd Division represented France on the occasion of a visit to Genoa by the King and Queen of Italy. During the same year wireless transmission (W/T) equipment was fitted in *Bouvet*.

On 31 March 1906, a hawser parted when *Bouvet* was attempting to moor to a buoy, injuring three men, one seriously. At 0110 on 11 April the 2nd Division put to sea in a hurry on a mission to give humanitarian assistance to the inhabitants of Naples following the eruption of Vesuvius. In May *Bouvet* participated in manœuvres off the coast of North Africa, with visits to the customary ports. The 2nd Division then sailed for Palma de Mallorca for the marriage ceremony of King Alfonso XIII. The year ended with the customary cruise to Corsica.

On 15 February 1907 *Bouvet* was placed in 'normal reserve' at Toulon and underwent modifications which aimed to lighten the tophamper. This refit was of short duration as, following the explosion which destroyed the battleship *Iéna* on 12 March, *Bouvet* was assigned for the next few months to the 3rd Battle Division as her replacement. On 16 August she hoisted the flag of Rear Admiral Marin-Darbel, who had been appointed to the command of the 4th Division, formerly the Training Division (*Division d'instruction*). In this new role *Bouvet* would engage in only limited activites for the next two years. In July 1908, during a firing exercise, the gunnery officer made a recognition error and opened fire on the destroyer *Arbalète*, which was towing the target. Luckily the destroyer did not sustain any damage, and as soon as the error was realised fire was shifted to the raft which constituted the target.

From the Middle East to the Northern Mists

On 16 August 1909, Rear Admiral Berryer took command of the 3rd Division (ex-4th Division). Soon, with the entry into service of the 'pre-dreadnoughts' of the *Patrie* class, the older batteships would be assigned to the 2nd Division of the 2nd Squadron, which was due to sail on a cruise of North Africa, with visits to Algiers 17–22 December, then Mers el-Kebir until 4 January 1910. *Bouvet* and her division then sailed for the Atlantic, dropping anchor at Quiberon on 23 February. The ship was at Cherbourg from 13 March to 9 May, where she was docked in April. The 2nd Division sailed

Bouvet moored off the coast of Brittany. She has now been repainted in a blue-grey livery overall. (DR)

Bouvet and *Suffren* in the Toulon roads in 1909. (Author's collection)

on 9 May, calling in at Mers el-Kebir before returning to Toulon on the 30th.

The three battleships left for their new home port of Brest on 6 August, and *Bouvet* returned to Cherbourg from 25 to 28 October. On 21 December, Admiral Berryer was succeeded in command of the division by Rear Admiral Adam. From 18 February 1911, *Bouvet* was docked for a short period for repairs to the port-side submerged torpedo tube. She would remain based in that port until late October in order to have some of her boiler pipework replaced. On 1st August the 2nd Squadron became the 3rd Squadron. On 11 November Vice Admiral Marolles hoisted his flag in *Bouvet* and would remain in command of the squadron until 15 April 1912.

After farewell visits to Dunkirk and Le Havre, the 3rd Squadron finally departed Brest on 16 October 1912. It was to join the *Armée Navale* of Admiral Boué de Lapeyrère at Toulon via Mers el-Kebir and Algiers, where it arrived on 9 November.

In early 1913 *Bouvet* undertook visits to Villefranche, Les Salins d'Hyères and a number of other ports in the region. On 18 March she sailed unaccompanied for Cherbourg, where she was due to undergo a major refit which took from 28 March to 19 August. During this refit the boilers were refurbished, and a propeller and some of the bollards used for mooring were replaced. *Bouvet* returned to Toulon on 27 August. She visited Marseille in October and Corsica in November. On 11 November she was assigned to the Reserve Division (*Division de complément*) along with *Gaulois* and *Saint-Louis*, the latter ship flying the flag of Rear Admiral Ramey de Sugny.

From 15 to 27 January 1914, *Bouvet* was at Villefranche, and on 7 March she conducted gunnery exercises not far from the Cerbicale archipelago in the south of Corsica. In June there were visits to Ajaccio, Bizerte, Port la Nouvelle and Golfe Juan.

With the German declaration of war on France on 3 August, the *Armée Navale* sailed from Toulon at 0315 and headed for North Africa. *Bouvet* put in to Mers el-Kebir then escorted a troop convoy to Toulon, returning on the 8th. From that date *Bouvet* and *Jauréguiberry* were to form the 1st Division of the Special Squadron (*Escadre Spéciale*). On the 9th the 2nd Division of the 2nd Battle Squadron was formed with *Bouvet*, *Jauréguiberry* and *Charlemagne* under the command of Rear Admiral Darrieux. Three days earlier, a naval convention had been signed in London stating that:

> … The French Navy will secure the protection of British and French commerce in the Mediterranean. In particular, it will act against the naval forces of Austria-Hungary and will prevent these forces from leaving the Adriatic. It will also monitor the exit from the Suez Canal and the Strait of Gibraltar.

That same day, 9 August, the 2nd Battle Division sailed from Toulon as escort for a convoy to Ajaccio. It returned on the 11th and then escorted another convoy between Marseille and Ajaccio. From the 14th *Bouvet* and *Jauréguiberry* were providing cover for the trans-

Bouvet at Toulon in 1910/1911. (DR)

Bouvet underway off Toulon for a sortie of the *Armée Navale* in the spring of 1914. (DR)

ports, but were also conducting patrols off Barcelona to monitor sea traffic. From 8 September the battleships were cruising between Cap Corse, the northern tip of Corsica, and Villefranche-sur-mer on the French coast. The two battleships then headed for Malta to escort a troop convoy of 25,000 men from Port Said to Marseille. From 17 October *Bouvet* and *Jauréguiberry* resumed their patrols in the Gulf of Genoa. During this period, 57 ships were inspected and six were seized. *Bouvet* then returned to Toulon for a period of leave. She would be docked from 23 November to 1st December and, the following day, sailed on a mission to patrol the Strait of Messina in company with the large protected cruiser *D'Entrecasteaux* and the auxiliary cruisers *Provence II* and *Lorraine II*.

During this period, the principal threat in the Mediterranean was represented by the presence of the German battlecruiser *Goeben* and the light cruiser *Breslau*. Having conducted a bombardment of Bône and Philippeville on 4 August, these two ships headed east at speed and passed through the Dardanelles on the 10th. From that moment it became necessary to mount a blockade to prevent these ships conducting operations against maritime traffic in the Aegean. On 11 August the British Vice Admiral Milne was ordered to secure this blockade with the battlecruisers *Indomitable* and *Indefatigable*, supported by the light cruiser *Gloucester*.

On the 16th, on board *Goeben*, the Kaiser's ensign was hauled down and replaced by that of the Ottoman Empire. *Goeben* and *Breslau* had now been ceded to the Turks, who had agreed to join the war on the side of the Germans.

On 26 September it was the turn of the French battleships *Suffren* and *Vérité*, under the command of Rear Admiral Guépratte, to join up with the Allied naval forces at the Dardanelles, which would from this point be commanded by Milne's replacement, Vice Admiral Carden. The main anchorage for the ships was the island of Tenedos, and the battleship *Gaulois* would arrive on 16 November, followed by her sisters *Charlemagne* and *Saint-Louis* on the 26th. For her part, *Bouvet* left Malta on 16 December, coaled at Port Sigri (Lesbos) on the 19th and arrived at Tenedos on the 20th.

The Dardanelles

In the meantime, on 3 November 1914 an intervention took place involving *Indomitable*, *Indefatigable*, *Suffren* and *Vérité*, which fired for 11 minutes on the Turkish fortifications at the entrance to the strait. This action was not followed up, so all hope of surprise was lost. It soon became clear that if the Allies wished to take Constantinople and to sink *Goeben*, they would have to force the straits. The Turks, advised by the Germans,

Suffren, left, and *Bouvet* at anchor at Mudros during early 1915. (DR)

The senior officers of *Bouvet*, a few days before the drama at the Dardanelles. Second from the left is CF Jean Autric, then Captain Rageot de la Touche with his military decorations and, in the centre CF Cosmao Dumanoir. (DR)

profited from the pause in the bombardments to reinforce their defences.

From 1st February 1915 the Bay of Mudros, on the island of Lemnos, was fitted out to receive the Franco-British fleet, and *Bouvet* dropped anchor there for the first time on 3 March.

Finally, at about 1000 on 19 February, a first attack was undertaken with a bombardment of the forts at the entrance to the straits. The British battleships *Vengeance* (VA de Robeck), *Triumph* and *Cornwallis*, and the French *Suffren* (CA Guépratte), *Gaulois* and *Bouvet*, together with the battlecruiser *Inflexible* (VA Carden) and the cruisers *Amethyst* and *Dublin* participated in this operation, during which Admiral Guépratte distin-

guished himself by coming to the aid of HMS *Vengeance*, which was in difficulties off Seddul-Bahr.

A second offensive operation took place successfully on the 25th, and a number of officers were convinced that the forts had sustained servere damage. A further bombardment was to have taken place the following day but was cancelled due to an unfavourable weather forecast. On 25 February, then on 1st, 6 and 7 March there were further bombardments of the forts on either side of the Dardanelles, none of which made much impact on the Turkish defences. Between 2 and 3 March *Bouvet* conducted a reconnaissance of the Gulf of Saros, on the north side of the Dardanelles.

The last sortie of *Bouvet* on 18 March 1915. Note the zebra-stripe camouflage on the funnels; it was intended to confuse enemy rangefinders.

The Attack on the Forts
18 March 1915

N

Gallipoli
Peninsula

Kilid Bahr

Chanak

Aegean
Sea

Kephez
Point

Kephez
Bay

Anatolia

Seddul-
Bahr

Queen
Elizabeth

Prince
George

Agamemnon

Nelson

Majestic

Cape
Helles

Gaulois

Inflexible

Irresistible strikes mine

Ocean strikes
mine: abandoned

Vengeance

Charlemagne

line of
20 mines

Irresistible

Bouvet

Albion

Suffren

Triumph

Bouvet
sunk by mine

Ocean

Kumkale

Swiftsure

Eren Keui
Bay

Key:

○ searchlight position

⊡ main battery

■ light gun position

...... mines

0   25km

© John Jordan 2015

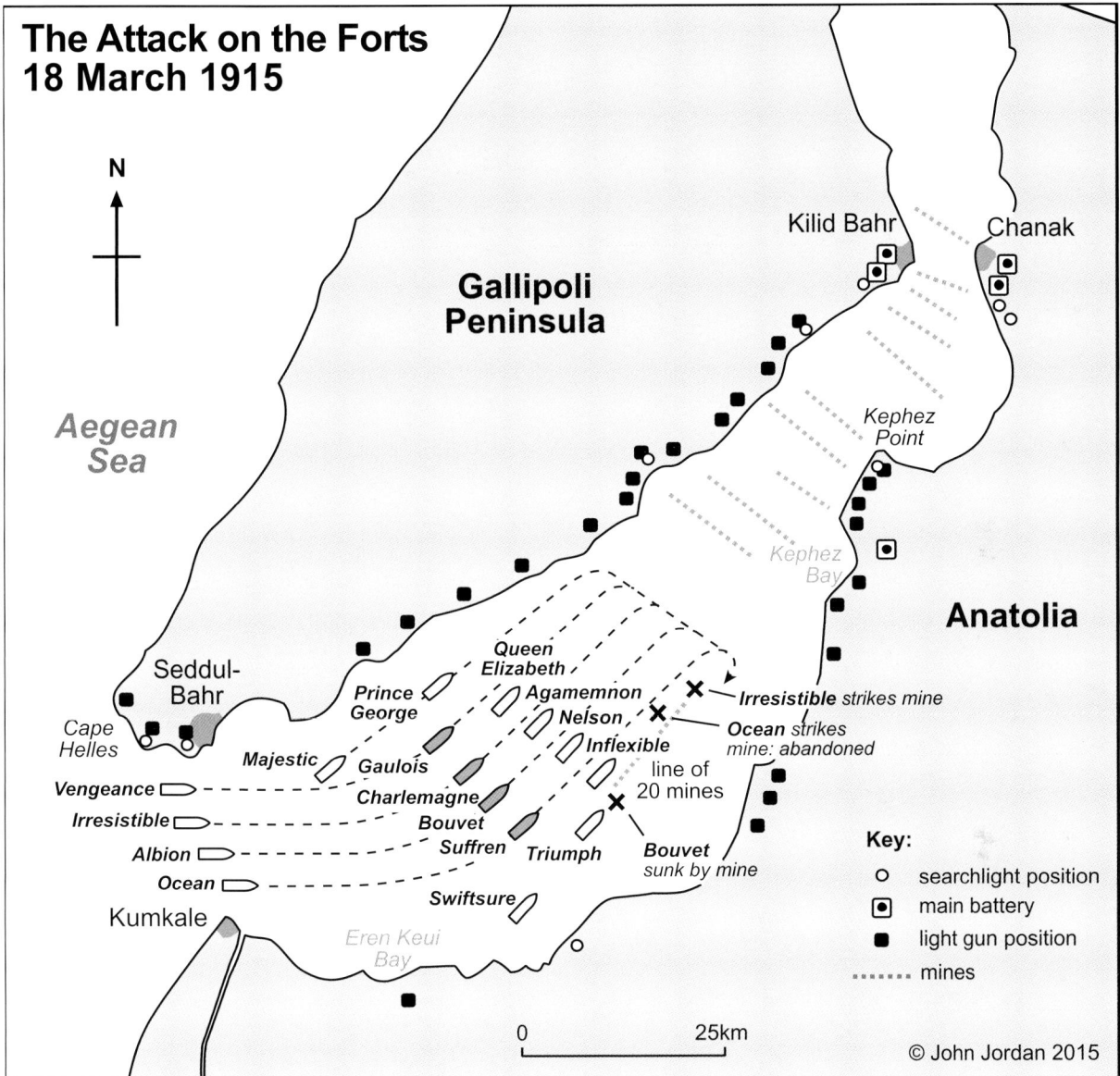

On the 16th there was development which caused consternation among the Allied officers: Admiral Carden had resigned and would be replaced by Vice Admiral de Robeck. Despite this, a major operation had been planned for the 18th. For this new attack on the forts, Admiral Guépratte had requested and been granted the honour of being at the head of the assault force. Two days before the attack, Guépratte received the commanding officer of *Bouvet*, CV Rageot de la Touche, to inform him that his elderly battleship would not take part in the operation. Guépratte subsequently reported:

The response of this valiant officer was of a silent eloquence, and all the more moving because of that: not a word, not a gesture, simply two big tears which ran silently down his long, pale cheeks that expressed his utter devastation at being forced to sit on the sidelines. These

generous tears were effectively a death sentence for the CO and his fellow officers as, deeply moved, I took back the order I had just signed from his hands and threw it in the bin.

At 0900 in the morning, on 18 March, the French division left the anchorage in Tenedos. It was formed into two sections: *Suffren-Bouvet* and *Gaulois-Charlemagne*. In line with custom and practice, the band of the flagship, lined up on the quarterdeck, struck up a traditional repertoire of anthems and patriotic songs.

At 1100 the British battleships *Lord Nelson*, *Agamemnon*, *Majestic* and *Triumph*, supported by the dreadnought *Queen Elizabeth*, opened up at long range against the Chanak forts. At midday the command was given for the French squadron to take up its position in the van. Actions stations was sounded and the musicians made their way below. At 1230 *Suffren*, followed by

141

Two views of the sinking of *Bouvet*, taken 30 seconds apart. The photos were taken by Chief Armourer Butler on board the British battleship HMS *Agamemnon*. (DR)

Bouvet, entered the straits and the two battleships made their way along the coast of Asia, while *Gaulois* and *Charlemagne* proceeded along the European shore. Each of these groups was reinforced by the presence of the British *Swiftsure* and *Prince George*, which followed at a distance.

The battleships proceeded until they were 9,400 metres from Yeni-Medjidieh without provoking any reaction except for a few isolated shots fired from Aren Koi. Closing to about 5,000 metres from the great forts they halted, ready to open fire, according to their orders. At that moment, at 1230, all the batteries on land opened fire simultaneously. The flashes from the four big guns of the Dardanus fort could be seen less than 4,500 metres away. The flashes from the two 150mm guns of Souan-Dere and the six big guns of Medjidieh could also be made out. *Suffren* was keeping up a rapid fire from all her guns. While Guépratte's flagship was conducting a methodical bombardment, *Bouvet* took on the field

batteries and the Souan-Dere fort. The current was running at 3 knots, and *Suffren* was allowed to drift, giving way to her partner. At 1300 *Bouvet* opened fire on the Chanak fort, but the Marbec breech flushing apparatus of the forward turret failed, and the turret was filled with noxious powder combustion gases. Several gunners quickly lost consciousness, while at the same time the Turks landed two shells on their target, putting two of the 138.6mm guns out of action. The ship drifted on the current, and at 1310 again yielded her place to *Suffren*.

Suffren was then taken under a heavy fire from the port side, putting casemate gun No 10 and turret No 8 out of action. Only the quick action of a quartermaster prevented a magazine explosion. *Bouvet*, seeing the flagship in difficulty, manoeuvred at speed to give support and opened a rapid and sustained fire on the Namazieh fort.

At 1337, Admiral de Robeck ordered Guépratte to

The survivors of *Bouvet*, gathered together on the quarterdeck of the *Agamemnon*. (DR)

withdraw and to give way to the British battleships. *Suffren* hoisted flag *trapèze 4*, signalling withdrawal. *Charlemagne* and *Gaulois* hoisted an acknowledgement, but *Bouvet* remained where she was. Despite the continued immobilisation of the forward 305mm turret, her CO Rageot de la Touche was reluctant to break off the action, despite the signals officer pointing out on no fewer than three occasions that Guépratte had ordered a withdrawal. Seeing that *Bouvet* had remained in position, *Suffren* turned 180 degrees and placed herself less than 30 metres from the elderly battleship and fired a blank round as a reminder that she needed to conform to the admiral's instructions. Finally *Bouvet* began to move, raising her speed to 12 knots and profiting from the turn to unleash a broadside against the Erenkeui heights. She had barely completed this manoeuvre when, at about 1400, there was a massive explosion abreast the wing 274mm turret to starboard. The ship heeled to port for a few seconds, then rolled over to starboard until she was on her beam ends. At 120 degrees the turrets came loose and sank, and the ship capsized, her propellers still turning.

Bouvet had sunk in 50 seconds, her lower hull ripped open by a mine. She took with her her commanding officer, 23 officers and 619 men. There was not a single survivor from the machinery spaces or the magazines below. Seeing his ship in her death throes, Rageot de la Touche sealed himself inside the conning tower, which would be his tomb, and *Bouvet* sank with her ensign still flying. In the damage control post the executive officer, aware that the ship had struck a mine, went below to assess the situation and was never seen again. Some of the gunnery personnal remained at their posts as the ship sank. Only six officers and 60 men survived the disaster. They were picked up by the steam pinnaces charged with locating mines.

Several cables away Admiral Guépratte saluted those who died and, impassively, ascribed the loss of *Bouvet* to her poor stability – a criticism that could have been applied to any of the other ships in his squadron. The CO of his flagship *Suffren* immediately summoned the bosun and ordered that all available boats be launched to rescue any survivors. He was informed that all the boats on board had been holed by shell splinters with the sole exception of the admiral's whaler. This was then launched on the orders of Guépratte.

Unfortunately, this drama was not the only one during this fateful day of 18 March 1915. *Gaulois* had suffered a number of hits by large-calibre shell and avoided sinking only by being deliberately grounded in the Rabbit Islands, eight miles southwest of the straits. *Suffren* herself had been struck by no fewer than

Commanding Officers

CV Noël	15 Jul 1897 – 5 Aug 1899
CV Leygues	5 Aug 1899 – 5 Aug 1901
CV Dufayot de la Maisonneuve	5 Aug-1901 – 20 Mar 1903
CV Guillou	20 Mar 1903 – 20 Mar 1905
CV De Faubournet de Montferrand	20 Mar 1905 – 16 Aug 1907
CV Papaix	16 Aug 1907 – 16 Aug 1909
CV Charlier	16 Aug 1909 – 20 Dec 1910
CV Fournier	20 Dec 1910 – 21 Jul 1911
CV Degouy	21 Jul 1911 – 20 Oct 1911
CV Fouet	20 Oct 1911 – 30 Mar 1912
CV Jochaud du Plessis	30 Mar 1912 – 10 Oct 1912
CV Rageot de la Touche	10 Oct 1912 – 18 Mar 1915

Flag Officers

VA Gervais	20 Jun – 21 Jul 1900
VA Gervais	27 Jun – 1 Aug 1901
VA Gervais	7 Jul 1907 – 8 Aug 1907
CA Marin Darbel	16 Aug 1907 – 15 Aug 1909
CA Berryer	16 Aug 1909 – 20 Dec 1910
CA Adam	21 Dec 1910 – 21 Jul 1911
VA de Marolles	11 Nov 1911 – 15 Apr 1912

Notes:

VA	*Vice-Amiral*	Vice Admiral
CA	*Contre-Amiral*	Rear Admiral
CV	*Capitaine de Vaisseau*	Captain

14 shells of all calibres. The Royal Navy likewise suffered grievous losses on that day, losing the battleships *Ocean* and *Irresistible*. *Bouvet* received a citation on 6 August 1915.

These days the wreck of *Bouvet* remains at a depth of 72 metres; the keel is uppermost, and the bow is deeply buried in the muddy bottom. The hole made by the explosion of the mine, on her starboard side, is still visible but, more surprisingly, there is a second large hole close to the stem. A prominent member of the French Académie de Marine has advanced the theory that a large-calibre shell struck *Bouvet* at almost the same time as the ship struck the mine, and that this contributed to the rapid capsizing of the ship. This theory has been confirmed by other marine experts who have examined the wreck. It also appears that the submerged torpedo tubes were removed from the wreck during the 1960s, so unresolved mysteries remain.

Editor's Note:

This article was translated from the French by the Editor. The drawings are based on the official plans of *Bouvet* held at the Centre d'Archives de l'Armement.

MUSSOLINI'S CAPRICES: THE ITALIAN MIDGET SUBMARINES AND ELEKTROBOOTE OF 1934–1943

Enrico Cernuschi looks at the development of the single-engine submarine in the *Regia Marina* between the wars.

When the Italian parliament elected Benito Mussolini Prime Minister on 17 November 1922 his knowledge of naval affairs was modest. He demonstrated this during his first international crisis. On 31 August 1923 the Italian battlefleet landed troops on the Greek island of Corfu in the Ionian Sea and conducted a symbolic bombardment of an old Venetian fort, killing sixteen civilians in the process. This event followed the 27 August murder on Greek territory of an Italian general and his staff, members of a League of Nations boundary commission, and Greece's subsequent failure to meet Italian demands for a formal apology and reparations for the crime, whose authors (possibly petty criminals) were never discovered.

Tensions with Greece over the status of the Dodecanese, the Aegean islands occupied by Italy in

Mussolini on the cruiser *Pola* at Tripoli on 18 March 1937 with Marshal Italo Balbo, the secretary of the Fascist party Achille Starace and, observing behind them, Admiral Domenico Cavagnari. (Gardin Archive, Author's collection)

1912 following its war with the Ottoman Empire, were high, all the more so following the new Turkish republic's decisive defeat of Greek forces in 1922. The Italian Navy, led by the Minister Admiral Paolo Thaon di Revel, advocated the occupation of a small Greek island[1] as a politically manageable pawn to be bartered for the right to build and fortify a naval base on Leros in the Dodecanese.

On 29 August 1923, Mussolini advised Thaon di Revel that he had secured French support for an Italian move in the new crisis, and he ordered a peaceful occupation of Corfu. The island had almost no defences, but the Governor refused to surrender so the Italian battlefleet shelled the town fortress, causing the casualties noted above. The British reacted immediately, and the Mediterranean Fleet sailed from Malta while London demanded the return of the *status quo ante*. The combative Mussolini, still confident of Paris's support, was confronted, on the morning of 1 September 1923 morning, by Thaon di Revel and his staff.

The admirals immediately told their prime minister that the Italian dreadnoughts could not match the British super-dreadnoughts, explaining that their 12in guns could not pierce the British battleships' armour while the British 15in guns would sink any Italian vessel with a few direct hits. Mussolini countered that Italy had MAS boats that had sunk two Austro-Hungarian battleships during the Great War. Thaon di Revel patiently explained that Italy had a coastline more than 7,500km long. The 40-odd MAS boats available, like the 30 submarines in commission, could only ambush a target that happened to cross within a few hundred yards from their bow: at night for the fast coastal forces, during the day for submerged boats creeping slowly at 4–5 knots. The Air Force had no torpedo bombers, and the small bombs available could not harm a battleship. With his back to the wall the future dictator[2] asked, reluctantly, how long Italy could stand against Britain. The blunt reply was '72 hours!'. The ball therefore passed into the hands of the diplomatic corps and, within a month and with French support behind the scenes, Athens was induced to accept Mussolini's demands, thereby saving his face, and the Italian troops departed Corfu.

After this episode Mussolini studied Sea Power and even wrote a booklet about Rome's Sea Power assisted by a ghostwriter, from which he often quoted, then and later.

Following suggestions by Admiral Costanzo Ciano (father of the better-known Galeazzo, future son-in-law of the now-Duce and the only former Italian Navy man who retained an important role under Fascism), Mussolini approved and financed, within the limits of the tight Italian budgets, a modern force comprising large, fast heavy cruisers, and powerfully armed scouts and destroyers. He then concluded, in 1927, that modern fast battleships were necessary, but deferred any order for capital ships until the early 1930s for financial reasons. Mussolini always opposed carriers, as his ideological faith in the future progress of the aeroplane and in the true fascist nature of the *Regia Aeronautica* he created in

Corfu, 31 August 1923. The Italian battleship *Cavour*. On her deck are 65/17 howitzers which will be landed on the island. (Author's collection)

March 1923, absorbing the Navy's *Forza Aerea* in the process, were a mantra for his political movement, based culturally and for propaganda purposes on the futuristic visions of speed and modernism.

At the heart of his naval thinking was the traditional combination of the submarine and the MAS boat. The *Regia Marina* continued to order both improved designs and experimental types; some proved successful, others not. Between late 1926 and 1931, with Italy's most likely adversaries being France and Yugoslavia and with the UK as a benevolent neutral, this defensive naval programme was sound. However, when this policy was replaced, following the introduction in 1931–32 of the British Imperial Preference tariffs, by a new Franco-Italian collaboration against the UK, the *Regia Marina* was seriously under-gunned and Mussolini had to adopt a different overarching naval plan. This reversal brought a change in the Navy's leadership. The Duce claimed for himself the position of Navy Minister, removing Admiral Sirianni, who had vigorously opposed planning for conflict with Britain and, in November 1933, nominated as Secretary (and effectively Minister) of the Navy, on Costanzo Ciano's advice, the now-Admiral Domenico Cavagnari. Because the Navy's Chief of Staff, Gino Ducci, was openly opposed to any anti-British policy, his desk was taken over by Cavagnari himself on 1st June 1934.

By that time the new Italian naval programme already included two modern 35,000-ton battleships, *Littorio* and *Vittorio Veneto*, marking the abandonment of several abortive studies for smaller ships similar to the French *Dunkerque*. The new capital ships would not be commissioned before 1940, and Cavagnari was the first to forecast a crisis with Britain before that time. It therefore became necessary to find a stop-gap solution for the period to 1937, when the two modernised *Cavour*s would be completed, or, better, until 1940. The solution that found favour with Mussolini was the revival of an old Italian Navy dream, the *sottomarino*.

A Shortcut to Sea Power

Since the late 19th century the Italian dictionary has featured two words for submarine: *sommergibile* and *sottomarino*. The first refers to the classic boat with two types of engine: one for surface navigation and the other for underwater operation. Petrol, steam or diesel were (and remain up to the present) the solution for the surface, while in the submerged condition the boat relied only on electric motors powered by batteries.

The *sottomarino* was a single-engine boat and this was the path the *Regia Marina* favoured from the outset, as only such a boat could be a realistic (and relatively cheap) alternative to the few costly battleships Italy could commission. As the electric batteries of the day did not allow for the long-range patrols and the high underwater speed that such a strategic submarine required, *the torpediniera sommergibile* ('submarine torpedo boat', the classic submarine since the French *Narval* of 1899) was the only realistic option for all navies of the period. Before 1914 Italy attempted to develop a true single-engine boat based on an efficient close-circuit diesel fed by oxygen cylinders. There were trials before and during the Great War of an anaerobic FIAT 500hp engine on the oil tanker *Acheronte*, built from scratch for this purpose, while studies continued on the hull-form of the desired future *sottomarino* with two secret semi-submersible boats of reduced scale named *Alfa* and *Beta* in Venice. The conflict against Austria-Hungary and Germany and the ultimate failure of the engine trials, which concluded with an explosion on board the *Acheronte* – the tanker was paid off in 1920 – put an end to this ambitious programme.

In 1934, however, the threat of a confrontation with Britain induced Cavagnari to accept a new proposal for a single-engine boat advanced that year by Professor Pericle Ferretti, the father of the Italian snorkel. The principle, perfected by Ferretti, was essentially the same as that conceived in 1912, and the first land-based trials of the engine in mid-1935 were successful.[3] The possibility of developing a viable larger engine appeared, however, remote from the outset while, in the meantime, the pending crisis with Britain in East Africa, which reared its head in the second half of 1934, would become a clear and present danger. General (GN: Naval Engineering Corps) Eugenio Minisini, President from 1934 of

Silurificio Italiano, projected a midget submarine built around the prototype of the new single engine, and an order was placed in January 1936 for a secret 10-tonne experimental submarine driven by screws ahead, designated *SI*. The submarine was laid down in the torpedo factory belonging to the latter company at Baia, near Naples, not in one of the three established Italian yards for submarines, which were located at Monfalcone, Muggiano (La Spezia) and Taranto.

The operational concept was as follows: once the enemy had been sighted, during daylight and in favourable weather conditions, each of the specially-fitted Italian cruisers would launch at sea, some 30,000 metres from the enemy, two of the tiny new boats. The midgets, running at 30 knots towards the enemy warships, would be able to locate their targets after 30 minutes, using a 3-metre long periscope located within a streamlined fin that would generate only a small wake. They would slowly gain a favourable launch position before retreating at flank speed, in the hope of being recovered by an Italian warship or reaching a friendly harbour. Despite their small size the midgets were technically complex, and the proposed Ferretti engine remained experimental. On 24 January 1936 Cavagnari therefore decided to hedge his bets.

Midget Submarines

The concept of the midget submarine had been popular with the Italian Navy since the turn of the century. As the 'punch' of such a boat was similar to that of a larger one the midget submarine was attractive from an economic point of view. Moreover, it would be a true (if short-legged) *sottomarino* with a single electric motor and a purely defensive role, allowing the classic *sommergibili* to be employed for offensive patrols only. The first proposal was made in October 1908 by the famous engineer Cesare Laurenti,[4] by far the most successful Italian submarine naval constructor of his time, whose boats and designs were purchased, both before and during the Great War, by a number of foreign navies: four for each of Britain, the United States, Portugal and Japan, 15 for Sweden, nine for Russia, three for each of Spain and Brazil, and one for Germany; a total of 38 Laurenti boats were also built for the *Regia Marina*.[5] The proposal was for an all-electric 50-tonne submarine 19 metres long able to conduct a single night patrol off harbours; it would have a flank speed of 10 knots and carry two 450mm torpedoes.

The project had been conceived as a consequence of the growing Bosnian crisis between Italy and Austria-Hungary and was renewed, now by General (GN) Edgardo Ferrati, in 1915. The first six boats of the 'A' class (*A1-A6*; 31.2 tonnes, 13.5 metres long; powered by a single-shaft electric engine capable of 6.8/5 knots and armed with two 450 mm torpedoes) were completed between December 1915 and March 1916. They were followed by an order for six 'B'-class boats, slightly larger (46 tonnes / 15.1 metres) with a petrol engine for

The midget submarine *A1* being 'launched' at La Spezia on 17 October 1915; the submarines of the 'A' class were designed to be transportable by rail. (USMM)

surface navigation that gave them a reasonable operating range (128nm vs 12nm). All these boats were, like the 18-metre MAS of the SVAN types, built within the limits of the Sagoma Ferroviaria, and could therefore be transported by rail to the desired area of employment – an arrangement that was deemed a failure by early 1917. The 'A' had poor endurance and seaworthiness and the 'B' was little better; *B4–6* were not laid down, while *B3* was converted as an attack craft intended to penetrate Austro-Hungarian harbours, where it would release two divers equipped with the *Cappuccio Belloni*, a sort of ancestor of the Davis Submerged Escape Apparatus. However, the war ended before this ambitious program could be completed.

After the Great War there were no further midget submarines until 1934, when the British and Italian confrontation in East Africa brought both countries to the verge of war. The Colonial Office, having taken control in 1933 of the Yemenite Hadhramaut region (which was as large as Czechoslovakia) in the face of opposition from the League of Nations, now had its sights set on the Haud territory bordering British Somaliland. That large area (about the size of Finland) was thought to have considerable mineral resources, and the British Cabinet offered the Ethiopian Emperor the small harbour of Zeila in British Somaliland, linked by a corridor to Ethiopia, in exchange. However, the contra-

dictory treaties signed between Rome, London, Paris and Addis Ababa between 1906 and 1928 had (not by chance) created a sort of 'No Man's Land' in that corner of the larger Ogaden region, and the Haud had been occupied by the Italians in 1925.

In 1934 a British military mission was sent to survey the new boundary between British Somaliland and Ethiopia. This provoked a two-day clash with Italian colonial troops at the Walwal Pits,[6] sparking the Ethiopian Crisis of 1935–36.

The naval confrontation between the *Regia Marina* and the Royal Navy promised to be an unequal contest. The former had in commission only two elderly dreadnoughts, *Doria* and *Duilio*, which had been despatched to Naples and Augusta respectively to augment the local coast defence.[7] The British Mediterranean Fleet of 1 January 1936. on the other hand, comprised seven battleships and two carriers distributed between Alexandria and Gibraltar. Both Cavagnari and Mussolini (again Minister of the Navy between 8 November 1933 and 25 July 1943) therefore welcomed any ideas to correct this imbalance. The first attack craft of the new generation, the *Siluri a Lenta Corsa* (SLC, the so called 'pigs') and the *Motoscafo Aviotrasportato da Turismo* (MAT, an explosive motor boat) were developed from 1935. Meanwhile, Mussolini personally ordered that the number of submarines be doubled from the 60 currently available, and the number of MAS boats tripled, from 30 to 100 by 1940. A shortage of steel prevented an increase in the construction of major surface vessels except for the 600-tonne torpedo boats of the *Spica* class, the number of which increased from six to 32 between 1935 and 1938; Mussolini practised a degree of self-deception in referring to these as 'destroyers' in his personal notes, while the Navy always considered them coastal units.

This policy was confirmed by the dictator's opting to pursue all ideas to increase fast (and cheap) new 'secret' naval weapons, as the crisis surrounding the import of raw materials during the Ethiopian Crisis was severe. One of his own ideas was to resuscitate the coastal defence midget submarine project during a speech to prominent Italian industrialists during the summer of 1935. The first response to his request was a proposal by the Breda enterprise for a revamped 'B' boat. Meanwhile, Colonel (GN) Franco Spinelli, the prominent naval salvage expert, put forward a more advanced project, named 'Progetto Sigma', which met with the approval of General (GN) Umberto Pugliese, the designer of the *Littorio*s and President of the *Regia Marina*'s Comitato Progetto Navi (Constructor Corps). Spinelli's proposal was for a two-man, 10-tonne boat 7.6 metres long (and secretly designated *S1*), designed around a pair of small 450mm Norwegian air-launched torpedoes manufactured by the Horten Shipyard and positioned under the hull. A few weapons of this type had been sold to the Italian Air force in 1932 for experiments with S 55 floatplanes, which however demonstrated that the qualities of these torpedoes failed to match the manufacturer's claims.

The original project was amended during the Winter of 1935–36 at the request of the Comitato in favour of a larger *S2* model, and then to a definitive *S3* boat, which had a length of 8.73 metres. By February 1936 the dearth of raw materials caused by the Ethiopian Crisis was severe. Count Giovanni Caproni, a famous aviation pioneer, designer and industrialist, presented a further project for a 6-tonne midget submarine, and there was a new proposal shortly afterwards by engineer Vincenzo Goeta for a completely different design designated 'Progetto Gamma'. General Pugliese decided, in March 1936, to merge *S3* with the aeronautical series production techniques of Goeta's solution. The view of the *Regia Marina*, from Cavagnari down, was that *if* (a big 'if' after the experience of the Great War) these 'toys' of

Mussolini really worked, they would be manufactured by the dozen, as the small submarines would have a short operational life similar to aircraft. Mass production was clearly the right option for what Count Caproni called '*i cacciabombardieri del mare*' (sea fighter-bombers). Moreover, the demand for raw materials for these unconventional programmes was so small that the Italian Navy could afford the related efforts, hopefully gaining the favour of the Duce for more important programmes, primarily the carriers and torpedo bombers the Italian Air Force was currently opposing.

The end of the war in Ethiopia and the crisis with Britain in July 1936 slowed the entire small submarine coast defence programme except for the continued preoccupation with development of the Ferretti closed-cycle engine.

The CRDA Electric Boat

By late 1935 the Italian Navy had conceived two different types of midget submarine: the *Sottomarino d'attacco* (fast-attack single-engine submarine) and a coast defence type, the *Sommergibile di agguato* ('ambush submarine', a conventional diesel/electric patrol type).

By late January 1936 development of the *Sottomarino d'attacco* was following two paths: the single-engine Ferretti (which could be stern-launched from an overhanging ramp) and an all-electric boat to be lowered into the water by a crane to prevent damage to the batteries. These approaches were the work of two private inventors, Engineer Mazzei and Shipmaster Di Vicino, and would have replaced the Ferretti-Minisini model if the new engine failed to confirm its promise. In April 1936 the original Mazzei-Di Vicino scheme had been finally judged as impratical by General Pugliese but, given that the programe was still accorded top priority by Mussolini – who had just decided to abandon any further compromise with the Ethiopian Emperor and Britain by ordering the capture of Addis Ababa – the two inventors were despatched to the CRDA (Cantieri Riuniti dell'Adriatico) shipyard at Monfalcone in the hope that a specialist submarine builder could produce an efficient midget boat. They were fortunate in that an engineer in that yard had, by chance, made a private (and unsupported) study of such a craft, and all that was needed was the exact type of electric battery being developed by CRDA and which would later be perfected by the Tecnomasio Brown Boveri enterprise of Milan. The final product of their efforts was a 17.4-tonne boat, presented here for the first time using the original plans, which are dated May 1937 and which were recovered some years ago by the late naval architect Franco Harrauer, a close friend of the author. Franco's drawings based on the plans are reproduced here.

The *Sottomarino d'attacco Mazzei–Di Vinicio–CRDA* was discussed in Rome on 20 and 25 May 1937, and the secret report related that, according to the new project, the boat would be able to reach a top submerged speed

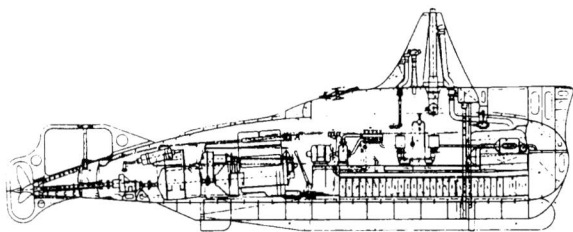

The midget submarine *S3*. (Drawn by Gino Chesi, Associazione Navimodellisti Bolognesi)

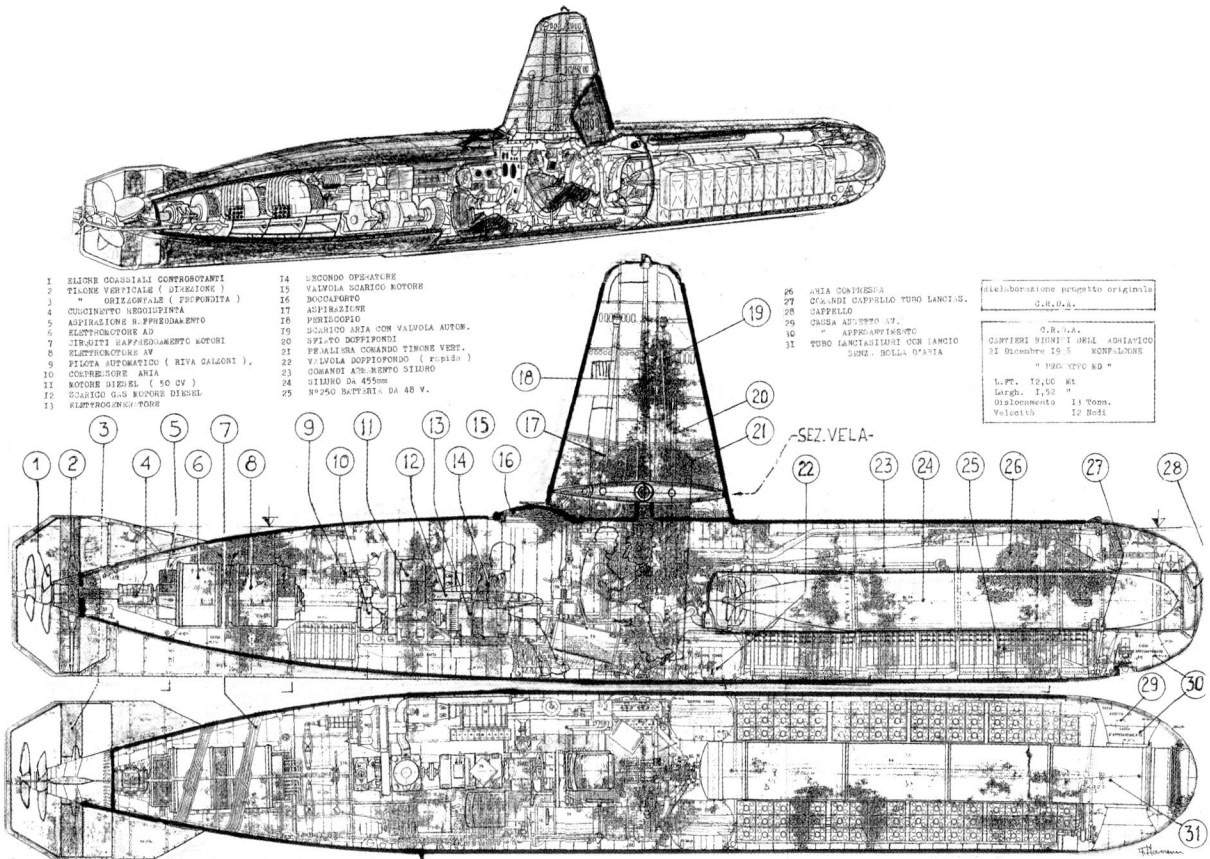

The CRDA *Sottomarino d'attacco*. Drawn by Franco Harrauer from the original 1/10 scale *Progetto esecutivo* dated 21 December 1938. (Author's collection)

The CRDA electric boat: artist's impression by Franco Harrauer, showing the two-man crew at the controls and the single torpedo tube forward, between the battery banks. (Author's collection)

of 16.5 knots with a twin coaxial screw. It was considered too small compared to the *SI*, the sea trials of which were due to begin the following summer and which had a projected range of 25nm. The CRDA boat had a range of 120nm at 6 knots in the submerged condition and, with a small diesel engine, could run at 4 knots on the surface for 340nm. Autonomy in real terms, after one hour at full speed with the electric engine, was calculated as 50nm, but even such a modest range would allow the boat to return to port. Diving depth was 30 metres, and the gyro compass would be located in the conning tower. The real advantage of the design, however, was the decision to carry only a single 450mm torpedo, the launch of which, using the weapon's own screws once the tube had been flooded, did not generate bubbles or compromise trim. This was important because loss of trim would have caused the boat to surface in an almost uncontrollable vertical position, resulting in the release of poisonous chlorine gas from the batteries. As the instability problem characteristic of all the very small midget boats projected in Italy in 1935–36 had yet to be fully experienced during the Froude tank model trials undertaken at la Spezia and in Rome, the unusual CRDA solution of what was effectively a torpedo launching a smaller torpedo was greatly appreciated. The CRDA boat's metacentric height was

particularly modest: 9cm surfaced and 8.7cm submerged (*vice* 12cm for the *SI*), and the dangers when the boat was running at high speed in the submerged condition remained a real threat to the safety of the crews. The final decision was to wait for the imminent trials of *S3* and the *SI* before continuing with the CRDA midget boat.

The first midget to be tested in the summer of 1937

La Spezia, January 1941. The first trials at sea of *CB1* and *CB2*. The 'ball' on the deck is an experimental infra-red projector to detect an enemy submarine running on the surface. Conscious that the *Regia Marina* could not hope to commission more than one hundred modern submersibles, in 1939 Mussolini believed it would be possible to double that total with the new midget submarines. (Author's collection)

I sommergibili tascabili tipo « C.A. » 25

CA 1-2

2 siluri da 450 mm

CA 2

CA 1

8 cariche da 100 Kg.

D.F.

cariche

accumulatori

I "C.A." della 1ª serie nelle versioni "silurante" e "d'assalto"

CA1 and CA2: the original design (above) and the 1941 attack craft version. (*Rivista Marittima*)

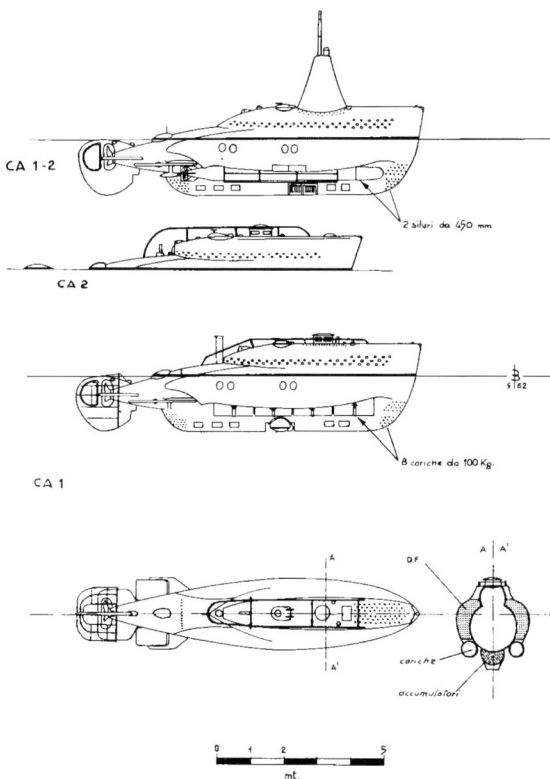

Bordeaux, September 1942: the 'mother' submarine *Da Vinci* with the attack craft *CA2* on deck. (*Rivista Marittima*)

was *S3* in Lake Iseo, where Count Caproni had a test centre for his floatplanes that provided the necessary security. There were stability problems from the outset but the boat proved to be mechanically reliable. It was therefore decided to halt the building of the *SI* (whose trials were scheduled for the following autumn) and to await the secret construction of two slightly enlarged midgets (*CA1* and *CA2*, 13.5 tonnes, 10 metres long) which were to be an improved version of *S3*; the latter boat was to be paid off and scrapped to recover some items for the pair of newly-built submarines. The cancel-

lation of the *SI* in turn compromised the CRDA project, which had been modified before being abandoned in December 1938. Following the Ethiopian and Spanish Civil War confrontations with Britain in October 1935 and August 1937, Mussolini's appreciation (quite different from that of the pessimistic Cavagnari) was that the Conservative cabinet in London could not afford a further risk of war in the Mediterranean, as such a move would have jeopardised the Commonwealth's much greater interests in the Far East. There was therefore no urgency to develop alternative naval weapon systems

Artist's impression of an Italian destroyer of the 'Soldati' class ferrying a *sottomarino d'attacco* by Franco Harrauer. (Author's collection)

while the classic naval programme approved in December 1937 for the new 35,000 tons battleships *Impero* and *Roma* was advancing according to schedule; the latter were due to enter service in 1942.

The two 'CA' boats were completed in April 1938 and,

despite modifications that included the lengthening of the periscope from 2.5 to 4.5 metres, their trials – first in Venice then, in February 1939, at La Spezia – confirmed they that were too small and unseaworthy, and that the solution of locating the torpedoes beneath the hull was

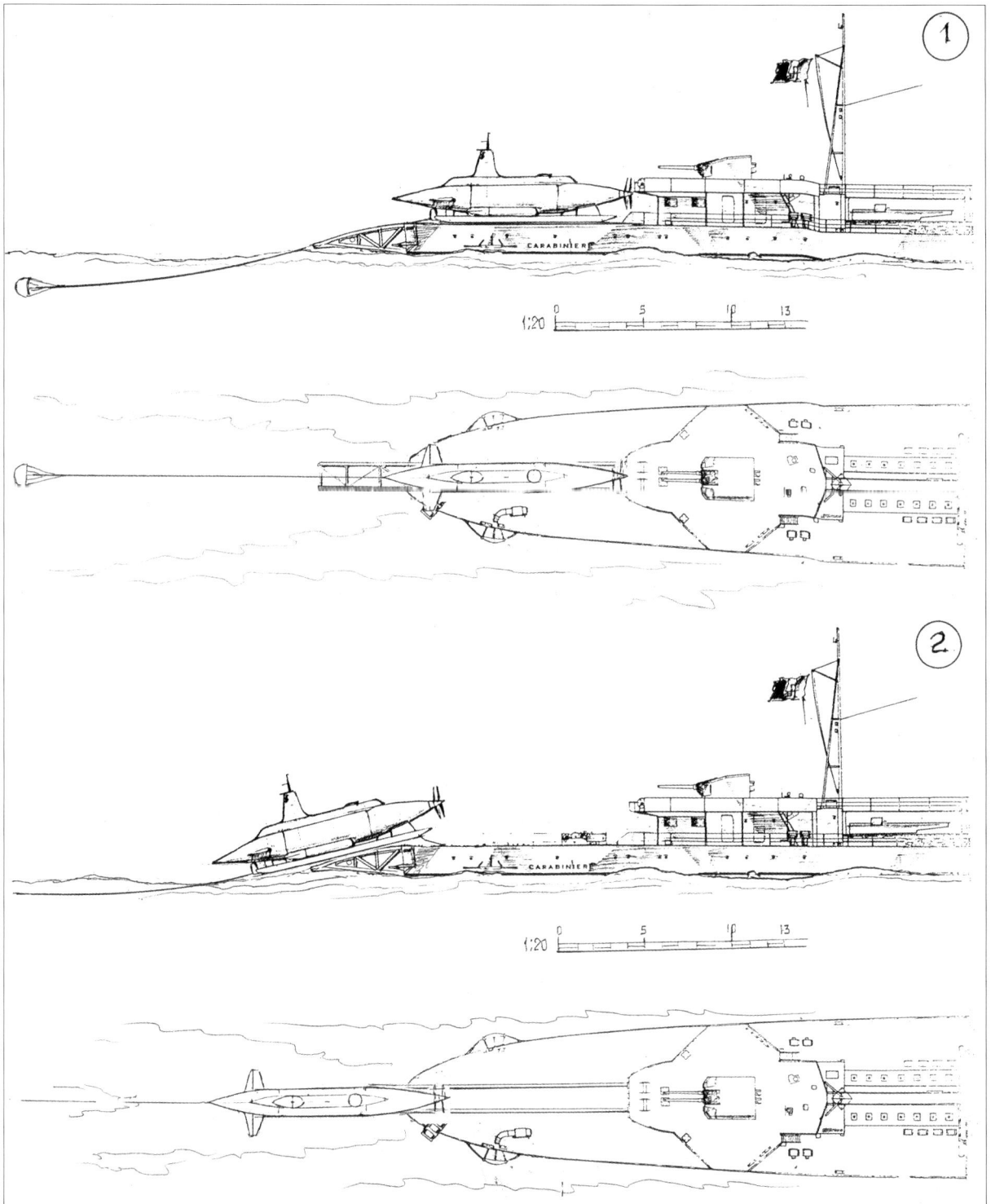

The three phases of the system conceived to launch a 'SA'-type midget submarine from a destroyer, drawn by Franco Harrauer. (Author's collection)

misguided. They were laid up, and recovered in 1941 to be radically modified as attack craft with only the electric engine; the midgets were to be transported across the Atlantic Ocean on the deck of a submarine.

By the end of 1938 Engineer Goeta had begun to study a further enlarged version, this time with a three-man crew, the 16-tonne *CA3*, but in March 1939 this programme was abandoned in favour of a project by Spinelli designated *S4* (28.5 tonnes, 14.0 metres long). Two months later a further revised design, *S5*, evolved, with the participation of a team from Caproni industries led by engineer Costa, who would later become better known. The 'CB' class displaced 36 tonnes and was 15 metres long. Completed and trialled in January 1941 these last midget submarines, which were designed from the outset as anti-submarine boats, gave a good account of themselves in 1942–45 in the Black Sea as coastal attack boats, and in the Adriatic, ferrying saboteurs and agents behind the Allied lines in Italy.

The old Minisini-Ferretti *S1* project was revived in September 1940, after Mussolini admitted he had been wrong in his original forecast stating that the war would be over before the autumn of that year. Two boats were completed the following year as *SA1* (SA = *Sottomarino d'attacco*) and *SA2*. While a complicated procedure using a floating anchor was developed to launch one of these midgets from a destroyer, the heel of the tiny submarine when launching its torpedoes remained a serious problem. This induced Minisini to radically modify his design, creating, in 1943, the revolutionary *SA3* which finally resolved the instability problem by launching her two 450mm torpedoes from aft-facing tubes; the boat

Baia, 1943: a photo of the newly-completed *SA3* midget attack submarine. (Author's collection, courtesy of Franco Harrauer)

was tested successfully off Baia, but was not yet operational when the 8 September 1943 armistice was announced.

Cloaks and Daggers

Following the sea trials of *CA1* and *CA2* in the lagoon of Venice there was little hope of a successful outcome pending a new test at La Spezia scheduled for January 1939. The new cycle of tests had a different purpose. As these two midgets had no military value – they were nicknamed 'Type Topolino' ('Mickey Mouse' Type) – Admiral Alberto Lais, the Chief of Italian Naval Intelligence, decided in 1938 to infiltrate an Italian

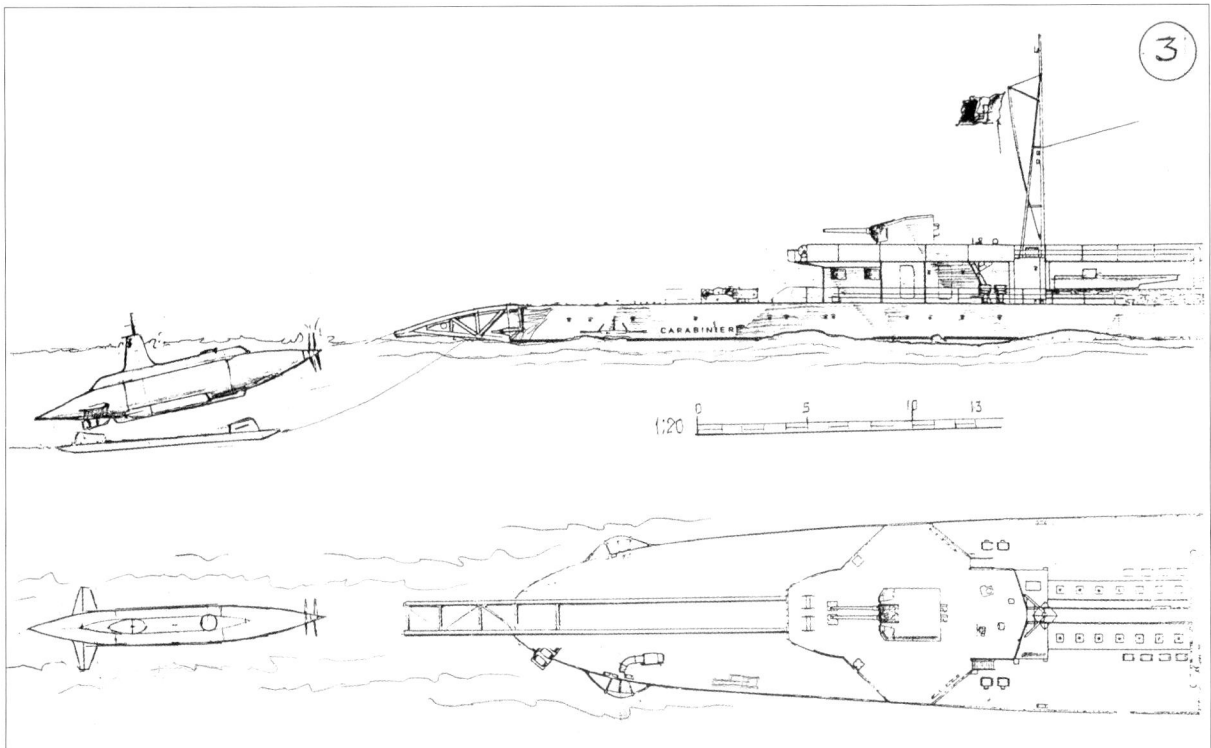

Artist's impressions of *SA2* by Franco Harrauer. Each of the two torpedoes was carried in tubes outside the casing. (Author's collection)

Two views of the
Baia salvage
operations on the
wrecks of *SA1*
and *SA2* by the
US Navy on
October 1943.
(NARA)

officer, Lieutenant Goliardo Zanfranceschi, of the
submarine *Anfitrite*, into the main French spy network
active in Italy. Zanfrancheschi sold some photos of those
boats to the *Deuxième Bureau*, thereby confirming his
trustworthiness. The idea worked well, the more so

because in January 1939 the young Rear Admirals
Fioravanzo and Sansonetti let slip, during a public
conference in Milan, a false news item stating that the
new *Littorio*s would carry two midget submarines which
could be launched before going into action for a torpedo

attack on enemy surface forces. This was the scheme for the *SI* and CRDA boats that had actually been dropped in 1938 (and recovered, for the more stable *SI* midget only, in the autumn of 1940). The young Aldo Fraccaroli, then a university freshman and future father of the modern generation of Italian naval scholars, was present at the conference and noted down this information, reporting it in good faith many years later.[8]

Thus, despite the technical failure of these midget submarines, the Italians were able to infiltrate the entire French spying network, most of whom were arrested at the beginning of the hostilities in 1940, with the remainder being employed as double agents until 1943.[9] One amusing sidebar of this sequence of events is that the only photos of *CA1* and *CA2* in 1938 that have survived are the ones from the French archives!

Conclusion

Experience has taught us that midgets can be successfully employed as special attack craft but make poor patrol submarines. Their dubious sea-keeping qualities present a hazard for their crews, who are always overworked. The CRDA *Elektroboot* was an original design and her engine and batteries had some interesting features, but the entire programme was simply one of Mussolini's many unrealistic 'bright ideas' as Navy Minister.

The similarities between the Italian and Japanese programmes are striking, even though they were developed independently. Like the *Regia Marina*'s *sottomarino d'attacco*, the better-known IJN 45-tonne midget submarines of the 'A' Type, designed from 1934, were to be used in a fleet action after being launched from fast seaplane carriers and specially-fitted cruisers. The real difference was the Italians were able to develop an effective attack craft doctrine, while by 1945 the Japanese were resorting to weapons of desperation in the form of the 8.2-tonne *Kaiten* manned suicide torpedoes.

Sources:

Archivio Ufficio Storico della Marina Militare, Rome, Fondo Supermarina Mezzi d'assalto, Archivio I, Cartella V, Documentazione varia. Fondo Invenzioni Busta 3 Fascicolo 2 & Fascicolo 27.

Vincent P O'Hara and Enrico Cernuschi, 'A Century-long Dream: Single Purpose Engine Submarines of the Italian Navy', *Warship 2004*.

Enrico Cernuschi, *Il sottomarino italiano*, supplemento *Rivista Marittima*, April 1999.

Enrico Cernuschi, 'Il progetto Sigma', *Storia militare*, October 1995.

Enrico Cernuschi, 'I sommergibili forzatori di porti', *Storia militare*, March 1997.

Enrico Cernuschi, 'Il Regio Sottomarino Sandokan', *Rivista Marittima*, Aug/Sep 2002.

Enrico Cernuschi, 'Sandokan alla riscossa', *Rivista Marittima*, April 2004.

Enrico Cernuschi, 'Il ritorno di Sandokan', *Rivista Marittima*, September 2005.

Alessandro Turrini, 'Il minisommergibile: un miraggio storico', Bollettino d'Archivio dell'Ufficio Storico della Marina Militare, March 1998.

Erminio Bagnasco, 'I sommergibili tascabili tipo C.A.', *Rivista Marittima*, September 1962.

Associazione Navimodellisti Bolognesi, Tecnica e storia attraverso i piani costruttivi navali, Bologna, 2001.

Endnotes:

1 On 4 December 1912 Greek troops landed on the small island of Saseno, which controls the Bay of Valona in Albania. Italian gunboat diplomacy induced Athens to retire that garrison the following year. HP Willmott, *The Last Century of Sea Power*, Vol I, Indiana University Press (Bloomington, 2009), 179.

2 Italy remained a parliamentary democracy. It would steadily transition to a dictatorship between November 1926 and March 1929.

3 Gino Galuppini, *Lo Schnorchel italiano*, USMM, (Rome, 1986), 127.

4 Alessandro Turrini, *L'opera di Cesare Laurenti: realizzazioni e progetti*, USMM (Rome, 2002), 298–300.

5 *I sommergibili italiani*, USMM (Rome, 1971), 15.

6 *Colonel Esmond Humphrey Miller Clifford (1895–1970) Summary Guide*, Kings College London, Liddell Hart Centre for Military History.

7 Emilia Chiavarelli, *L'opera della Marina italiana nella guerra italo-etiopica*, Giuffrè (Milan, 1969), 74.

8 Aldo Fraccaroli, 'The Littorio Class', *Warship* No 3, 12.

9 Marc'Antonio Bragadin, 'Risposta del SIS ai francesi: come arrivammo ai segreti di Palazzo Farnese', *Storia Illustrata*, Novembre 1969. Mimmo Franzinelli, *Guerra di Spie*, Mondadori (Milan, 2004), 236.

FIT FOR PURPOSE? THE ROYAL NAVY'S FISHERY PROTECTION SQUADRON, 1883–2023

Jon Wise looks at the history of fishery protection in the Royal Navy from the late 19th century to the present day.

The Royal Navy claims that the duties undertaken by the Fishery Protection Squadron are the oldest in its history, dating from the 14th century. Several name changes have been applied over the centuries: currently fishery protection (FP) is subsumed rather obscurely under the 'umbrella' title 'The Royal Navy Overseas Patrol Squadron'.

This article, covering the modern phase of a long history, begins in May 1882 with the signing of the International Convention for Regulating the Police of the North Sea Fisheries Outside Territorial Waters by six countries bordering the North Sea, including the UK. The intention was to encourage conformity and regulation in a 'fish-rich' area of the West European continental shelf that could be applied to international waters outside the agreed three-mile territorial delimitation of individual countries. The British Sea Fisheries Act followed in 1883 with arrangements made for 'gunboats and cruisers', selected by the Admiral Superintendent of Naval Reserves and under the command of a senior officer, to be in charge of all matters concerning the North Sea fishery. The remit was to deter attempts by foreign vessels to poach or make incursions within territorial waters, to dampen down disputes between fishermen, and to provide help to a fishing vessel in need.

Fresh sea fish was a luxury enjoyed only by the well-off until the second half of the 19th century. In contrast, its salted equivalent had been a staple part of the diet for the vast majority of the populace over several centuries. Thus, starting in the late medieval period, the protection of the country's territorial waters and the convoying of the commodity needed to be secured, as '… the sea was widely perceived as a lawless realm beyond the frontiers of all nations, where neither law nor truce nor treaty ran.'[1] The task became more demanding when fishermen began to exploit what were termed 'middle' and 'distant' waters – particularly in the Arctic and on the northeast American littoral.

Comparisons with the duties of a policeman on the beat have frequently been made in relation to fishery protection. In order to be effective all year round, an FP vessel has needed to be durable enough to cope with most sea states. Since the age of steam it has also required both good manoeuvrability and a margin of speed to out-run recalcitrant fishing boats. Lastly, although only permitted to be used as a final resort, the vessel needed to be armed with a light-calibre weapon. As this article will demonstrate, suitability for the task has been a perennial problem because, understandably, assigning vessels to fishery protection duties at the expense of front-line, warfare-related tasks has never been a priority.

The Late Victorian Period

Details of operations undertaken by designated RN fishery protection vessels in the late Victorian period, following the signing of the North Sea Convention, reveal a variety of ships assigned to the task. HMS *Hearty*, for example, first commissioned in September 1886, was employed in an assortment of roles in addition to Fishery Protection. Her distinctive cutaway stern gave her a tug-like appearance, a duty in fact she undertook on several occasions during a long career.

HMS *Galatea*, on the other hand, despatched in 1899 to investigate harassment of British trawlers off Iceland and the Faroe Islands, was a very different proposition. On that occasion, the use of the 5,620-ton *Orlando*-class armoured cruiser, armed with 9.2 inch guns, was an overt example of 'gunboat diplomacy' in action. The ruling

HMS *Hearty*, a fishery protection vessel, photographed in the 1890s. (Author's collection)

power Denmark, despite being a North Sea Convention signatory, had had the temerity to declare a 13-mile delimitation around its overseas territories![2]

In 1905 Lt-Cdr Rooke, as 'Senior Naval Officer, North Sea Fisheries', took command of the torpedo gunboat HMS *Halcyon*, assisted by five other assorted gunboats and destroyers. *Halcyon* was one of five torpedo gunboats of the *Dryad* class and was first commissioned in 1895. She displaced 1,070 tons, and was armed with two 4.7in quick firing guns and five 18in torpedo tubes. The *Dryad* class had an unusual profile with the two funnels set wide apart. *Halycon* shared the inherent weakness of the earlier *Alarm* class, being fitted with unreliable locomotive boilers, which may well account for the fact that she was allocated the FP task, in which the radius of operation was limited to the parameters of the North Sea. The torpedo gunboats, sometimes called 'catchers' in reference to their role as counters to the numerous torpedo boats of the French and Russian navies, were not a great success as a generic design. In addition to defective boilers, they struggled to keep the sea even in favourable weather conditions such as during summer manoeuvres.[3]

The pre-First World War squadron also included HM Ships *Skipjack* and *Spanker* of the thirteen-strong *Sharpshooter* class. These torpedo gunboats were smaller than the *Dryads*, displacing only 735 tons, but were marginally faster. HMS *Leda*, of the above-mentioned *Alarm* class, was also assigned to the flotilla, but in 1913 HMS *Wear* was employed as a temporary replacement. *Wear*, a 'River'-class destroyer, differed from the other early members of the squadron: the destroyers, which were gradually replacing the torpedo gunboats, enjoyed a considerably faster maximum speed of 25 knots. The sturdiness of this Admiralty design allowed the ships to maintain this speed in most sea conditions, making them appear (at least on paper) more suited to the protection task.

Problems Enforcing International Law

It would be inaccurate to assert that the 1905 Fishery Protection Flotilla represented a settled, coherent unit during the decade leading up to the Great War. HMS *Skipjack*, for instance, had been brought forward from the Reserve Fleet in 1905, and had then reverted to an RNR drill ship for a short period before undergoing a further refit for FP duties. In 1907 the ship was converted to a minesweeper, having her torpedo tubes removed in the process.

In 1910 *Skipjack* was instrumental in the arrest of the French trawler *G599* in territorial waters in the Thames Estuary. This operation was routine, passing off without incident. On the other hand, two years later the Belgian sailing trawler *0.100* was discovered fishing off Southwold inside the three-mile limit by *Skipjack*'s sister-ship HMS *Spanker*. The RN torpedo gunboat despatched her ship's boat across to the trawler, which immediately drew in her nets and set off home, nearly swamping *Spanker*'s boat in the process. The gunboat pursued the culprit and closed her, but the Belgian boat then contrived to get her mizzen boom entangled in the starboard after back-stay of *Spanker*'s fore-topmast.

The trawler fled again. This time *Spanker* did not pursue because the CO believed he had enough evidence to secure a conviction. He was wrong, because the incident had started *within* territorial waters, thus exposing a loophole in the 1882 Convention.[4] This example, coupled with the obvious problem of ensuring that fishery protection vessels were in the right place at the right time, and the rare likelihood of transgressions occurring in daylight with pin-sharp visibility, were examples of the day-to-day problems faced by the flotilla at a time when ship-to-ship and ship-to-shore communications were still primitive.

Responsibility

Responsibility for Fishery Protection was the subject of a Government Interdepartmental Conference in 1907. The Admiralty stated tangentially that the vessels currently used for FP were rapidly becoming obsolescent and unfit for 'war purposes'. The main thrust of its argument then turned inevitably to the matter of finance, questioning whether the Navy should continue to undertake full responsibility for fishery protection and thus be charged on the Navy Vote.[5] It turned out that the Treasury was the stumbling block in what appeared to be the Navy's covert plan to rid itself of fishery protection. The Admiralty claimed the Treasury had 'misunderstood … the recommendations of the Conference'; the Treasury refuted the accusation and the matter lay unresolved until it was resurrected in 1919.

Both the First and the Second World Wars were remarkable for the fact that the trawlers and drifters, together with their crews, became integral parts of the war effort, and peacetime fishery protection arrangements were largely set aside. The design of the fishing boats, their gear and the experience of the fishermen proved an invaluable resource, particularly with respect to minesweeping and antisubmarine warfare. Large numbers of vessels were requisitioned, and in both conflicts Admiralty-designed boats were constructed to supplement the commercially-built craft. Details of these ironic circumstances, where in effect 'the protected became protectors', is beyond the scope of this narrative.

The unfinished business from the 1907 Interdepartmental Conference for fishery protection (see above) was resurrected after the Great War, and the outcome took the form of a report entitled 'Admiralty Responsibility for Protection of Fisheries' composed by a senior naval officer, Captain Dugmore. Among other issues, it examined the existing structure of the service, and questioned the Royal Navy's continuing participation, the types and numbers of vessels required, and the sea areas under its jurisdiction.

Dugmore addressed the thorny issue of the Admiralty's

own involvement, particularly with regard to finance. He stated that the maintenance of the prewar torpedo gunboats then in use was unnecessarily expensive: the annual cost of HM ships *Halcyon*, *Skipjack*, *Spanker* and *Leda* amounted to £53,701 out of a total of £102,249 allocated to Fishery Protection in 1913. He admitted that if Fishery Protection were assigned to another department, it would need to have the powers to exercise the same authority to deal with foreign vessels outside territorial waters as currently possessed by ships flying the white ensign, which included being armed. The important and related matter of 'showing the flag' was addressed: Dugmore remarks, tellingly, 'it [the RN] certainly exercises a more effective influence on the fishing population than any other service would be likely to exercise'.[6]

A sub-committee considered the most appropriate design of vessel to meet the needs of the flotilla. It was concluded that it should be not too distinctive in appearance, have sufficient speed, and be capable of keeping the sea in bad weather. The prewar torpedo gunboats had failed to meet these criteria, being found unsuitable for navigatation around the islands and headlands encountered in the Shetlands, for example, which in turn inhibited them from apprehending trawlers that were breaking the law. Moreover, their distinctive appearance made them easily recognisable miles away on shore, where there were reported to be 'spies' on watch to observe and report the movement of the protection vessels.

Indeed, 'appearance' seems to have been a key factor in determining suitability. The report suggested employing the same 'Fleet Sweeping Vessels' or 'Sloops' of a generic 'Flower Class' group. They were recommended because they resembled tramp steamers and therefore would be less obviously RN-manned. Unfortunately, these ships were found woefully lacking when used in the Barents Sea some three years later.

A retired officer, Commodore Ellison, was consulted and he remarked, prophetically as it turned out: 'I recommended, and still recommend that the proper type of fishery protection vessel is a specially built trawler, considerably larger than ordinary fishing trawlers, with decent accommodation for officers and men; a vessel that would keep the sea in all weathers and would be more or less comfortable.' It took half a century for Ellison's words to be heeded.[7]

However, the report concluded that fishery protection should remain in the hands of the Royal Navy, although Dugmore advised that the cost should be shared by other Government departments, including the Board of Trade and HM Customs. Nevertheless, one detects in the subtext of the report a shared feeling within the Navy that the task was an unnecessary burden, that HM ships should not be called upon to undertake these kinds of routine constabulary-type duties. One might attribute such sentiments to the fact that those involved had just experienced a long and difficult war. The same arguments were to be revisited again and again over the course of the next century.

Suitability for Task

The matter of settling on a satisfactory, generic design for a fishery protection ship never became a priority in the interwar years. Three examples follow which are illustrative of this shortcoming.

Britain did not formally recognise the Soviet Union until 1924. Prior to that there had been arrests of British trawlers off the Murman Coast east of the Norwegian border. The Soviets had adopted the same 12–mile delimitation advocated during the preceding Czarist Regime. In March 1922, in response to the arrest of the trawlers *Magneta* and *St Hubert*, HMS *Harebell* (*Anchusa* class) and *Godetia* (*Arabis c*lass) were reported as being readied for deployment to the Barents Sea. Both were considered to be better seaboats with a greater radius than the twin-screw minesweepers they had replaced.

Soon after the outbreak of the First World War, the Admiralty had recognised a requirement for auxiliary-type vessels capable of undertaking a variety of duties that included minesweeping, towing and the transportation of servicemen and stores. A design produced by the Director of Naval Construction, using mercantile scantlings, aimed to deliver vessels that were simple, robust and capable of being built in non-specialist yards. The resulting coal-burning, single-screw ships had a displacement of 1,200–1,250 tons, depending on the individual class, and carried one or two single 4in guns. Being single-screw ships meant that they had large turning circles and so required a 'steadying sail' to assist in keeping the head to the wind. They were euphemistically described as 'very lively ships'.[8]

Harebell and *Godetia* returned to northern Russian waters in 1923, having been withdrawn at the end of the previous year due to what were termed 'technical difficulties'. One might have assumed that this referred to the design problems mentioned above, but it turned out that this explanation was only partly true. The Squadron's Senior Officer Captain Aylmer subsequently reported to the Admiralty on a visit he had made to HMS *Godetia* at Sheerness following her patrol in late 1922. *Godetia's* Commanding Officer had found his ship's company 'thoroughly discontented', although nothing had occurred that might have incurred disciplinary action. Aylmer's subsequent investigation discovered that the basis of the grievances centred on the 'harsh and uncomfortable conditions' experienced during such deployments, the absence of any hard-lying money that the crew felt they were owed, and an unofficial report which had been circulated that stated that the ship was unfit for service in Arctic waters, and that the chances of *Godetia* making a safe return were poor.

Captain Aylmer instigated further enquiries. The unofficial report had originated from one of the trawlers with whom *Godetia's* crew had fraternised at the port of Honningsvåg in Norway. He deduced that what he described as the 'nervous condition' of the ship's company had originated mainly from their experiences during the return voyage to the UK in late 1922. They

HMS *Godetia*: her poor sea-keeping qualities in northern waters caused unrest among the crew in the 1920s. (World Ship Society Photographic Library)

had sailed through the Inner Lead off Norway in darkness and driving snow, and *Godetia's* CO had admitted that he had 'nearly lost his ship' on three occasions despite the presence on board of two Norwegian pilots. Furthermore, Alymer discovered from talking to a trawler owner in Hull that all the skippers who had observed the two sloops during the winter months considered them to be most unsuitable for the work owing to their light draught and high upperworks.[9]

The difficulties associated with keeping the seas in harsh northern waters were, of course, not confined to the Barents Sea: stormy waters were encountered perennially around the coasts of the British Isles, particularly in winter. On Monday 25 January 1937, Captain Fishery Protection and Minesweeping, Captain VAC Crutchley, received a signal from the District Inspector of Fisheries that information had reached Hull that the trawler S/T *Amethyst* was in difficulties about 120 miles east of Kinnaird Head in northeastern Scotland. Signals at approximately 22.00 hours the previous night had stated: 'Engine trouble, cracked cylinder head, position 120 miles East of Kinnaird Head', and then later, ominously, 'Boiler explosion, heavy list to Port, can do nothing'. The District Inspector reported that the *Amethyst's* owners had asked for assistance as they

considered their vessel might still be afloat and drifting helplessly. S/T *Amethyst* was comparatively modern, having been built in 1928; she displaced 357 tons and had a triple expansion engine and a single boiler.

HMS *Kingfisher*, the ship assigned exclusively to work in Scottish waters as part of a historic 1815 Statute, was immediately ordered to sea from Invergordon and a wider, international call for assistance was broadcast through the Ministry of Agriculture and Food (MAF) in case other ships were in the locality. *Kingfisher* was the lead ship of a new class of sloops or patrol vessels, laid down in 1934 by Fairfield and completed the following year. The ships of the *Kingfisher* class were intended to operate as coastal escorts, suitable for a range of tasks to include fishery protection and peacetime antisubmarine warfare training.

The 530-ton sloop responded within two hours, proceeding to sea at 16 knots initially, then 14 knots as 'the ship was bumping badly, weather conditions at the time being Wind E.S.E., force 8, sea and swell 45'. The ensuing report stated that *Kingfisher's* speed had to be progressively reduced, until during the Monday evening she was making only three knots with the propellers regularly out of the water. The sloop was hove to a short while later and unable to make any headway. Captain

Crutchley was forced to order her to abandon the search.

Meanwhile, HMS *Harebell*, his command, had sailed north from her base at Portland to Leith. Crutchley was of the opinion that, unless the *Amethyst* had succeeded in riding some form of sea anchor 'such as her trawl gear shot forward', she had undoubtedly foundered. The trawler's owners disagreed, however, arguing that their boat was quite likely still to be afloat. Accordingly, the Admiralty ordered a further search. Having coaled, *Harebell* resumed the search nearly five days after *Amethyst's* signal had been received.

The sea state had increased to such an extent that by mid-afternoon on 30 January it became obvious to Captain Crutchley that he would lose both his whalers if the ship remained on course and the weather deteriorated further. He decided to heave to with revolutions for a speed of seven knots so as to maintain some steerage way. Late in the evening a particularly heavy wave submerged the port whaler causing the falls and gripes to part. The wind was measured as Force 11 and the sea state and swell as 79 on the Douglas Scale.

Meanwhile, other ships had joined the search. HMS *Kingfisher* was unable to leave Leith until 2 February as it was decided that she would make little headway in the prevailing weather conditions. Finally, late on the evening of 1 February, the wind had abated sufficiently for *Harebell* properly to resume her search. Sadly, despite best efforts, there was no trace found of the *Amethyst* and the search was officially abandoned by the Navy on 5 February. An 'Admiralty regrets …' notice was issued three days later.

HMS *Kingfisher*, despite being only two years old and built with fishery protection duties in mind, had proved inadequate to cope with the savagery of the winter weather to the north of the British mainland. Inexplicably though, by way of replacement, the CO and ship's company were transferred to HMS *Sheldrake*, another unit of the *Kingfisher* class, while *Kingfisher* herself was assigned to the admittedly less stormy Channel Area.[10]

Robustness was one issue, but fishery protection

HMS *Kingfisher*, a general-purpose large patrol vessel, photographed in the 1930s. *Kingfisher* took part in the abortive search for the Hull trawler *Amethyst* in 1937. (Author's Collection)

vessels also required a sufficient margin of speed in order to catch recalcitrant fishing vessels. In the early afternoon of 14 November, 1933, the *Mersey*-class trawler HMS *Doon*, patrolling near the southern reaches of the Outer Hebrides, encountered a group of trawlers apparently working illegally. The first boat approached was the Fleetwood trawler *Lucida*, and her skipper Bert Jinks was promptly charged with illegal trawling in contravention of the Herring Fisheries Act of Scotland, 1899. After dealing with the other trawlers in the group, which were also working illegally, HMS *Doon*'s CO, Lt-Cdr Dallison, returned to S/T *Lucida* and, leaving Leading Seaman Forrest aboard as the official boarding party, ordered Jinks to proceed to Stornoway.

However, as soon as *Doon*'s sea boat had cast off, the trawler made off southward with the FP vessel in pursuit. Repeated orders to stop were issued via *Doon*'s search-light and three blank rounds were fired. These finally elicited a reply via the *Lucida*'s whistle, 'Bound for Fleetwood'. It was reported that Jinks had admitted earlier that he had been twice convicted for working inside the Scottish fishing limit and that when he had heard *Doon* firing her gun while apprehending another trawler, he had reassured his crew with, 'Go on, don't mind that f——g gun, he daren't hit yer, it's only a blank'.

The chase south continued all day. By dusk it became apparent that Bert Jinks was not returning to Fleetwood. He was spotted by HMS *Doon* making his way up the Lune Estuary early the next morning. A hair-raising pursuit ensued with both vessels dodging through a crowded waterway before *Lucida* was finally apprehended. Frustratingly, the intricacies of Scottish bye-laws prevented Skipper Jinks from receiving the maximum fine, but it was reported later the same year that the owners of S/T *Lucida* had sold the ship, while Bert Jinks was said to have been out of employment and in receipt of public assistance.

On paper, HMS *Doon* and S/T *Lucida* were of similar size, displacement and age, and both were classified as trawlers. Officially, the RN vessel had a maximum speed of 11 knots, the *Lucida* a slightly slower 9 knots. HMS *Doon*'s inability to overhaul *Lucida* during the race southwards was remarked upon in a subsequent exchange in the House of Commons. It was disclosed that *Doon* had superior speed but, rather conveniently to avoid further scrutiny of the matter, it was announced that this could not be divulged for 'security reasons'.[11]

Postwar Reorganisation

Shortly after VE Day, as had happened after the end of the Great War, there was a firm commitment made by the Admiralty to reorganise its fishery protection fleet now hostilities had ceased; this was in anticipation of a rush back to the fishing grounds. In November 1945 The Fishery Protection and Minesweeping Flotilla consisted of the sloop HMS *Fleetwood* (Senior Officer command), two 'Castle'-class corvettes, *Allington Castle* and *Bamborough Castle*, together with two trawlers and initially three

Harbour Defence Motor Launches (HDML) for inshore work. Geographically, the flotilla was distributed across three sea areas: the North Sea, the Channel and the Irish Sea, with HMS *Fleetwood* (later *Stork*) based at Lowestoft. The presence of the sloop and the two 'Castle'-class corvettes meant that the backbone of the squadron all had recent experience of the harsh weather conditions of the North Atlantic, and a sufficient margin of speed to be effective in any stern chase. This constituted an improvement on the prewar state of affairs.

The shore organisation was also streamlined over the next decade, with the Civilian Fishery Inspectors providing 'intelligence' for the Navy and working more closely and directly with the Senior Officer Fishery Protection. The importance of the additional minesweeping task was initially lowered but then raised again following the Korean War, when the threat imposed by modern, highly sensitive magnetic mines rendered the steel-built ocean minesweepers largely obsolete and heralded the building of the very numerous, wooden-hulled 'Ton' class.

A final, important step was taken in the late 1950s towards presenting a coherent approach to fishery protection with the allocation of 'Ton'-class coastal minesweepers for inshore waters and the employment of

Type 14 frigates for distant waters support. This decision was made shortly before the start of the 'First Cod War' in 1958. Although none of the ships was actually purpose-built for the task, all were modern having been constructed in the last decade. This contrasted sharply with previous practice, where availability rather than suitability had been the watchword. However, although this decision looked positive on paper, the reality was not entirely satisfactory.

The Type 14 (*Blackwood* class) was conceived as a 'second rate' frigate, a successor to the 'Flower'- and 'Castle'-class corvettes of the Second World War, and was relatively cheap and quick to build. In point of fact, the actual complexities of the resulting design meant that it took on average three years to construct each ship, while the overall aim of increasing numbers of hulls at no extra cost resulted in a very austere ship not well suited to a peacetime navy. In theory, the Type 14's high prow enabled the ships to ride up and over a wave without it submerging the forecastle. Structurally, however, the hull proved frail and subsequently needed considerable strengthening for fishery protection duties in stormy northern waters. Moreover, accommodation was sparse, and 50 per cent 'hard lying' money had to be paid to the crews while they were assigned to the FPS.[12]

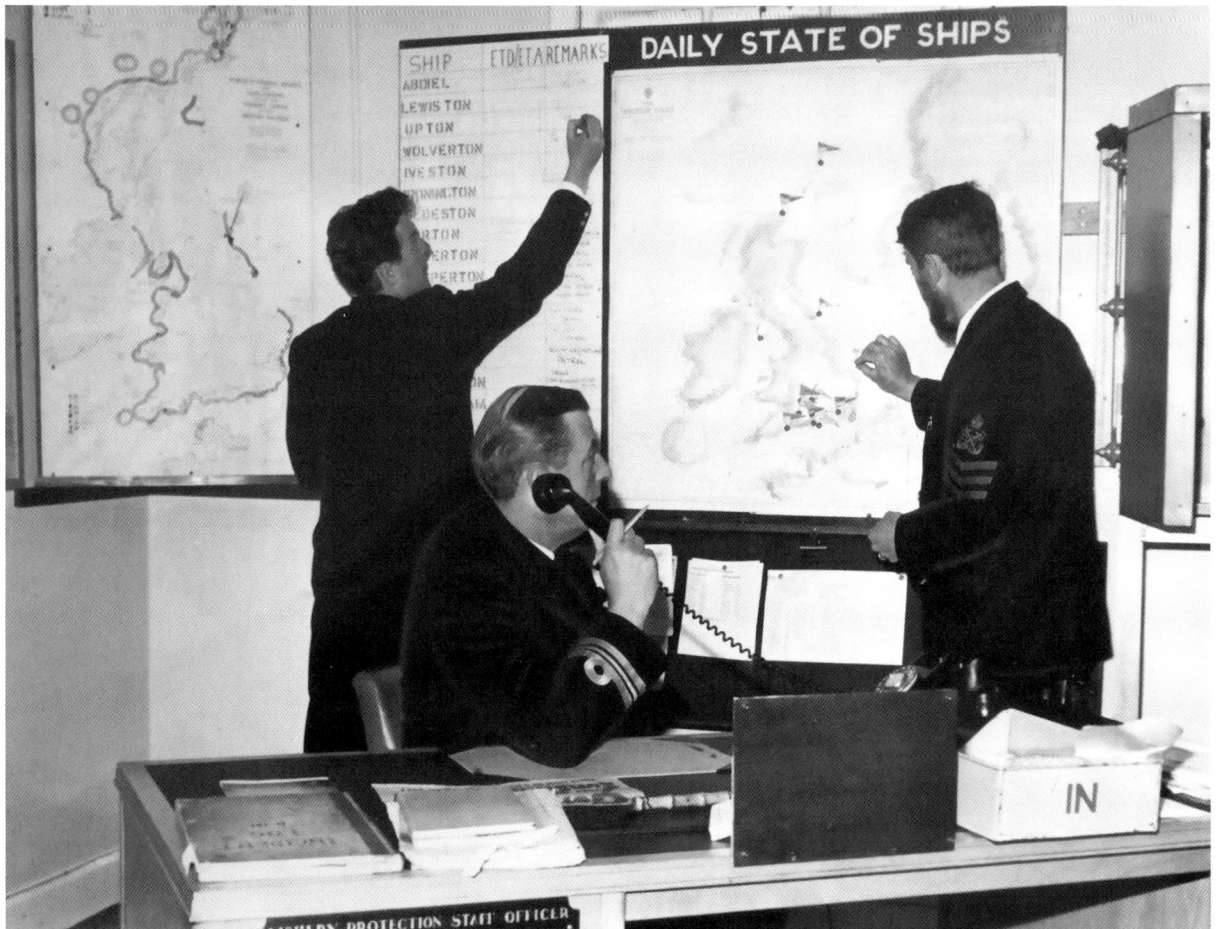

A view of the Fishery Protection Squadron Headquarters Operations Room at Port Edgar, Scotland, probably in the late 1960s. (MoD)

The slim hull of the Type 14 (*Blackwood* class) frigate is evident in this photograph of HMS *Murray* (F91) taken in May 1969 in the English Channel. Her sister-ship *Russell* broached in a heavy following sea in the Pentland Firth while returning from patrol off Iceland during the First Cod War, and was saved from disaster only by the quick thinking of her navigation officer. (MoD)

An improved *Algerine* class fleet minesweeper, with a large 1,500-ton displacement, had been announced as part of the 1951 'War Programme', but the magnetic mine threat mentioned above put paid to that design. The perceived requirement for suitable mine countermeasures vessels was urgent, and the much smaller 360-ton coastal sweeper ('Ton' class) emerged as part of the solution. In its primary role the 'Ton', with its non-magnetic hull of wood planking over aluminium frames, together with a reduced pressure signature due to its modest displace-

ment, was a very successful design. No fewer than 118 were built for the RN alone; this excessive number proved to be something of an embarrassment for the naval planners, and some orders were subsequently cancelled.

However, fishery protection benefited. Initially, four 'Tons' were allocated to what was described as 'The Home Division' of the FPS with responsibility for the entire coastline of the British Isles with the exception of the English Channel. An inexplicable MoD decision resulted in the so-called 'Fish Tons' being allocated the open-bridge variant while the ships with an enclosed bridge were sent to tropical Singapore. These minesweepers had a maximum speed of 15 knots, which was sufficient to outpace most modern trawlers. Their small size enabled them to refuel in small harbours, allowing opportunities to meet the local inhabitants including fishermen who, in turn, provided up-to-date intelligence about the migration of fish shoals and thus the concentrations of the fishing fleets.

Declining Power

The three disputes with Iceland 1958–61, 1972–73 and 1975–76 represented a turning point for the British distant water fishing industry and, in the broader sense, a further indication of the country's diminishing ability to influence international events on the world stage. The long-running arguments with Denmark, Norway and the

'Ton' Class Fishery Protection Vessel

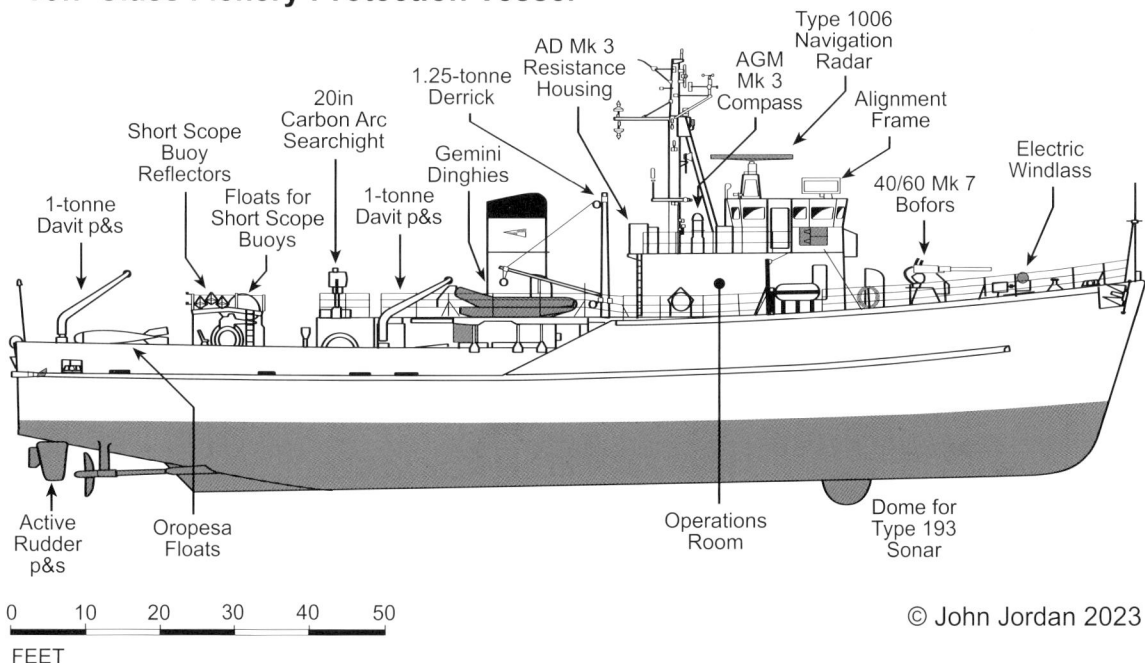

© John Jordan 2023

The generic 'Ton' or *Coniston* class of minesweepers were constructed during the 1950s in response to the perceived threat from the Warsaw Pact countries' manufacture of large numbers of highly sensitive magnetic mines. In order to minimise the magnetic signature the 'Ton'-class hulls were constructed of wood (double diagonal mahogany planking) on an aluminium frame. The superstructure was also wholly aluminium. No fewer than 281 vessels were built in a wide variety of shipyards as part of a coastal build programme. The 'Ton' class was successfully employed on fishery protection tasking for 30 years.

HMS *Wasperton*, one of the so-called 'Fish Tons'. Note the open bridge and the 20-inch carbon arc searchlight abaft the funnel, dubbed 'death ray' by the crew. (Courtesy of The Ton Class Association)

Soviet Union over the UK's adherence to the historic three-mile territorial water delimitation since the end of the Great War ended with Norway's success in 1951 in obtaining approval for extending its territorial waters when the matter was brought before the International Court of Justice in the Hague. This victory was followed by similar concessions made to the Soviet Union during the same decade. In the meantime, the postwar Truman Declaration on ownership of the continental shelf, together with similar claims by South American countries, showed that international opinion was shifting, rendering the old imperial power ever more isolated.

A boarding party from HMS *Ashton* undertaking the routine task of checking a fishing boat's net mesh size. (Courtesy of The Ton Class Association)

The three 'Cod Wars' resulted in progressive retreat and inevitable defeat. Britain extended her own exclusive fishing zone from three to twelve miles in 1964. By the time of the last 'War', in which Britain disputed Iceland's claim to a 200-mile delimitation, the country was already a member of the EU and about to become a signatory to the Common Fisheries Policy. Six months after the conflict ended on 1 June 1976, Britain declared its own 200 nautical mile extended fisheries zone (EFZ) to complement the similarly-sized exclusive economic zone (EEZ). By March 1976, the Government and the Fishing Industry were already discussing the consequences of the impending collapse of the country's distant water fishing; the writing was already clearly on the wall.

The FPS, in the shape of the Type 14s, played a significant role during the first Cod War, their numbers supplemented by various destroyers and frigates drawn from routine Cold War duties. The abrasive nature of the second and third conflicts resulted in almost the entire Royal Navy contribution being provided by modern front-line frigates whose speed, in particular, was needed to counter the sturdy and highly manoeuvrable Icelandic gunboats. Unlike the 1958–61 conflict, when the main Icelandic tactic was to attempt to board and arrest individual trawlers, the latter two wars were characterised by the gunboats seeking to cut across the fishing boats' sterns in order to sever the nets with the aid of a 'warp cutter'. In order to succeed they had to out-manoeuvre the RN frigates shielding the trawlers.[13] This led to highly dangerous close-quarter encounters and, inevitably, to collisions.

A tense moment during the Third Cod War with HMS *Mermaid* (F76) and ICGV *Baldur* just feet apart prior to a collision on 6 May 1976. (MoD)

HMS *Salisbury*, seen here in 1956, was the oldest RN frigate to take part in the Third Cod War (1975–76). Her CO, Commander Hugo White, used his ship's diesel engines to good effect to counter the 'violent sternboard and accelerating manoeuvres' of the similarly-propelled ICGV gunboats. (MoD)

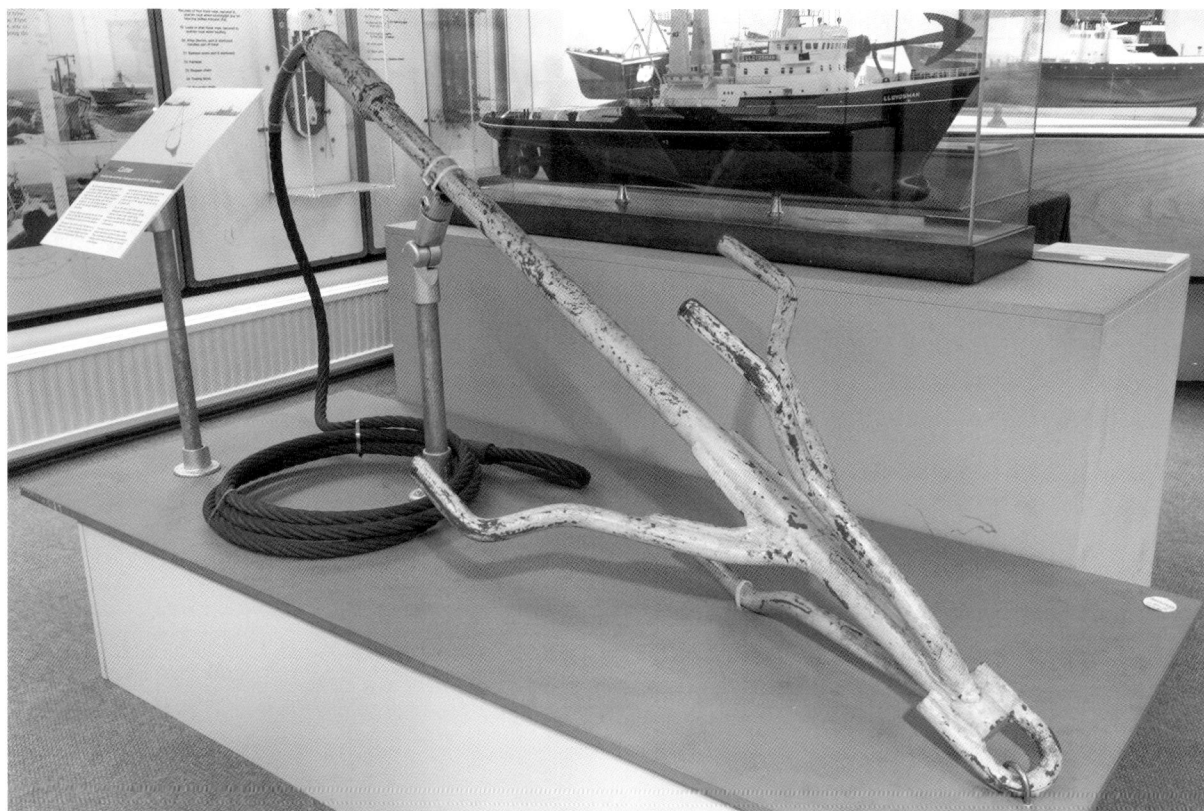

The highly effective Icelandic secret weapon, the 'Trawl Cutter'. Devised in the late 1950s by two Icelandic blacksmiths, it was fitted on all the Icelandic gunboats but not used until the Second Cod War (1972–73). This example was gifted to Hull Maritime Museum by The Reykjavík Maritime Museum in 2017. (Courtesy of Maritime Museum, Hull)

The Off-Shore Tapestry

Although routine FP duties had been undertaken during the Icelandic disputes, Britain's membership of the European Community and its vastly extended exclusion zone required a thorough reappraisal of security arrangements during the early 1970s. The catch-all term 'The Off-Shore Tapestry' entered the vocabulary. The distant-water fishing industry was in its death-throes as ever more countries were declaring exclusion zones. However, at the same time Britain's much-enlarged EEZ itself required policing. A new threat had emerged when Warsaw Pact countries began to despatch large modern trawlers supported by 'mother-ships' to British waters, where they were permitted to operate under licence. Reports of Proceedings of the time show how much attention was being paid by the COs of FP vessels to the whereabouts of the mother-ships and their 'brood' and whether their activities were within the law.

Although the 'Ton' class still constituted the backbone of the coastal protection force, by the mid-1970s these vessels were ageing, and various replacement options were explored. The expensive 'Large Patrol Craft' of the 'Bird' class were described as 'vile' in heavy seaways, while plans to employ the private-venture fast attack boat *Tenacity* and the new hydrofoils were discarded as the age-old need for a robust design capable of dealing with most sea states remained paramount.[14]

The remit of the fishery protection task became more complex with the additional responsibilities associated with the security of the burgeoning oil and gas industry, which was mainly concentrated in the North Sea. The term 'Offshore Patrol Vessel' (OPV) started to be used instead of the narrower fishery protection vessel nomenclature. Clearly there was an urgent requirement for a class of ship capable of coping with this new task. By early 1976 there proved to be sufficient unspent funds available from the previous year's estimates for the MoD to order five craft based on an 'off-the-shelf' design.[15]

The result was the 'Island' class, a diesel-engined 925-ton ship with a speed of 16 knots and a theoretical range of 7,000 miles at 12 knots. The design was closely based on the Scottish Fisheries Department vessel *Jura*, which was subsequently loaned to and trialled by the Royal Navy in 1975–76. The design of *Jura*, in turn, had been influenced by the excellent sea-keeping qualities demonstrated by the 573-ton *Switha* that had provided prolonged service to the Scottish Department in the immediate postwar period.

Owing to the importance of the oil and gas rigs to the economy in the face of the difficult balance of payments situation the UK faced in the mid-1970s, the 'Island' Class was subjected to prolonged, mostly negative criticism in the media and even in Parliament. One backbencher was notably dismissive: 'These vessels are certainly slow, having a maximum speed of 16 knots. They are equipped with only one Bofors gun. I should describe them as Donkey class ships.'[16]

Most of the opprobrium centred on their lack of speed. It was argued in some quarters that the 'Islands' could be outstripped by modern Warsaw Pact trawlers. Despite the defence of the 'Island' class mounted by the MoD, that speed was not *the* determining factor and that endurance and sea-keeping qualities counted for more, these objections seemed to stick. Likewise, the lack of any airborne facilities continued to be perceived as a shortcoming. Eric Grove observed: '… it was recognised that the "Islands" were something of a "quick fix", and in the late 1970s Naval Staff Target (NST) 7040 was drawn up for a more capable and flexible vessel with helicopter facilities that would displace almost 1,500 tons.'[17]

The policing of the new extended fishing limit was described in 1977 as a 'two tier system'. Officially, it was conceded that this reflected more a recognition of the limitations of the available vessels than an acceptance that it was the best way of achieving the RN's current and future protection patrol duties. Therefore, in the immediate future, the 'Tons' would be used exclusively within the twelve-mile inshore zone while the 'Islands' would be assigned to the further reaches of the EEZ. Describing the general unsuitability of the old minesweepers for fishery protection, it was stated: 'In addition to some erosion of their War Role efficiency caused by this type of employment, these ships are barely adequate for the tasks because their designed endurance and seakeeping qualities were not intended for prolonged open water patrol activities'.[18] It was concluded that, in the longer term, a 'one-tier' system was the better option, with a single type undertaking the total patrol task. This would result in greater operational flexibility, thus (crucially) leading to a reduction in the number of vessels required.

HMS *Orkney* dressed overall in Portsmouth Naval Base in June 1995. The 'Island'-class off-shore patrol vessels (OPVs) were the first purpose-built fishery protection ships. The design was based on the Scottish Fishery Board vessel *Jura* and was criticised for being a hastily arranged 'off the shelf' response to the demands of policing the hugely expanded economic exclusion zone of the 1970s. (Author's collection)

The chosen option to meet the one-tier requirement was the 'Castle' class (OPV Mk 2). It was initially envisaged that six ships would be ordered for the Royal Navy, and the design was also aimed at serving the export market. Although publicised as having been developed by Hall Russell and Co Ltd, who had been responsible for the 'Island' class, the detailed ship design was in fact undertaken by a team from the Royal Corps of Naval Constructors under the leadership of David K Brown. Brown researched his subject thoroughly. He concluded that it was unprofitable to put effort into making improvements to the 'Island' design. Consultations with Navy personnel convinced his team that high speed should not be the dominant factor; in reality '… the required success rate in catching poachers was lower than that in full-scale naval warfare. A moderate number of arrests and convictions would make illegal fishing unprofitable'. Also, the number of trawlers capable of speeds exceeding 12 knots was very small. Although there was pressure to deliver a 25-knot ship, in the end 19.5 knots was accepted.

Great emphasis was placed on tackling the problems of discomfort and nausea caused by the rough seas in which the 'Castles' were expected to operate. By producing a longer ship – 266 feet (81m) as opposed to 195 feet (59.5m) for the 'Island' class, Brown was able to incorporate a flight deck aft capable of accommodating a large helicopter and to bring the living and working spaces to the centre, where ship motion was least pronounced. The resulting profile was distinctive and unusual, with a long, flared prow, high freeboard and superstructure concentrated amidships.[19]

Despite frequent references made during the period 1980–85 to further orders for 'Castle'-class hulls Nos 3-6, none was forthcoming. Disagreements about capital cost sharing between the MoD and the other Government departments were at least partially caused by prevarication. In the meantime, HM Ships *Leeds Castle* and *Dumbarton Castle* were both sent to the South Atlantic in 1982 as part of 'Operation Corporate', where they successfully undertook the role of dispatch vessels, and one was later retained as the permanent guardship. A rather bald statement made in the House of Commons in June 1985 brought to an end plans for the 'one-tier' fishery protection ship as a long-term replacement for the 'Ton' and, eventually, the 'Island' classes: 'OPV3 has not secured a place in the forward defence programme when set against competing priorities for the Royal Navy and for the other services'.[20]

In 1984 Michael Jopling, now Minister of Agriculture, Fisheries *and* Food (MAFF), mounted a powerful argument in favour of privatisation. He proposed that on the grounds of cost-effectiveness, the entire fisheries protection operation should be taken over by a commercial enterprise. 'Civilianisation' would, he argued, provide up to 200 additional merchant marine jobs in the private sector and more than £27.5 million of prospective orders for the shipbuilding industry, as opposed to roughly half that if MoD naval contracts were awarded instead. It was estimated that roughly five and a half million pounds would be saved.

Jopling lost his campaign on that occasion; the Government of the day concluded that the seaborne role should remain with the Royal Navy. However, it was decided that the RAF contribution of Nimrod LRMP aircraft would be replaced by lighter, less expensive civilian aircraft.

During the remaining years of the 20th century the problem remained of the inability of the RN to meet its schedule of providing sufficient ship days on fishery protection duty to achieve an agreed inter-departmental target. For example, it was pointed out that in 1995 a total of 229 ship days of patrolling had been lost due to bad weather, caused either by the FP vessels being unable to leave port or being at reduced effectiveness, which made boardings impossible. The 'Ton' class MCM replacements, the 'Hunt' Class, were considered slightly more susceptible to bad weather than the 'Islands' and generally performed less well than the OPVs, but both were considered superior to the newly-built fleet minesweepers of the 'River' class.

The demands of the Off-Shore Tapestry had been a struggle for the Royal Navy throughout the last quarter of the century. The situation was exacerbated by the requirement to spend less money on defence owing both to the parlous state of the UK economy in the early years

HMS *Blackwater*, a 'River'-class minesweeper, photographed here off Northfleet in Kent in 1993, served briefly with the Fishery Protection Squadron. The 'Rivers' proved less successful than the 'Hunt' class in this role. (Author's Collection)

of the period, and later by the effects of the 'peace dividend' at the end of the Cold War and the subsequent cutbacks in defence spending. Sea fishing was also an industry in decline; a watershed had been reached in 1984, when the UK became a net importer of fish, but in fact the downward spiral had been evident since the end of the Second World War.

A fine view of HMS *Leeds Castle* entering Portsmouth Harbour in April 2005. Arguably the most innovative indigenous design of her kind, neither *Leeds Castle* nor her sister *Dumbarton Castle* saw much service in the home waters OPV role, being employed instead on Falkland Islands guardship duties. (John Jordan)

Tyne (2003)

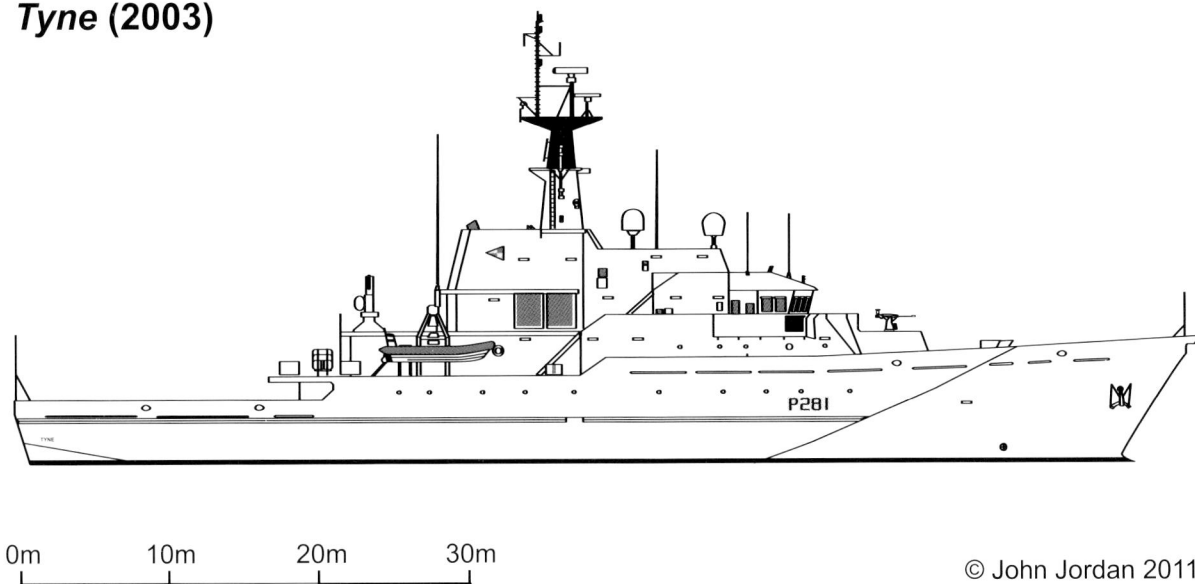

0m   10m   20m   30m

© John Jordan 2011

In 2000 Vosper Thornycroft won a competitive tender to finance the design and build of what became known as the Batch 1 'River' class. The fine bow with its prominent bulbous forefoot and a larger full-load displacement (1,700 tons as opposed to 1,250 tons in the 'Island' Class) were both aimed at improving durability in heavy weather. These factors took precedence over the need for an aircraft hangar, increased armament or speed. The three Batch 1 ships are still in service.

Shrinking Responsibility for an Industry in Decline

By 2000 the ageing and apparently increasingly unreliable 'Island'-class OPVs were in the process of being sold, principally to the Bangladeshi Navy. Vosper Thornycroft (VT), in May 2001, won a circa £60 million contract to construct, lease and support three new 'River' Class OPVs, to be named HMS *Tyne*, *Severn* and *Mersey*, to replace the 'Islands'. The contract introduced a novel leasing arrangement whereby the shipbuilder financed the design and build of the new ships and chartered them to the Navy for an initial five-year period. A second and key element, negotiated under a separate contract, was a Contractor Logistic Support package guaranteeing a minimum level of operational availability. 'This arrangement effectively committed VT to making each River Class vessel available for 320 days each year compared with around 160 days for the ships they were replacing'. MAFF's major bone of contention regarding its contract with the Royal Navy at the end of the previous century had been the latter's inability to achieve its target number of days on FP duties due to vessel breakdowns and shipyard maintenance overruns. The new plan also serves to explain why it was possible to reduce the number required to just three ships.[21]

The resulting 'River' class was a simple vessel focused on low-intensity constabulary missions. In keeping with the original requirement to confine the duties to coastal waters, the ships were constructed to commercial shipbuilding standards with, as Conrad Waters describes, 'a minimal overlay of Ministry of Defence specifications in areas such as damage control and stability'. The 'Rivers'

were fitted with a fine bow with a prominent bulbous forefoot and had a higher full-load displacement in comparison with the 'Islands' (1,700 tons as opposed to 1,250 tons), both features being aimed at improved durability in heavy weather.

Interestingly, there are no marked differences between the machinery fit of the 'River' class and that of its predecessor. Both were fitted with MAN (formally Ruston) diesels, albeit with different specifications, that provided sustained speeds of 16 knots in the 'Islands' and a marginally faster 16.5 knots in the three 'River'-class ships. It is of note that the range of the 'Island' class at 12 knots was 11,000 nautical miles whereas the 'River' Class attained a much-reduced 7,800nm at the same speed. The bow thruster fitted to the three VT boats was aimed at improved manoeuvrability. A major criticism levelled at the 'Island' class related to its minimal weaponry. However, the philosophy had not changed in the intervening 25 years; the 'River' class was fitted with a single 20mm gun, although provision was made for the installation of a medium-calibre gun if that was deemed necessary at some future stage.[22]

If the Batch 1 'River' class represented the more austere end of the OPV 'spectrum', the same cannot be said of the Batch 2 ships which followed, the initial contract for which was signed by BAE Systems in August 2014. The five 'River 2' ships are faster and have a considerably higher displacement than their predecessors: 2,000 as opposed to 1,677 tonnes. The increased length, just shy of 300 feet (90m), reflects their capability to operate a large helicopter from the flight deck aft.

It was originally proposed that these vessels would replace the Batch 1s on a one-for-one basis.

HMS *Mersey* photographed in 2005. Vosper Thornycroft (VT) was contracted in May 2001 to construct, lease and support three new OPVs of the 'River' class. A package was agreed that included an all-important Contractor Logistic Support arrangement guaranteeing a minimum level of operational availability, something the RN had been unable to achieve in the 1990s with its previous generation of protection vessels. (John Jordan)

Consequently, in May 2018, HMS *Tyne* entered Portsmouth ahead of her pre-planned decommissioning but, confusingly, by July the ship was reportedly still flying the White Ensign and therefore still in active service. Subsequently, an official announcement of the retention of the three Batch 1 'River' OPVs was made towards the end of 2018.

The Batch 2s are a reflection of the changed priorities brought about principally by Britain's decision to leave the European Union. The political argument is that the resulting independence enables the country, through its navy, to exercise a more visible presence and influence across the world that will, in turn, serve to boost the economy. This means that these ships are not currently engaged in patrol or fishery protection duties in home waters. Further, it explains the retention of the Batch 1 vessels.

The decline in the UK fishing industry has meant that fishery protection now forms only a part of the duties performed by the RN ships assigned to the task. In October 2021, the Office for National Statistics calculated that the fishing industry contributed around 0.03 per cent of the total UK economic output. Additionally, the FP task is now shared with a civilian body, the Maritime Marine Organisation (MMO), which contributes its own personnel, ships and aircraft. The MMO also provides a vessel monitoring service through its headquarters in Newcastle. The Royal Navy is therefore reduced to a subsidiary role, and a proportion of its time is spent on other constabulary-related duties such as border and smuggling patrols and routine shadowing of foreign naval vessels.

Currently, it is forecast that the Batch 2 'River' class will continue its overseas service for a further five to six years before being withdrawn to home waters, being replaced possibly by the new Type 31 frigate; this will in turn lead to the Batch 1 'Rivers' being withdrawn. It has long been argued that the Royal Navy's ability to undertake the fishery protection role cannot be satisfactorily replicated by a civilian organisation. Whether this view will withstand the inevitable financial pressures that will be exerted on defence spending during the next decade and allow the Batch 2 'River' class to continue the centuries-old duty of protecting the nation's fishery is another matter.

Editor's Note:

Jon Wise's book *The Royal Navy and Fishery Protection: From the Fourteenth Century to the Present* was published by Seaforth in August 2023; price £25.00.

HMS *Forth* 2020

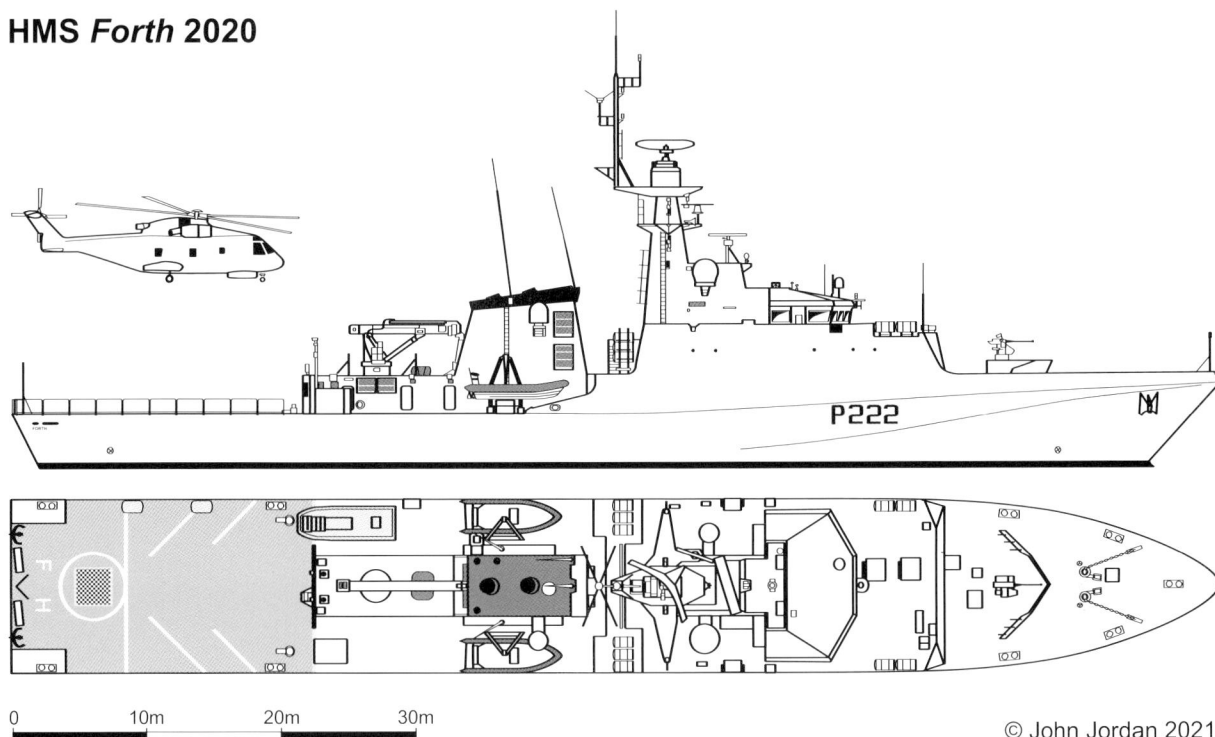

0   10m   20m   30m

© John Jordan 2021

The OPVs of the Batch 2 'River' class were originally intended to replace the Batch 1 vessels in a rapidly diminishing fishery protection task that would be supplemented by other 'constabulary'-type duties. The design is derived from an original 2007 VT 90-metre OPV design and the Brazilian Navy's *Amazonas* class, built by VT Shipbuilding (now BAE Systems) at their Portsmouth facility. *Forth* (P222) and her four sisters were contracted to BAE Systems and built on the Clyde.

By the time the five Batch 2 'River' class ships, including HMS *Tamar* pictured here, had commissioned in mid-2021, times and circumstances had changed, and these vessels are now allocated to what might be described as 'colonial-era sloop' duties in various overseas locations as part of the Government's global 'presence' mission in a post-Brexit world. At present, the Batch 2 'Rivers' are scheduled to take over what will remain of the Royal Navy's historic fishery protection task during the latter years of the 2020s. (Conrad Waters)

Appendix

A comparison of the characteristics of some of the vessels mentioned.

HMS *Galatea* (date mentioned in text: 1896): *Orlando* class armoured cruiser
Builder: Robert Napier, Govan; *Completed* 1887; *Displacement* 5,620 tons; *Dimensions*: 300 x 56 x 24ft (to nearest foot); *Machinery*: 2 triple-expansion steam engines, coal fired; *Speed* 18 knots; *Complement* 484; *Principal Armament*: 2 x 9.2in, 10 x 6in guns.

HMS *Skipjack* (1905): *Sharpshooter* class torpedo gunboat
Chatham Dockyard; 1891; 747 tons; 252 x 27 x 11ft; 2 x triple expansion; 19 knots; 91 crew; 2 x 4.7in.

HMS *Godetia* (1923): 'Flower' type sloop
Connell, Scotstoun; 1916; 1,250 tons; 268 x 33 x 11ft; 1 x 4-cylinder triple expansion; 16.5 knots; 85–116 crew; 1 x 4in.

HMS *Hastings* (1937): *Hastings* class sloop
Swan Hunter, Tyneside; 1930; 1,045 tons; 266 x 34 x 12ft; Parsons geared turbines (oil-fired); 16 knots; 100 crew; 2 x 4in.

HMS *Allington Castle* (1946): 'Castle' class corvette
Fleming & Ferguson, Clyde; 1944; 1,060 tons; 252 x 37 x 13ft; 1-shaft VTE; 16½ knots; 120 crew; 1 x 4in.

HMS *Hound* (1958): *Algerine* class minesweeper
Lobnitz, Clyde; 1942; 950 tons; 225 x 35 x 15ft; 2-shaft VTE; 16½ knots; 85 crew; 1 x 4in.

HMS *Palliser* (1963): Type 14/*Blackwood* class frigate
Stephen, Clyde; 1957; 1,180 tons; 310 x 35 x 15ft; 1-shaft geared steam turbines; 27 knots; 140 crew; 3 x 40mm.

HMS *Wotton* (1976): 'Ton' class coastal minesweeper
Philip, Dartmouth; 1957; 360 tons; 140 x 29 x 8ft; 18-cylinder Napier Deltic diesel; 15 knots; 29 crew; 1 x 40mm.

HMS *Leeds Castle* (1981): 'Castle' class offshore patrol vessel
Hall Russell, Aberdeen; 1981; 1,427 tons; 265 x 38 x 11ft; 2 x Paxman diesels; 20 knots; 40 crew; 1 x 40mm.

HMS *Tyne* (2002): Batch I 'River' class offshore patrol vessel
Vosper Thornycroft, Southampton; 2002; 1,700 tons; 262 x 45 x 23ft; 2 x MAN 12RK 270 diesels; 20 knots; 30 crew, 1 x 20mm.

HMS *Tamar* (2021) Batch 2 'River' class offshore patrol vessel
BAE Systems, Clyde; 2020; 2,000 tons; 297 x 43 x 12ft; 2 x MAN 16V 28/33D diesels; 25 knots; 50 crew; 1 x 30mm.

Endnotes:

[1] 'The Naval Service of the Cinque Ports', in NAM Rodger, *Essays in Naval History, From Medieval to Modern*, Ashgate Publishing (Abingdon, 2009), 646.

[2] The National Archives (TNA), ADM116/100, Faroese and Icelandic Fisheries, 1899. HMS Galatea – remarks by Captain Cross, 19 June 1899.

[3] Robert Gardiner (ed), *Conway's All the World's Fighting Ships, 1860–1905*, Conway Maritime Press (London, 1979), 87, 89 & 113.

[4] *Hansard House of Commons Debates*, Vol 2, 24 March 1892: 'Destruction of Nets on The North Sea Fishing Grounds'; TNA, TS 32/63 Fishery Protection Vessels: Use of Force, 1912–1925.

[5] TNA, T1/11012: Report: Interdepartmental Conference: Protection of Sea Fisheries, 1908.

[6] TNA, ADM1/8553/74: Paper by Captain Dugmore RN, 1919.

[7] *Ibid*.

[8] Randal Gray (ed), *Conway's All the World's Fighting Ships 1906–1921*, Conway Maritime Press (London, 1985), 94; O Parkes & F McMurtie (eds), *Jane's Fighting Ships 1924*, David and Charles Reprints (Newton Abbott, 1973), 81.

[9] TNA, ADM 116/2178: Remarks on Recent Patrol of HMS "Godetia".

[10] TNA, ADM 116/3361: Report: Search for the missing trawler "Amethyst" H.455, 6 February 1937.

[11] *Hansard, House of Commons Debates*, Vol 283, 23 November 1933: 'Fishery Patrol Vessel (Speed)'.

[12] DK Brown & George Moore, *Rebuilding the Royal Navy: Warship Design since 1945*, Chatham Publishing (London, 2003), 81.

[13] The 'warp cutter' proved very effective: a large number of trawlers lost their gear as a consequence. HMS *Salisbury*'s Admiralty Standard Range 1 diesels made her unique among the other RN frigates employed off Iceland; see TNA,

ADM 330/110, HMS Salisbury and Tyr, 5 incidents, 1 April, 1976; Commander HM White: Report of Proceedings, 9 April, 1976: 'With her diesel propulsion and high minimum shaft revolutions, SALISBURY has good brakes and was able on nearly all occasions to prevent TYR from crossing her stern. However TYR's subsequent accelerations and use of wheel was more than SALISBURY could match, and this movement had to be contained by anticipation'.

[14] Capt JE Moore, *Warships of the Royal Navy*, Macdonald and Jane's, (London, 1979), 84–86.

[15] *Ibid*, 84.

[16] Hansard, House of Commons Debates, Vol 373, 12 May 1976: 'Exclusive Economic Zone'.

[17] Eric J Grove, *Vanguard to Trident: British Naval Policy since World War II*, The Bodley Head (London, 1987), 333.

[18] MAF 452/9 Fishery Protection: replacement of Royal Navy Ton Class minesweepers, 1977–81. Memo: Conduct of Operations (The One/Two Tier System) 1977.

[19] Utilising his extensive knowledge and experience, David Brown later became widely known for his books and articles on Royal Navy ship design. He wrote about the 'Castle' class design in Brown & Moore, *Rebuilding the Royal Navy*, 13–37, and 'The Design of the "Castle" Class: a Personal View' in John Jordan (ed), *Warship 2006*, Conway Maritime Press (London, 2006), 78–85.

[20] Hansard, House of Commons Debates, Vol 80, Defence Estimates 1985, 13 June 1985.

[21] Cdre Stephen Saunders RN (ed), *Jane's Fighting Ships 2002–2003*, Jane's Information Group Ltd (Coulsdon, 2002), 784; see also Conrad Waters, 'British-Built Offshore Patrol Vessels: Balancing Cost and Capability', in Conrad Waters (ed), *Seaforth World Naval Review 2012*, Seaforth Publishing (Barnsley, 2011), 109.

[22] *Ibid*, 110–112; see also Robert Gardiner (ed), *Conway's All the World's Fighting Ships, 1947–1995*, Conway Maritime Press (London, 1983), 537.

FROM *ORËL* TO *IWAMI*

After the new Russian battleship *Orël* surrendered to the Japanese following the Battle of Tsushima, she was extensively modified and eventually entered service with the IJN as *Iwami*. However, standard reference works offer conflicting information regarding her characteristics. **Stephen McLaughlin** attempts to sort out the discrepancies using contemporary reports from the American naval attachés in Tokyo, recent publications and other sources.

Anyone looking for accurate information about the Japanese 'predreadnought' *Iwami*, ex-Russian *Orël*, is soon confronted by a series of contradictions. Table 1 summarises the data given in the major published sources; while almost all agree on the ship's displacement, there are considerable differences in her other parameters. It is possible, however, by using a variety of sources to get a clearer picture of *Iwami*'s particulars. However, first some background information on the ship is in order.

Tsushima

The keel of the Russian predreadnought *Orël* (pronounced OrYOL, 'Eagle') was laid at the Galernyi Island Shipyard in St Petersburg on 7 November 1899.[1] She was one of five battleships of the *Borodino* class intended to bolster Russia's naval forces in the Far East to counter the growing Japanese fleet. She was launched on 19 July 1902, and in May 1904 was transferred to Kronshtadt, the island naval base about 14nm west of St Petersburg, for fitting out. By that time Russia was at war with Japan, and *Orël* was earmarked for the squadron being formed to reinforce Russia's besieged Port Arthur Squadron. Her completion and trials were rushed, and she departed with Vice Admiral Z P Rozhestvenskii's Second Pacific Squadron on 15 October 1904.

The sad tale of Rozhestvenskii's voyage to the Pacific is well known: after various trials and tribulations his squadron arrived at the Tsushima Strait on 27 May 1905, where the Japanese Combined Fleet under Vice Admiral Tōgō Heihachirō lay in wait. The four battleships of the *Borodino* class made up the First Battleship Detachment (the fifth sister, *Slava*, could not be readied in time to sail with Rozhestvenskii's squadron). The flagship *Kniaz Suvorov* led the battleline, followed by her sister-ships *Imperator Aleksandr III*, *Borodino* and finally *Orël*, in that order.[2] One by one *Orël*'s sisters fell out of line, wrecked by the concentrated fire of Tōgō's battleships; all three eventually sank. Despite sustaining serious damage *Orël* was still afloat as evening drew on; she joined the remnants of the Russian squadron, now led by Rear Admiral N I Nebogatov, and set course for Vladivostok. However, the next day Nebogatov's five ships were intercepted by the almost completely intact Japanese fleet and, after a brief exchange of fire, the Russians surrendered – except for the small cruiser *Izumrud*, which used her speed to escape but later ran aground on Russia's Pacific coast, where she was destroyed by her crew to prevent her capture.

Commander Tōgō Kichitarō of the battleship *Asahi* led the Japanese prize crew that boarded *Orël*. They found her topsides a mass of wreckage.[3] Their survey estimated that the ship had been struck by approximately 55 shells of various calibres between 6in and 12in; the muzzle of the forward turret's left-hand 12in gun had been chopped off by a shell hit, the roof of the after turret had been dented, preventing one gun from elevating, most of the 6in twin turrets had been damaged to one degree or another, almost all of the light guns were out of commission, the ship's unarmoured side was holed in several places, and many shipboard systems had broken down. The ship had also taken on a considerable amount of

Table 1: **Comparison of Characteristics**

Source:	*Naval Annual*	*Jane's*	Gordon & Watts	Jentschura *et al*	Lengerer & Ahlberg
Date/page(s):	1914/p236	1914/p224	1971/p33	1977/p21	2019/pp274–276
Displacement:	13,516 tons	13,566 tons	13,516 tons	13,516 tons	13,516 tons
Main battery:	12in	12in/45	12in/45	12in/45	12in/40
	N/A	Obukhov	Obukhov	Armstrong	Obukhov
Torpedo tubes:	3 (2 submerged)	2 x 18in submerged	2 x 18in submerged	2 x 18in submerged	3 x 15in (2 submerged)*
Belt armour (max):	9in	7.5in	7.5in	9in	7.64in
Boilers:	Miyabara	Belleville	Belleville	Miyabara	Belleville

*Replaced by 2 x 18in submerged in 1911.

Orël in Kronshtadt roadstead; the photo was probably taken in September 1904, shortly before Rozhestvenskii's squadron departed for the Far East. Note the lofty superstructures, the tall funnels, the fighting tops on the masts, the embrasures for 75mm guns below the forward 12in turret, and the casemate for the 75mm battery close to waterline amidships – all features that the Japanese would eliminate. (Naval History and Heritage Command, NH 45853)

Orël after the Battle of Tsushima; the photo was taken at Maizuru, Japan, on 3 June 1905. Some of the damage she received during the battle can be seen in this photograph: the truncated left-hand 12in gun of the forward turret and holes in her unarmoured sides. Less visible is the internal damage: the 75mm guns in the forward and after casemates were all out of action, and most of the 6in turrets were jammed in train or otherwise damaged. (NHHC, NH 84789)

water, but her machinery and steering gear were intact, and she was able to steam to Maizuru (although at one point *Asahi* had to take her in tow), arriving on 30 May. The work there was intended only to make her seaworthy enough to reach the Kure Navy Yard safely; on 6 June, while still at Maizuru, she was renamed *Iwami* after a province in western Honshu. She arrived at Kure on 29 July 1905.[4]

A Japanese sentry patrols near *Orël's* forward 12in turret. The muzzle of the left-hand gun was chopped off by a hit, probably by an 8in shell. The broken-off piece reportedly was found on the starboard bridge wing, lodged in a signal locker. (NHHC, NH 66269)

This hole was just forward of *Orël*'s starboard midships 6in turret in the unarmoured side. As was the case with many of the Japanese shells, it exploded on impact, tearing ragged-edged openings in the plating. (NHHC, NH 66262)

Typical wreckage caused by the extreme violence of the Japanese Shimose explosive. This photograph shows the port side of *Orël*'s forward bridge. In the right foreground is the breech of a 47mm (3pdr) gun, and behind it is another; almost all these guns had been knocked out of action by the end of the battle. (NHHC, NH 66265)

The American Reports

It is at Kure that the American naval attachés first make their appearance. At this point Japanese-American relations were still on a friendly basis – the United States had been largely sympathetic toward Japan during the war and, as Evans and Peattie noted, 'there was in 1907 no clash of fundamental interests between Japan and the United States nor any indication that either the Japanese or the American government desired confrontation'.[5] It was only later that Japanese-American relations would be poisoned by distrust and hostility.

Thanks to this generally cordial atmosphere two American naval attachés, Commander Frank Marble and Commander John Allen Dougherty, were given the

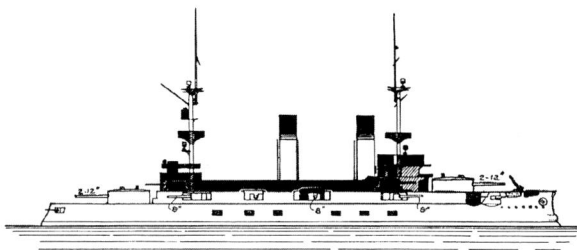

A crude but helpful contemporary sketch of *Iwami* published in the US journal *Scientific American*, 4 April 1908. The parts of the ship rendered in black show the structures removed during reconstruction. The caption for the original reads: 'Before her capture from the Russians the "Iwami" was a high-freeboard ship with most of her guns carried 32 to 34 feet above the water …. After her capture the Japanese, as the result of their experiences in the war, cut down the freeboard amidships by 8 feet; abolished six of the lofty turrets; lowered the secondary battery to the main deck; cut off 20 feet of the smokestacks, and removed the flying bridges, the fighting tops, and the whole of the superstructure'. [In fact the funnels had been shortened by only about 6ft.]

opportunity to inspect *Iwami*.[6] In their reports quoted below, slight changes have been made to punctuation, etc, for clarity, while additional information and corrections have been inserted in square brackets.

Marble visited *Iwami* on 1 October 1905 and submitted a report on his impressions, dated 11 October; he noted that:

The [Japanese] Admiralty have decided to remodel the *Orel* (which I shall hereafter call by her Japanese name *Iwami*), in order to increase her metacentric height and improve the distribution of her main battery and armor, judging that since she is a comparatively new ship it is worth the expense. This of course will prolong the original estimate of three months to recommission her.[7]

The 'original estimate of three months' suggests that initially the IJN planned to make minimal changes to the ship. But now a much more extensive reconstruction was underway. Marble went on to report that:

The change in [*Iwami's*] appearance, now that she is cleaned and stripped, is extraordinary. All four 12" guns have been taken out (the turret tops were removed and then laid back), and all the 6" guns, and all the 6" turrets except the port forward one [have been removed]. Most of the superstructure has been cut away, and all boats etc hoisted out, and she is nearly empty of coal. The ship now floats at approximately her designed draft! The marks of barnacles on her sides show that when she went into action she was at least 4ft deeper in the water. The waterline came just below the bow and stern torpedo tubes. By the marks on the stem she drew 30ft forward. The main armor-belt was entirely and the upper belt half submerged. Looking at her side now, it is easy to see how her

A close-up view of the damaged left-hand 12in gun of the forward turret. The inner 'A' tube and its shrunk-on supporting tubes have been sheered about 7ft 6in from the muzzle, making it almost impossible to repair. It was replaced by a similar gun from another captured Russian battleship, *Retvizan*, which entered service in the IJN as *Hizen*. (NHHC, NH 66268)

sister ship was sunk by gunfire penetrating above the armor. [This is a reference to *Borodino*, sunk at the very end of the daylight action by a shell hit that triggered an internal explosion.]

The plan for remodelling is approximately as follows: The 12" guns will be replaced. Instead of the twelve 6" guns there will be six 8" mounted singly in the broadside turrets. The after pair of broadside turrets will be brought down one deck, to the same level as the midship pair; this will of course interfere with the theoretical end-on fire

astern of the midship turrets. The whole two-storeyed superstructure amidships will be cut away, together with most of the top-hamper about the masts – particularly the foremast – leaving the upper deck clear from the stern to about the forward pair of broadside turrets, to which the topgallant forecastle will extend. [...] The change will make a much more roomy upper deck .... There will of course be some sort of a superstructure or light boat deck amidships, but it will be at least one deck lower than the present one ....

The foregoing description was given me, a little at a time, in conversation, with the remark that the design was not fully worked out – or decided yet. It may, therefore, not be entirely accurate. The object of the reconstruction of course is to reduce weights and lower the centre of gravity. The *Iwami* will undoubtedly be a better looking as well as a more efficient ship than her step-mother the *Orel*.

The next report in the file was submitted by Dougherty, Marble's replacement as attaché, and is dated 9 July 1907.[8] He had requested information about the ship on 4 May, and finally received a detailed description from 'the Imperial Japanese Navy Dept' on 1 July. As he noted, 'Several apologies were made for the delay in furnishing this information', and that it 'is to be considered CONFIDENTIAL'. His report is quoted below with only minor deletions:

Armament

The two twelve-inch turrets are practically all that remains of the old battery. The six pairs of 6" guns in turrets are replaced by six 8" guns mounted singly, not in turrets; and both the forward and after pair are brought down one deck to the level of the midship pair – that is to say, all six

Iwami (ex-*Orël*) following her reconstruction. This broadside view clearly illustrates the changes in the ship's appearance: the reduced and simplified superstructures, the removal of the fighting tops, and the central 8in gun on a shielded pivot mounting amidships. Somewhat lost in the dark paint scheme are the 8in casemates fore and aft of the central gun. (NHHC, NH 45832)

6" [*sic* – in fact 8in] guns are on the upper deck. [*Orël's* 6in guns were used to replace those of the battleship *Tango* (ex-*Poltava*).[9]] The forward and after pair of 8" guns are protected by separate armored casemates, as shown in the plan, 6" thick outboard and 2" thick inboard. These guns also carry cylindrical shields, 3" thick, which turn in the port openings of the casemate. The midship pair of 8" guns is protected only by shields, which will be – as I remember it – 6" thick in front …. The Naval Constructor at Kure told me that this plan was adopted to save weight. I should have supposed that single turrets for the 8" guns would be lighter than casemates. All the 8" guns secure with their barrels inboard of the rail.

The 12pdr guns [pdr = pounder; these were 3in/76.2mm guns, although the original Russian guns were actually Canet-pattern 75mm guns] are all removed from the main deck, except the aftermost pair, in the Admiral's cabin. The Japanese officers said 'our experience has shown us that it is useless to carry guns so low'. The old main deck battery would have been 12ft above the water amidships, with the ship at her designed draft, as she will float now; it was 4ft lower when the Russians took her into action.

The twenty 12pdrs [*sic* – strictly 75mm] of the old battery are replaced by sixteen [3in], mounted as follows:

Two on the main deck, in the stern;
Two on the upper deck, in the bow (the foremost pair of the old);
Two on the upper deck amidships, abaft the after funnel;

Four on the roofs of the 8" gun casemates;
Two on the after corners of the forecastle;
Two on the after bridge;
Two on the forward upper bridge (on the level of the conning tower);
There are four 47mm on the rail amidships, and four Maxims on the bridges.

Armor
The main and upper belts remain the same as before. I noticed that the upper edge of the main belt amidships came just even with the 27ft draft mark – ie, it will be very little above water even at the designed draft. The upper belt is about 6ft wide – ie, it just reaches the edge of the main deck.

From the top of the upper belt to the level of the upper deck (or just below it), the side amidships – about from the foremast to the mainmast – will be protected by 3" armor. The old plates have been removed and new ones are being made at Kure without openings for gunports. The old armor here was 3" thick, continuous but for the gunports, as shown in Brassey, 1905; not detached casemates 2" thick, as given in Clowes, 1904. ['Brassey' refers to *The Naval Annual*, edited by Thomas, Earl Brassey; 'Clowes' is *The Naval Pocket-Book*, by Sir W Laird Clowes.]

The old 6" gun turret supports were (by actual measurement) 1¼" thick; not 5" or 6" as given by Clowes and Brassey. [Despite Dougherty's 'actual measurement', Russian construction drawings confirm that the thickness

Iwami at Kure, 2 November 1907. In this view the casemates for the 8in guns are clearly visible, as is the middle gun in its shield trained outboard. Several of the 12pdr (3in) guns can also be seen: in the embrasure below the forward 12in turret, atop the forward 8in casemate, and on the starboard bridge wing above. Another detail to note is the anchor-bed for a single anchor; originally it extended farther forward and accommodated two anchors. (Wikimedia Commons)

Iwami (ex-*Orel*)

Sketch of Rig

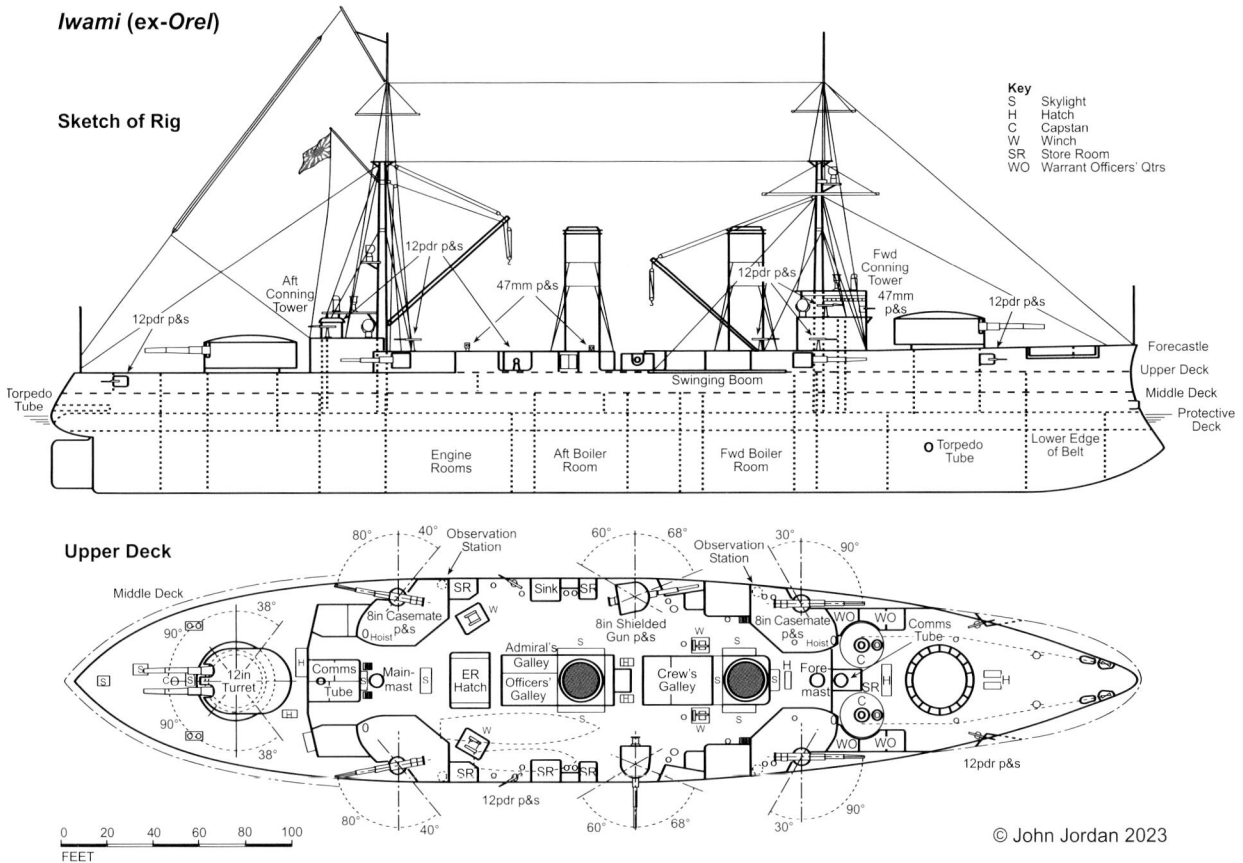

Key
S	Skylight
H	Hatch
C	Capstan
W	Winch
SR	Store Room
WO	Warrant Officers' Qtrs

Upper Deck

© John Jordan 2023

of the ammunition tubes for the 6in turrets was 6in, not 1.75in.] These of course remain, up to the level of the upper deck, as protection for the ammunition hoists of the 8" guns.

Clowes is also in error as to the thickness of decks and bulkheads. The main deck …, which is just above the upper armor belt, is actually 2½" thick, the protective deck, which is just above the water-line amidships and joins the lower edge of the main armor belt, is only 1½" thick on the flat. [In fact the main deck had a maximum thickness of 2in; on the other hand, the 1.5in thickness of the 'protective' (lower) deck is correct.] Why this is thus is a question for the Russian designers.

The fore-and-aft bulkheads below the protective deck to the turn of the bilge on each side are one inch thick (not 4" as stated by Clowes). [The longitudinal bulkheads were actually 1⁵⁄₁₆in thick, composed of two layers.] These will remain.

The conning tower, as shown in the profile, is the same as before. It was impossible to lower it, because it barely looks over the forward turret as it is. There is an after conning tower, 8ft in interior diameter.

Tophamper & Superstructure
Practically the whole superstructure has disappeared. The topgallant forecastle now extends aft to the foremast; from there to the mainmast is a straight bulwark, cut away or recessed for the [8"] gunports, continuing the

upper sheer-line of the ship. Formerly there was a solid deck on this level (topgallant forecastle) extending well abaft the mainmast, with a high superstructure above that. Now, on the upper deck level, abaft the break of the fore-castle, there is properly speaking no superstructure, but the 8" gun casemates and a series of light deck-houses amidships and along the sides, as shown in the plan. Across the top of the bulwarks (on the level of the forecastle deck) will be light skid beams for boats. That is, the boats in their cradles have been brought down more than the height of a full deck.

Both the forward and after bridges are one storey lower. Formerly the forward bridge was on top of the conning tower, with a pilot-house and flying bridge above that. Now the bridge is on the level of the base of the conning tower, with only a compass and range-finder platform on top.

The funnels have been shortened 6ft …. The military tops are removed, and the heavy yards replaced by lighter ones. The anchor-beds are made shorter, to stow only one anchor on either bow, instead of two.

As the Russian Naval Attaché said, when he visited the *Iwami* at Kure, 'I would hardly recognize the old *Orel*'. [The Russian attaché was Lieutenant A N Voskresenskii.[10]]

It is now estimated that it will take nearly a year longer (from the middle of October) to complete the remodelling of the *Iwami*. It could of course be done sooner, in an emergency; as it is, the work is going on steadily but not hastily, because of the congestion of work in the dockyards on repairs to other Russian prizes and on new ships building, and, most of all, because of lack of money.

The blueprints mentioned by Dougherty have been redrawn by the Editor, and these give a good overview of the ship's configuration when she entered service with the IJN. The final report in the file is a brief note on the ship's recent trials:

> The *Iwami* (ex-*Orel*) had a trial test of her guns off Suwo Nada September 26th [1907]. The tests were reported satisfactory. She has made 18 knots, and I was told at the Navy Department that it is hoped to get out of her 18.5 knots. She will be put in the 1st Squadron.[11]

Evaluating the Information

Dougherty's summary of the reconstruction – despite his errors regarding armour thicknesses – combined with an extensive list of particulars and the plans he obtained from the Japanese Naval Ministry, give us a fairly complete description of *Iwami*. Taken together with other sources, it is possible to fill in other blanks. Table 2 gives the ship's most likely characteristics before and after her reconstruction by the IJN.

One interesting point concerns the broken left-hand gun of the forward turret. Repair was impractical, given that the inner 'A' tube was shattered some distance from the muzzle. There were probably no spare 12in barrels at Port Arthur, since that base lacked significant repair facilities. Therefore the only replacements available, at least initially, were mounted in the other captured Russian battleships armed with the same model 12in/40 gun, *Poltava* and *Revizan*. These had been salvaged at Port Arthur and eventually entered service with the IJN as *Tango* and *Hizen* respectively. According to reports from Lieutenant Voskresenskii, the Russian naval attaché, *Iwami* received one of the guns from *Hizen*, which gun was probably replaced in turn by the least-worn of *Tango's* guns.[12] The latter ship was reportedly rearmed with Russian-pattern 12in guns manufactured at the Kure Naval Arsenal. These guns were still in place when the ship was sold back to Russia in 1916.[13]

Some confusion still surrounds *Iwami's* armour protection. Archival drawings show beyond doubt that the belt had a maximum thickness of 7.6in (194mm).[14] But while Lengerer and Ahlberg (following Ishibashi) agree that her original belt armour was 7.6in, they go on to say that the Japanese replaced it with 8.9in (225mm) plates; moreover, they state that the middle deck was increased from 2in to 4in, the conning tower from 8in to 11in, and the barbettes from 9in to 11in. These increases in protection seem unlikely given the expense and work (not to mention the extra weight) involved. The American attaché report from July 1907 – shortly before the reconstruction work was completed – states that the 'main and upper belts remain the same as before', and it does indeed seem most probable that *Iwami's* basic armour scheme remained unchanged – but there is as yet no indisputable evidence for this.

Another view of *Iwami*, probably taken shortly before the First World War, again emphasising the radical change in her appearance after her reconstruction. In the background beyond her bow can be seen part of the battleship *Settsu*. (NHHC, NH 101762)

Conclusion

The American reports and recent publications, as well as other sources previously unavailable, help clarify some of the uncertainties that have surrounded *Iwami*'s characteristics, but there is still room for doubt on several issues. One may hope that future research in Japanese sources will clear up the remaining points of contention.

Table 2: **Characteristics**

	Orël As designed & built	*Iwami* Naval Attaché Reports[1]
Displacement:	13,516 tons design (normal) 14,151 tons actual	13,280 tons (normal)
Length:	375ft 4in / 114.4m pp 389ft 5in / 118.7m wl 397ft / 121.0m oa	376ft / 114.6m pp – 397ft / 121.0m oa
Beam:	76ft 1in / 23.2m	76ft / 23.2m
Draught:	26ft / 7.9m (designed) 27–29ft 2in / 8.24–8.9m (actual)	25ft 6in / 7.8m (normal)
Armament:	Four 12in (305mm)/40 (2 x II; 60rpg) Twelve 6in (152mm)/45 (6 x II; 180rpg) Twenty 75mm/50 (20 x 1; 300rpg) Twenty 47mm/43 Two 37mm/23 Four 0.30 (7.62mm) MGs Two 2.5in (63.5mm) field guns Four 15in (381mm) torpedo tubes, Two above-water (bow & stern) Two submerged broadside	Four 12in/40 (2 x II) Six 8in (203mm)/45 (6 x I) Sixteen 3in (7.62cm)/40 (16 x I) Four 47mm/30 Four 6.5mm MGs Two 2in Three 15in (381mm) torpedo tubes, One above-water (stern) Two submerged broadside[2]
Protection:[3]		
waterline belt	7.6/5in[4] amidships, reducing to 6.5/4.8in, 5.8/4.4in and 5.7/4.3in fore & aft	Unchanged
upper belt	6in amidships, reduced to 5.2in, 4.5in, and 4in fore & aft	Unchanged
casemate	3in	
main battery	10in turret sides, 2in roofs, 9in–7in barbettes	Unchanged?
secondary battery	6in turret sides, 1.5in roofs, 6in–5in barbettes	6in casemates, 3in shields (corner guns) 6in shields (middle guns)
conning tower	8in sides, 1.5in roof, 6in tube to middle deck	Unchanged?
decks	2in middle deck reduced to 1.26in inside casemate; 1.56in lower deck (2 x 0.78in)	Unchanged
torpedo bulkhead	1.56in (2 x 0.78in)	Unchanged
Machinery:	Twenty Belleville boilers Two 4-cylinder triple-expansion engines, 15,800ihp	Unchanged Unchanged Unchanged
Speed:	18 knots (design)	18 knots
Endurance:	1,158 tons coal maximum 2,760nm at 10 knots[5]	600 tons normal, 1,200 tons full ?
Complement:	826 to 846 in service	793

Sources:

– McLaughlin, *Russian and Soviet Battleships*, 136–137.
– RG 38, O-12-b, 'Japanese Battleship "Iwami"', 9 July 1907.
– Lengerer & Ahlberg, *Armourclad Fusō to Kongō Class Battle Cruisers*, 274–276.

Notes:

[1] Some information taken from Lengerer and Ahlberg.

[2] Replaced by two 18in (450mm) submerged torpedo tubes in 1911.

[3] The armour protection is based on Vinogradov, *Bronenosets 'Slava'*, 50–52.

[4] The armour plates of the waterline belt tapered from the upper edge to the lower, hence the thicknesses are given in the form upper edge/lower edge.

[5] Coal capacity and endurance figures from Vingradov, *Bronenosets 'Slava'*, 54. Actual bunkerage seems to have varied slightly among the ships of the class.

Iwami's Career

Work began on *Iwami* at Kure on 4 August 1905, and on 12 December 1905 she was classified as a battleship (*senkan*). Her reconstruction was completed on 26 November 1907, and on 5 January 1908 she was assigned to the First Fleet. She underwent hull and machinery repairs at Sasebo in July–October 1908. She was again attached to the First Fleet on 11 May 1909. In February–March 1910 she operated in the South China Sea; on 1 December of that year she was placed in reserve. The following year her original 15in underwater torpedo tubes were replaced by two Armstrong-pattern 18in (45cm) tubes. In 1912 a 9ft rangefinder was fitted on her bridge, and on 28 August of that year she was re-rated as a first-class coast defence ship (*ittō kaibōkan*). On 18 August 1914 she was assigned to the Second Fleet.

On 23 August 1914 Japan declared war on Germany, and *Iwami*, along with other IJN ships, took part in the blockade of the German treaty port of Tsingtao, China. After the German colony surrendered on 7 November she returned to Yokosuka. In early 1915 she was given a major overhaul, which was completed only in October 1916. In January 1918 she was assigned to the Third Fleet's 5th Squadron, and served as the flagship of Rear Admiral Katō Kanji during operations at Vladivostok; these would continue on and off through 1922 as Japan tried to take advantage of the chaotic situation in Russia resulting from the revolution and civil war. After the Japanese withdrawal from Siberia, *Iwami* was demobilised at Kure on 15 May 1922 and stricken on 1 September, being reclassified as a miscellaneous harbour ship (*zatsuekisen*). She was subsequently converted into a target ship and arrived at Yokosuka on 24 March 1924. In July she was reclassified first as an 'abolished' ship (*haikan*), and then on 7 July as a target ship (*hyōtekisen*). On 9 July 1924 she was bombed and sunk west of the Miura Peninsula by aircraft of the Yokosuka Air Group.

Acknowledgements:

The author would like to express his thanks to Lars Ahlberg, who provided valuable information from Japanese-language sources; to Sergei Vinogradov, who sorted out the replacement of the damaged 12in gun; to John Jordan, who redrew the blueprints of the ship; and, as ever, to my wife Jan Torbet for her keen editorial skills.

Sources:

National Archives and Records Administration, Washington DC
Record Group 38, O-12-a: 'Remodeling of the "Iwami"', 11 October 1905.
Record Group 38, O-12-b:
—'Japanese Battleship "Iwami"', 9 July 1907.
—'Ships, Japan, Miscellaneous Notes', November–December 1907.

Published Sources
Alekseev, Mikhail, *Voennaia razvedka Rossii: Ot Riurika do Nikolaia II* (4 vols), Russkaia razvedka (Moscow, 1998).

Evans, David C & Peattie, Mark R, *Kaigun: Strategy, Tactics, and Technology in the Imperial Japanese Navy 1887–1941*, Naval Institute Press (Annapolis, 1997).
Gribovskii, V Iu, *Eskadrennye bronenostsy tipa 'Borodino'*, Gangut (St Petersburg, 2010).
(Viscount) Hythe & Leyland, John, *The Naval Annual, 1914*, William Clowes and Sons (London, 1914).
Ishibashi Takao, *Nihon Teikoku Kaigun zen Kansen 1868–1945*, Vol 1: *Senkan, Junyōkan* [All ships of the Imperial Japanese Navy 1868–1945, Vol 1: Battleships, Battle Cruisers], Namiki Shobō (Tokyo, 2007).
Jane, Fred T (Ed), *Jane's Fighting Ships 1914*, Sampson Low Marston, 1914; reprint edition by Arco Publishing Company (New York, 1969).
Jentschura, Hansgeorg, Yung, Dieter & Mickel, Peter, *Warships of the Imperial Japanese Navy, 1869–1945*, Arms and Armour Press (London, 1977).
Lengerer, Hans & Ahlberg, Lars, *Armourclad Fusō to Kongō Class Battle Cruisers*, Despot Infinitus (Zagreb, 2019).
McLaughlin, Stephen, 'Aboard the *Orël* at Tsushima', *Warship 2005*, Conway Maritime Press (London, 2005), 38–65.
'Sledovat' v Aleksandrovsk ...', *Gangut* No 15 (1998), 115–126.
Sudovoi spisok. 1904 goda, [Russian Naval Ministry?] (St Petersburg [?], 1904?)
Vinogradov, Sergei E, *Bronenosets 'Slava'. Nepobezhdennyi geroi Moonzunda*, Iauza; EKSMO (Moscow, 2011).
Watts, Anthony J & Gordon, Brian G, *The Imperial Japanese Navy*, Doubleday & Company (Garden City NY, 1971).

Endnotes:

1. All dates in this article are according to the modern Gregorian calendar; according to the Julian calendar then used in Russia she was laid down on 26 October.
2. For a detailed account of *Orël*'s part in the battle, see McLaughlin, 'Aboard *Orël* at Tsushima'.
3. The following description is based on Lengerer & Ahlberg, *op cit*, 277–287; see also McLaughlin, 'Aboard *Orël* at Tsushima', 45–58, 62–64.
4. These dates and subsequent information about *Iwami*'s career in the IJN are taken from Lengerer & Ahlberg, 310–312.
5. Evans & Peattie, *Kaigun*, 151.
6. The attachés were identified in the various editions of the *Register of the Commissioned and Warrant Officers of the United States Navy and Marine Corps* (Washington DC: Government Printing Office).
7. RG 38, O-12-a: 'Remodeling of the "Iwami"', 11 October 1905.
8. RG 38, O-12-b: 'Japanese Battleship "Iwami"', 9 July 1907.
9. 'Sledovat' v Aleksandrovsk ...', 120.
10. Alekseev, *Voennaia razvedka Rossii*, 1:524.
11. RG 38, O-12b: 'Ships, Japan, Miscellaneous Notes', November–December 1907.
12. Sergei Vinogradov, email to the author, received 3 April 2023.
13. 'Sledovat' v Aleksandrovsk ...', 120. Three more replacement 12in/40 guns of Russian pattern were ordered in December 1910 from the Japan Steel Works, probably to replace worn barrels; email, Ahlberg to McLaughlin, 26 February 2023.
14. These are reproduced in Gribovskii, *Eskadrennye bronenosty tipa 'Borodino'*, 52–54; the thickness is also confirmed by the official *Sudovoi spisok* (ship list) for 1904, 40–41.

WARSHIP NOTES

This section comprises a number of short articles and notes, generally highlighting little known aspects of warship history.

FROM *GRAF ZEPPELIN* TO *AQUILA*: THE ITALIAN NAVY'S ASSESSMENT OF THE GERMAN CARRIER, 1941–1942

Regular *Warship* contributor **Enrico Cernuschi** has provided new insights into the influence of German thinking on the *Regia Marina*.

This Note is based on two Italian naval missions to Germany made for the purpose of examining the German *Graf Zeppelin* and her arrangements for landing aircraft at a time when Italy had finally committed to the aircraft carrier as a type. These reports shed new light on Germany's intentions regarding carrier aviation in the crucial 1941–42 period, and they illuminate some of the reasons why neither Germany nor Italy ever deployed a carrier.

Things proved a lot harder than they at first seemed. The mechanics of getting a modern aircraft off the comparatively small flight deck of a moving ship and then back on were by no means straightforward. Moreoever, it is no coincidence that no nation with a large navy that maintained a monopolistic Air Force – although arguably only Germany and Italy belong to that club – succeding in deploying a carrier, despite strenuous efforts. Only navies that had their own aviation arm successfully operated carriers in the Second World War.

The Context

The history of the German carrier *Graf Zeppelin* is well known. Laid down on 26 December 1936, she was launched on 19 January 1939 and was never completed. The keel of a sister ship was laid in 1939, but construction was halted in 1939 and she was broken up the following year. Some of the catapults, lifts and arrester wires intended for the second carrier were completed as spares for *Graf Zeppelin*; they would be sold, as we will see later, to Italy in December 1941.

The Italian dictator Benito Mussolini had changed his mind in January 1941. After more than sixteen years of stubborn opposition to carriers for the Italian Navy – a consequence of the ideological favour he bestowed on the Italian Air Force, the Fascist service he had created in 1923 – he had revised his opinions following the loss of Bardia, Libya, when rain and mud put out of service the *Regia Aeronautica* airstrips in Cyrenaica while the aircraft of the carrier *Illustrious* continued to conduct the customary flight operations from her steel deck.

Despite this change of heart on the part of Mussolini, the *Regia Aeronautica* was able to delay the *Regia*

Marina's carrier programme until June 1941, when General Ugo Cavallero, the Chief of the General Staff, now back in Rome after the end of the Greek campaign, was able to impose his will on the Italian Air Force Chief of Staff, General Francesco Pricolo. The latter would finally be removed on 15 November 1941 and replaced by the less obstructive General Rino Corso Fougier.

The *Regia Aeronautica* was being totally honest when it stated, at the end of January 1941, that the service had no knowledge regarding the take-off techniques currently employed on foreign aircraft carriers, not to mention the arrester gear needed for landing on. Indeed, any study of these procedures had been forbidden from 1924, and the only information available was from American and British magazines!

This sorry state of technological understanding was responsible for the excessive flank speed of 38 knots requested by the Italian Navy from General GN (*Genio Navale* = Constructor Corps) Umberto Pugliese in 1935, when he was asked to upgrade the light carrier design drawn up in 1927–31 by his colleague Filiippo Bonfiglietti. It was estimated that only such a speed would allow the heavier modern aircraft to fly off under the ship's own power, with the ship giving them the maximum available combined wind and carrier speed and take-off distance.

The Italian Navy had been asking in vain for a carrier since 1922, and numerous projects had been drawn up and revised prior to 1941. Following the abandonment in July of a proposal to complete the battleship *Impero* as a carrier, the final decision was for a conversion of the liners *Roma* and *Augustus* into fast fleet carriers in line with a project upgraded between August 1940 and January 1941 by Generals (GN) Gustavo Bozzoni and Carlo Sigismondi.

The problems of operating aircraft from a flight deck were resolved in 1941 by engineer Giovanni Pegna, a former Navy GN officer and an aviation pioneer, who concluded that the former liner *Roma*, reconstructed with new machinery originally ordered for two of the cancelled light cruisers of the 'Capitani Romani' type, would have sufficient flight deck length to allow a take-off run and possibly landing on – trials would be required – with ten planes parked on deck. The aircraft were initially to be Re 2000 fighters, a modified version of the American Seversky P-35. In December 1941 the order was revised, replacing them with 50 of the Re 2001, which had improved performance.

The *Regia Marina* schedule involved the completion of

the conversion of the liner *Roma* within twelve months. Mussolini, by a stroke of pen, then ordered this to be reduced to eight months – a totally unrealistic demand. When finally, in July 1941, after six months lost due to the opposition of the Italian Air Force, the programme began, it was for a *'lanciaerei'*, a carrier able to launch but not recover aircraft. As the *Regia Marina* expressed a strong preference for a true carrier able to sail beyond 100 miles from the coast – the fighters would otherwise have to find, after a maximum 3-hour mission, an airstrip on which to land – the Italian Navy staff asked, in September 1941, for German assistance; with hindsight this proved to be a fatal mistake.

The full story of this developmental 'dead end' is now available through previously unpublished documents found in the *Archivio dell'Ufficio Storico della Marina Militare* (the Italian Navy Historical Branch) in Rome. These papers also cast a new light on the true story of *Graf Zeppelin*.

A Troubled Partnership
The Italian mission left for Germany on 28 October 1941. It was led by Major (GN) Aurelio Molesini, responsible for the fitting out of the Italian carrier at Genoa, which had just begun.

The first meeting was held on 30 October with Admiral Fuchs, Chief of Naval Construction in the *Kriegsmarine*. During the following three days there was an intense programme which covered: vortex turbulence on the landing area generated by the island, hangar arrangements, aircraft handling, avgas security, sprinkler systems against fires, landing gear and crash barriers. The landing gear for *Graf Zeppelin* was designed and manufactured by the DEMAG company and comprised four arrester wires, the first of which was 28 metres forward of the stern post with the remaining three at intervals of

18, 10 and 12 metres respectively, each with a brake winch able to sustain the weight of a 21-tonne aircraft; deceleration was 2g, and one landing was possible every two minutes. The crash barrier was of the American type formed by two steel wires; the number of wires was increased to four following a critical observation made by the *Kriegsmarine* about the lifts of the French experimental carrier *Béarn*.

On 3 November a three-hour visit to *Graf Zeppelin* took place at Stettin. The following day there was a long debate about the Deutsche Werke compressed air catapults (4.25g), which were fitted flush with the flight deck right forward, and the trolley and rail track system on which the aircraft were moved from the hangar to the catapults for launch. According to the Italian delegation, this operation took three minutes even with a well-drilled team, as the procedure was complex and the two types of trolley (one for use in the hangar and on the flight deck, the other a carriage used by the catapult) were very heavy. On 5 November the Germans admitted that their Fieseler Fi 167 biplane torpedo bomber, which had a maximum weight of 4.8 tons, had to begin its take-off run from the central area of the flight deck, abaft the after windbreak, to give it a 120-metre run without the assistance of catapults. They also expressed disapproval of Pegna's non-assisted launch proposals, stating that only catapults could launch the Italian fighters and the 2-seat Re 2003 reconnaissance aircraft and their successors at high all-up weights. The rate of launch would be one aircraft every 30 seconds for 4 minutes alternating between the two catapults; the compressed-air stowage allowed a total of nine launches by each catapult before a 50-minute pause required to refill the tanks.

On 6 November, at the Travemünde *Luftwaffe* experimental centre, not far from Lübeck, a Ju 87 with arrester hook was shown to the Italians and moved by trolley; the

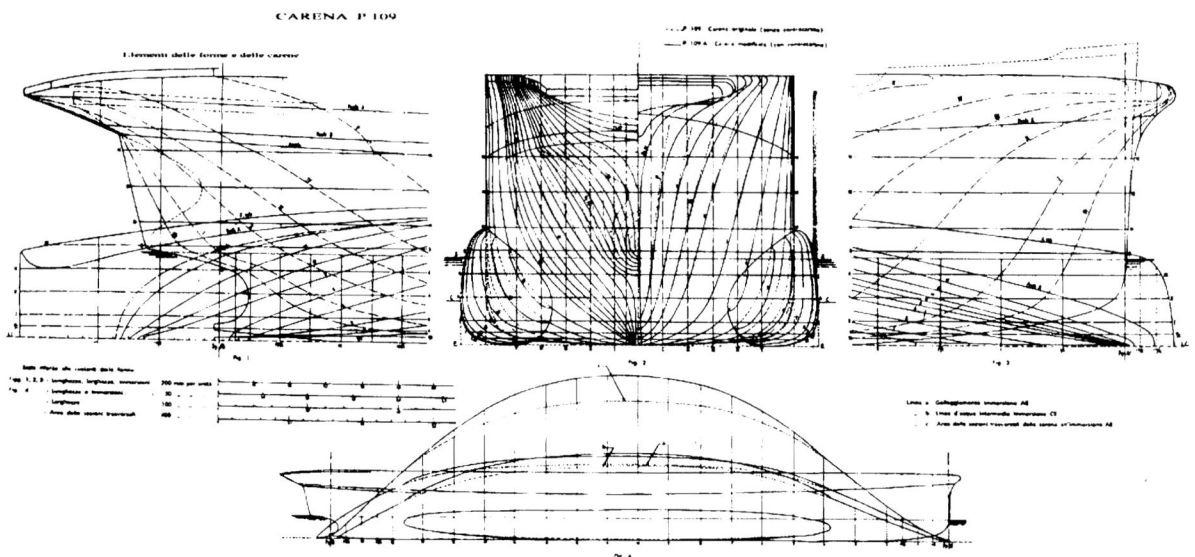

Body plans showing the bow and stern of the Italian carrier *Aquila* in their final configuration. The original project was amended to accommodate the German catapults, resulting in a delay until the late Spring of 1942. Note the new bulges. (Author's collection)

An aerial photo of the German carrier *Graf Zeppelin* taken at Gotenhafen (Gdynia) by an RAF reconnaissance Mosquito aircraft on 6 February 1942. North is to the bottom left of the picture. Amongst the other vessels visible in port are a three-masted sailing ship and at least one U-boat. (Naval Historical Heritage Command, NH 78307)

(manually) folding wings were also demonstrated. Other aircraft observed were an He 50 training biplane for simulated carrier deck landings, a Fi 167 and a Bf 109T fighter. A dozen landings were accomplished during the visit by all these aircraft except the Bf 109. On 9 November the mission was back in Italy.

The final choice made in November was to modify the Italian design to introduce the German lifts, catapults, landing gear and windbreaks. Only two lifts had so far been completed by the Germans, but the Italian project for a lift was still a paper study and was much criticised by the Germans who, in contrast, had tested their system in a Froude tank. As for the catapults, one was ready for installation, and construction of a second had been halted in 1940. The bow of the Italian ship had, accordingly, to be radically redesigned, and this task would require some months.

The Italians estimated that *Graf Zeppelin* was about 80 per cent complete. The Germans claimed that she would be combat ready in six months but no worker was

A close up of *Graf Zeppelin* enlarged from the same original image. (NHHC, NH 78306)

The stern of the former Italian liner *Roma* in dock at Genoa in July 1942 during her conversion into a fleet carrier. (Enrico Cernuschi collection)

seen on board. The machinery and boilers were almost ready, but the electrical cabling and related systems were still far from complete. There was no armament or fire control in place, and the wooden flight deck had sections missing.

The Second Round

By August 1942 the Italian carrier, which had been named *Aquila* in February, was at least one year behind the original schedule due to delays in the delivery of the German equipment. Mussolini was not amused, and his mood did not improve when he was informed that the planned second fleet carrier he wanted to be included in the 1943 program would not materialise, as the Germans were refusing to manufacture the lifts, catapults and arrester gear for that ship; all that was available to the

Italians was what had been begun in 1938 for the planned sister of *Graf Zeppelin*. It was therefore decided, in October 1942, to convert the liner *Augustus* into a slow escort carrier with her original diesel engines, equipped with STOL Ro 63 short-range reconnaissance aircraft to be used in the problematic anti-submarine role. This programme was cancelled by the end of December 1942 as the Ro 63 was powered by German-made Hirth HM 508D engines that Berlin was no longer willing to export; the engines were needed for the new series of Bf 108 trainers requested by the *Luftwaffe* to expand its fighter force to confront the increasingly-powerful USAAF Eighth Air Force based in Britain.

In the meantime a new visit to *Graf Zeppelin* was requested by the Italians on 8 September 1942. Its purpose was to find a solution for the various problems encountered in trying to develop a true carrier, in particular the landing-on procedures. The Germans were in no hurry to comply, and the new mission arrived in Berlin only on 7 December. *Graf Zeppelin* was visited in a floating dock at Kiel three days later, the visit involving just a two-hour ride. There was then a second visit to Travemünde, where fifteen trial flights were observed, most of them by an Arado 96, a single-engine, low-wing monoplane recently introduced for carrier training. On 18 December the Italian Navy party was back in Rome. Little had been accomplished beyond a follow-up visit that took place in February 1943 by the various types of German aircraft equipped with arrester hooks to the airport of Perugia, where a mock-up of the flight deck of *Aquila* had been made.

Contrary to the general claims by the *Kriegsmarine* that the fitting out of *Graf Zeppelin* had resumed for a period of seven months in 1942, the Italian appreciation

The Italian carrier *Aquila* at Genoa on 23 August 1943. (Aldo Fraccaroli, Enrico Cernuschi collection)

was that little progress had been made on her construction since November 1941. At best she was now 85 per cent complete. The new, asymmetrical bulges had been fitted, but there were as yet no weapons or fire control installations, and work on her seemed to have stalled; no workers were seen, only a skeleton maintenance crew. The Italians were able to ascertain from their *Wehrmacht* colleagues that the *Luftwaffe* had little enthusiasm for the carrier programme and that the experimental centre at Travemünde had encountered more difficulties than expected; the Fi 167 had been abandoned and the Ju 87 would have to be converted to a torpedo bomber and would need to be catapulted. On paper this was still possible, but it was a solution fraught with difficulty. The German Air Force blamed the Navy for the delay and *vice versa*.

For their part, the Germans welcomed Italian proposals to lighten and simplify the trolley and, above all, the new arrester hook adopted for the Re 2001. This was a masterstroke of reverse engineering by Pegna on the basis of a study he made in Egypt in 1941 of the wreck of a Fleet Air Arm Grumman Martlet shot down by a Fiat G 50 near Sollum. Pegna, who was Project Leader for the aviation side of the Italian carriers, had worked using just the holes in the fuselage, as the tail-hook had been removed before that aircraft, which belonged to a batch purchased originally by Belgium and later delivered to Britain, left the USA.

The Pegna hook was introduced by the *Luftwaffe* after the trials at Perugia. By July 1943, however, the Italian Navy and the Italian Air Force concluded, for once by mutual agreement, that the landing system for the Re 2001 was still far from operational despite the addition of a fifth arrester wire.

Conclusion

In 1945 a US Navy mission visited the never-completed carrier *Aquila* at Genoa. It concluded that the German arrester wire gear was unreliable and dangerous. Moreover, the *Kriegsmarine* catapult system was slow and complex compared to the American wire strop (or 'bridle) that connected the aircraft to the moving shuttle of the catapult – a solution also used by the Royal Navy on board escort carriers supplied under Lend-Lease from 1943.

The building and operation of a viable aircraft carrier was a long and difficult path, and there were no short-

The carrier *Aquila* at La Spezia in 1950 just before being broken up. The German rail tracks can be seen on the deck in the foreground. (Enrico Cernuschi collection)

cuts. Italy lost its last real opportunity to build two small 9,000-ton fast light carriers, known as the Vian design, in 1932 when the *Regia Aeronautica* Chief of Staff, Italo Balbo, halted that programme with the blessing of Mussolini. In September 1935 a design for the conversion of the liner *Roma* into a crude, slow emergency carrier was dropped by Il Duce, both for ideological and economic reasons.

Source:

Archivio dell'Ufficio Storico della Marina Militare, Rome. Fondo Nave *Aquila*, Busta 6, Titolario A 2: 'Relazione della missione in Germania per visita N.P.A.' and 'Relazione della missione eseguita in Germania per aggiornamento di argomenti sulle navi portaerei'.

THE *NIGER*, *LE MAGE* AND *FAIDHERBE*: A TALE OF RIVER GUNBOATS

Ian Sturton has provided an account of a little-known episode in European colonial history.

Between 1850 and 1900 a handful of dedicated Frenchmen explored, occupied and ultimately pacified a huge area of West Africa stretching from Dakar to Lake Chad and from the Sahara to the coast of Dahomey. Notable among them was Eugène Mage, the first French

naval officer to see the Niger. The British were also exploring, and to avert possible clashes the Anglo-French Convention of 5 August 1890 fixed the village of Say or Saay (Saï, now in the Republic of Niger) as the lower limit of French influence on the Niger.

The French penetration was slow; it was also brutal. The indigenous inhabitants were forced to be guides and porters. The desert *cafard* affected Europeans severely, causing madness, mutinies and murders. Malaria, typhoid fever and typhus were endemic; their causes and treatment

Portrait of LdV Eugène Mage (1837–69), the first French naval officer to explore and map the upper Niger. He drowned when his ship, the paddle corvette *La Gorgone*, was wrecked off Ushant with the loss of all hands, 19 December 1869. His body was never recovered. (Wikimedia commons)

were then largely unknown, leading to invalidity and more deaths. Frequent government and ministerial changes in Paris led to one military faction and then another taking control on the ground in West Africa. Very slow communications – letters to France took three months for a reply – and rivalry between Army and Navy meant orders could be and were ignored or varied with impunity; specious excuses and dumb insolence multiplied unchallenged.

Arrival of the Gunboats

The first gunboat (*chaloupe-canonnière*) on the Upper Niger was the iron-hulled sectional (*démontable*) *Niger*, built by the Société Génerale des Forges et Ateliers de St Denis to Claparède drawings and brought overland in 126 crates to Bamako (Bammako) for assembly. *Niger* was 18 metres long and designed to carry 10 days' provisions for 15 men, 24 hours' fuel (coal or wood) and a gun, a 37mm Hotchkiss revolver cannon. She was assembled by *Enseigne de Vaisseau* (EdV) Froger in 1884, running trials in September 1884.

In 1887 the naval authorities decided that a second gunboat was necessary for when *Niger* was refitting. For economic reasons the new unit would be built of wood at Bamako; only the machinery would come from France. The construction was entrusted to *Lieutenant de Vaisseau* (LdV) Caron.

In 1888 Caron published an illustrated article about his contributions to the Niger navy. The illustrations included accurate scale drawings of a wooden-hulled shallow-draught gunboat, the *Faidherbe* (ex-*Le Mage*), and of a wooden *chaland* (barge without machinery), the

Manambugu. The gunboat, much larger and roomier than *Niger*, was a strange-looking craft. She had two masts raked at different angles, no visible funnel or engine exhaust and three possible positions for the armament of two 37mm Hotchkiss revolver cannon: on the deck fore and aft and in a 'fighting top' on the foremast. The *chaland*, of 12 tonnes capacity, had dimensions of ten metres by 2.80 metres max beam (*largeur au maître*).

In 1886 Colonel Galliéni arrived as military commandant. He was very critical of the *Niger*, writing that she was more suited to pleasure trips on the Seine than rough work in Africa. To get a second active hull as quickly as possible, he insisted on a new France-built iron-hulled sectional gunboat; he was supplied with a duplicate of Claparède's *Niger* (notwithstanding his disapproval of the type), to be named *Le Mage*. At the same time he encouraged the progress of Caron's hull, also a *Le Mage*. On 4 April 1887 he both witnessed her ceremonial naming (see photo) and, in the same speech ordered her renamed *Faidherbe*, as the new *Le Mage* was *en route* from France. The *Faidherbe*'s machinery never reached her; although put afloat in June, she was allowed to deteriorate at Bamako, a monument to wasted time, effort and money.

Galliéni's *Le Mage*, fully assembled and in service by September 1887, was rebuilt in 1888 by LdV Davoust with a wooden outer hull about 80cm distant from the iron inner hull, the extra storage space increasing endurance from one to six months; fortuitously, the speed of the enlarged hull increased from 5.0 to 5.3 knots. Davoust, exhausted and mortally ill, died in Senegal, 31 December 1888/1 January 1889. LdV Jaime further modified *Le Mage* in 1889 at Koulikoro, fitting

A contemporary image of the Niger flotilla off Nyamina, 5 July 1887, at the start of Caron's expedition. The boats are (left to right) the *sharpée Titi*, the *chaland Manambugu* and the sectional gunboat *Niger*. Sectional gunboats were fabricated in France and assembled there for trials using bolts instead of rivets. They were then dismantled and packed in crates with the necessary riveting materials, tools, spare parts etc, for transport to West Africa and reassembly. Loss of crates could cause prolonged delays. The *sharpée*, a wooden one-man utility boat, was built by Caron at Bamako and named *Titi* after the chief there; she was about eight metres long. (from Jaime)

Faidherbe (ex-*Le Mage*)

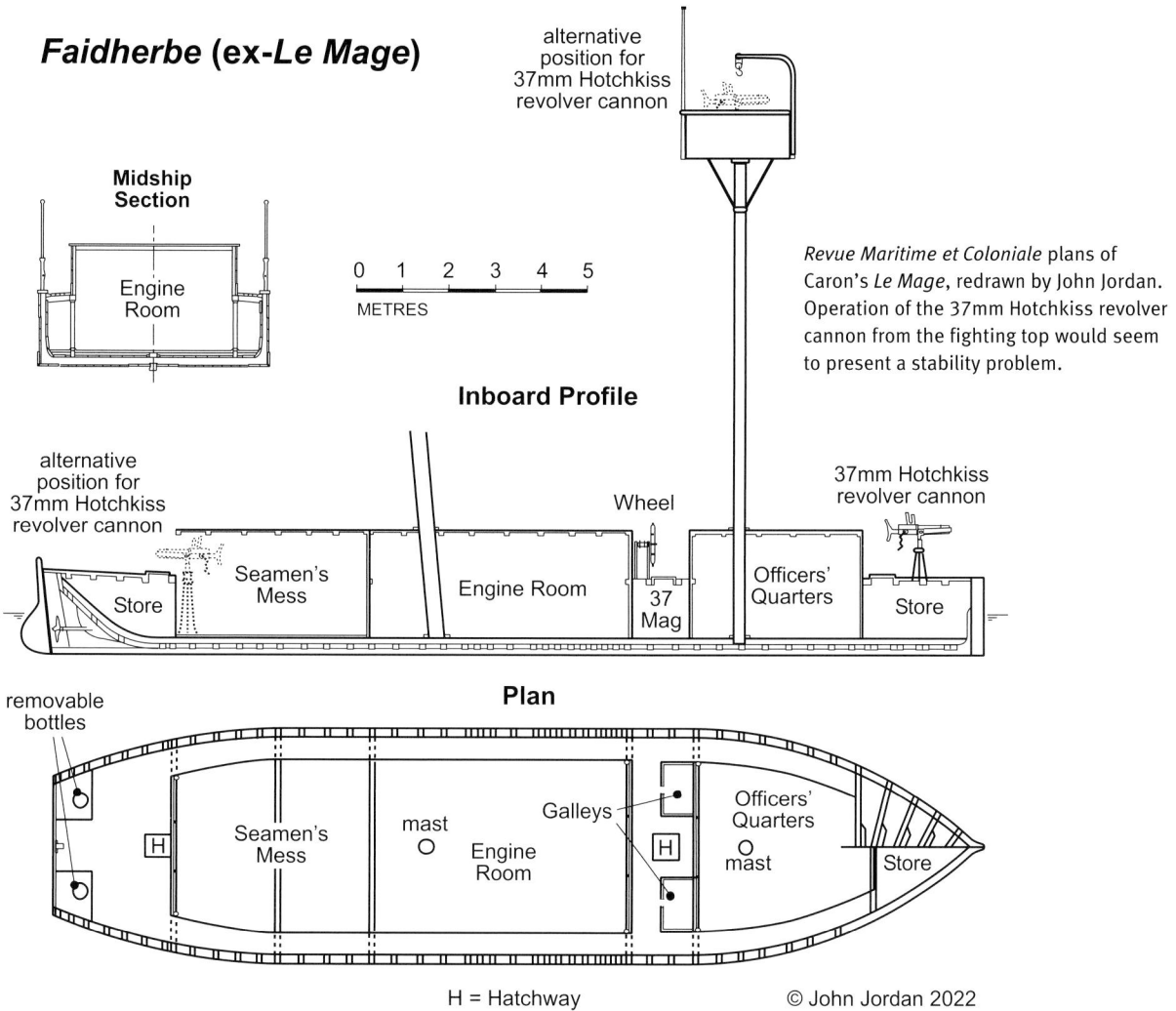

Midship Section

Engine Room

0 1 2 3 4 5
METRES

alternative position for 37mm Hotchkiss revolver cannon

Revue Maritime et Coloniale plans of Caron's *Le Mage*, redrawn by John Jordan. Operation of the 37mm Hotchkiss revolver cannon from the fighting top would seem to present a stability problem.

Inboard Profile

alternative position for 37mm Hotchkiss revolver cannon

Store | Seamen's Mess | Engine Room | Wheel | 37 Mag | Officers' Quarters | 37mm Hotchkiss revolver cannon | Store

Plan

removable bottles

Seamen's Mess | mast | Engine Room | Galleys | Officers' Quarters | mast | Store

H = Hatchway

© John Jordan 2022

Gunboat *Le Mage*

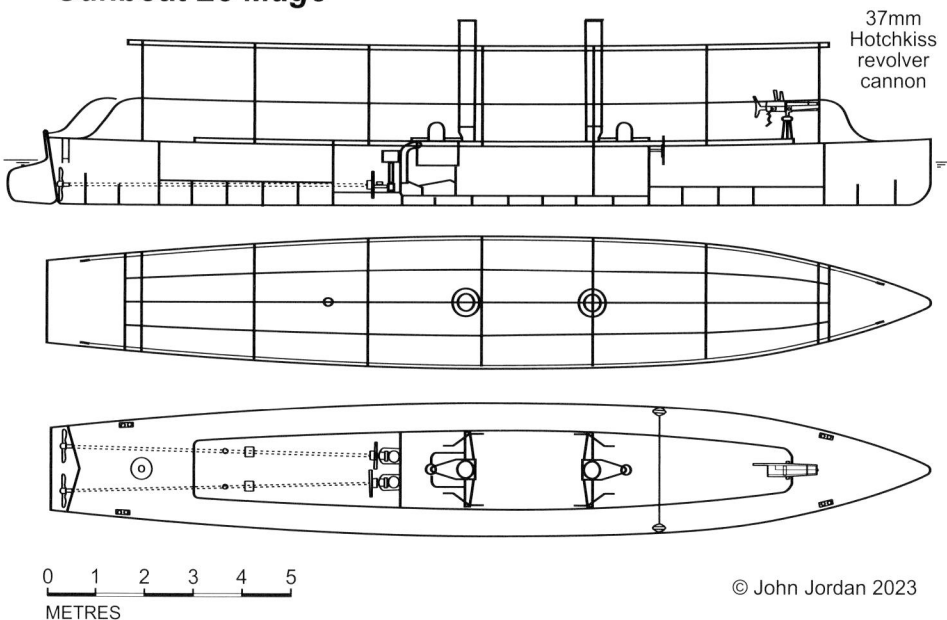

37mm Hotchkiss revolver cannon

Claparàde profile and plan of Galliéni's *Le Mage,* 8DD[1] series, redrawn by John Jordan; *Niger* was almost identical. Contemporary images of both gunboats show taller funnels. In retrospect, the design was most unsuitable for Upper Niger river conditions, lacking power, endurance and cargo capacity.

0 1 2 3 4 5
METRES

© John Jordan 2023

Barge *Le Manambugu*

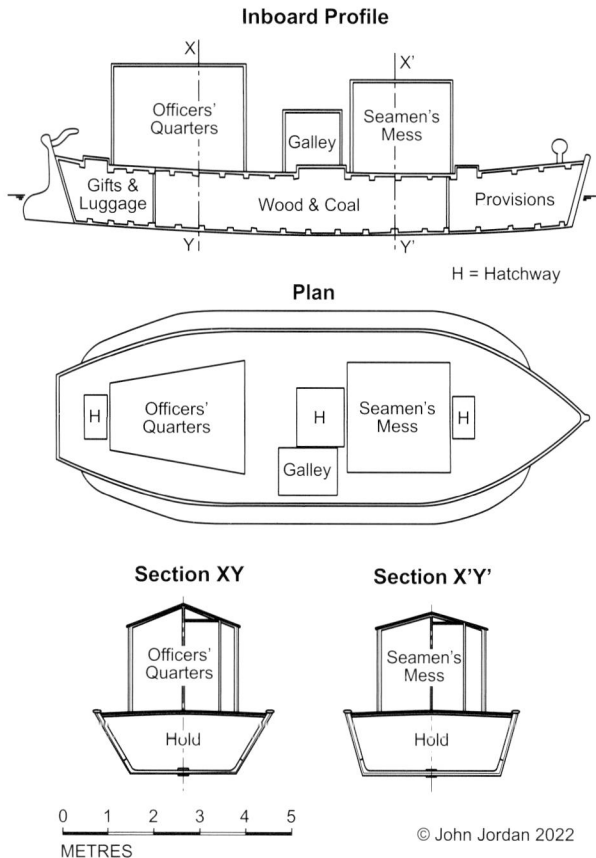

Inboard Profile

Officers' Quarters | Galley | Seamen's Mess
Gifts & Luggage | Wood & Coal | Provisions

X | X'
Y | Y'

H = Hatchway

Plan

Officers' Quarters | H | Galley | H | Seamen's Mess | H

Section XY   Section X'Y'

Officers' Quarters — Hold

Seamen's Mess — Hold

0 1 2 3 4 5
METRES

© John Jordan 2022

Revue Maritime et Coloniale drawings of Caron's *Manambugu*, redrawn by John Jordan. The slight curvature of the keel would have assisted towing over mud banks.

a Leclanché cell and a Gramme machine; their intense bursts of electricity operated a searchlight, the first on the river (see photo).

River Operations

An attempt at exploration was made in 1885 when the newly-completed *Niger*, commanded by Davoust, reached Diafarabé.

A contemporary image of the naming of Caron's *Le Mage*, shortly to be renamed *Faidherbe*, Bamako, 4 April 1887. Galliéni is presiding; note the general paucity of Europeans. (Courtesy of Stephen S Roberts)

The first major voyage was in 1887, when Caron was ordered to take *Niger* to Timbuktu with *Manambugu* and the *sharpée Titi* in tow. The outgoing journey started from Manambugu, some 60 kilometres below Bamako beyond the Sotuba rapids, on 1 July. *Niger* reached Koriumé, the river port for Timbuktu, on 18 August, but Caron could not make useful contacts with the Tuaregs controlling the area; without ground troops, his force was just too weak and unimpressive. After two or three days of futile talks he abandoned the attempt, struggling back upstream. Desperately short of fuel even after breaking up and burning the *chaland*, he reached Manambugu on 6 October. Over the next year, the gunboat base was gradually moved about 40km downriver to Koulikoro, healthier than Manambugu.

In 1889 Jaime's force for the second river expedition to Timbuktu consisted of *Niger* and the rebuilt *Le Mage*, each towing two *chalands*. It was nearly ready when on 31 August *Niger* broke loose in a storm and crushed one *chaland*. An 8-metre replacement with a capacity of four tonnes was constructed in just six days. Minor damage to one of *Niger*'s propellers was made good.

The flotilla finally left Koulikoro on 16 September, was at Nyamina on the 17th and reached Mopti on 21 September. As *Niger* lagged behind because of three ruptured boiler tubes, her commander, EdV Hourst, was

Data:

Contemporary accounts are short on precise data, preferring approximations, and are not always in agreement.

Niger
Roberts, *op cit*, 286: 18.75m wl x 2.60m moulded x 1.1m depth x 0.8m max draught; c28t; 30ihp (8nhp); 1–37mm Hotchkiss revolver cannon. Laid down 1884, launched 1884, completed Sep 1884.
Caron, *op cit*, 71: 18m x 3m, 5kts.

Le Mage
Roberts, *op cit*, 286: as *Niger*, but laid down and launched 1887, completed Sep 1887.
Jaime, *op cit*, 21: 18m x 2.7m; 22.75t full load; Jaime, 47: 18m x 4m x 80cm or 90cm depth (data after widening). 5kts (before widening), 5.5kts trials (estimated), 5.3kts actual, both after widening.

Faidherbe (ex-*Le Mage*)
Caron, *op cit*, 516: 20.25m length pp x 5m *largeur au maitre*; 28t (bare hull), 76.1t (fully loaded with 25 men, three months' provisions, 2–37mm Hotchkiss revolver cannon, 20t fuel); 50ihp for a designed trial speed of 7kts.

ordered to wait there for two months and return upriver alone if *Le Mage* did not return. *Le Mage* left Mopti towing two *chalands* laden with food and fuel on the 26th, and reached Koriumé on 3 October. After fruitless attempts at meaningful talks, Jaime left Koriumé four days later. As *Le Mage* had insufficient power to tow two *chalands* upstream, one was sacrificed as fuel, enabling her to reach Mopti on 11 October. Hourst and *Niger* were not there; the locals were silent on the subject, their response to enquiries being *'le mutisme le plus obstiné'*. Continuing upriver, Hourst was found at Diafarabé on 16 October, and Jaime moored at Koulikoro on 25 October, his epic journey complete. *Le Mage* had been underway for 34 of the previous 40 days, steaming an average of ten hours daily.

In terms of territory gained and treaties signed, he had been no more successful than Caron two years earlier. His voyage had been much better organised, and excellent survey and ethnological work had resulted, but the limitations of gunboat diplomacy had been exposed yet again. Galliéni continued his policy of slow, steady penetration by land, following the directions of least confrontation.

A more active policy and additional boots on the ground were needed; in late 1893 Major Joseph Joffre's soldiers seized Timbuktu, adding the legendary city to the French colonial empire.

The need for gunboats was ending. In July 1894 Hourst, promoted to LdV and commander of *Le Mage*, also headed the Niger river force. *Le Mage* was No 101 on the official list of 111 active naval units; *Niger* was no longer listed.

Both *Niger* and *Le Mage* were stricken on 1 June 1895 and used as barges.

Afterword

In 2023 former French West Africa is divided into nine independent, weak and quarrelsome republics. The stretch of the Niger between Bamako and Timbuktu is in the Republic of Mali, a country especially prone to *coups*

A contemporary image of Galliéni's *Le Mage* as rebuilt. LdV Jaime was proud of his searchlight, which he used to impress the Koulikoro locals on Bastille Day 1889. (from Jaime)

and insurrection, and small armed craft once more patrol the troubled waters of the great river.

Asknowledgements:

Grateful thanks to Iain Riley, Librarian, Glasgow Life in the Mitchell Library, for responding to a barely coherent initial inquiry, and to Dr Stephen Roberts for cheerfully searching his inimitable collection in answer to some very basic questions and introducing me to the 8DD[1] series of French warship drawings.

Sources:

E Caron, *La Revue Maritime et Coloniale*, Vol 99 Oct–Dec 1888 (Paris, 1888). His article, 'La Marine au Niger', is on pp504–534 + three unnumbered pages of illustrations.

E Caron, *De Saint-Louis au Port de Tombouktou*, Augustin Challamel (Paris, 1891).

J-S Galliéni, *Deux Campagnes au Soudan Français 1886–1888*, Librairie Hachette (Paris, 1891).

JGN Jaime, *De Koulikoro à Tombouctou sur la Canonnière Le Mage*, Les Libraires Associés (Paris, 1894, revised ed).

L Renard, *Carnet de l'Officier de Marine*, Berger-Levrault (Paris, 1894).

Stephen S Roberts, *French Warships in the Age of Steam 1859–1914*, Seaforth Publishing (Barnsley, 2021).

HMS *COSSACK* AND MR RAPLEY: A CAUTIONARY TALE

John Roberts investigates an account of an incident involving HMS *Cossack* during the last months of the Spanish Civil War that turns out to be seriously flawed.

In 1994 I began work on an Anatomy of the Ship volume on the 'Tribal' class destroyer *Cossack*. This project came to an end with the demise of Conway Maritime Press, although some of the research I had done resurfaced when I authored Seaforth's *Destroyer Cossack* (2020, reviewed in *Warship* 2022, 209-10), part of their series utilising the Admiralty's Builders Plans.

In the early stages of the original research I consulted, among other published works, 'The Tribals' by Martin Brice. The first few paragraphs of the chapter devoted to

Cossack covered her basic operational history from completion in 1938 to the outbreak of the Second World War. Within this was the following relatively detailed description of an incident on 22 August 1938, a few months before the end of the Spanish Civil War.

> *Cossack* was ordered to Barcelona to collect the British Consul there. As she could not enter the port because of the fighting, *Cossack* anchored off the neighbouring resort of Caldetas, sending her whaler to take the consul off the beach. Unfortunately, being a rather elderly gentleman, he slipped and broke his leg as he was getting into the boat. *Cossack* had to rush him to hospital in Marseilles, making 30 knots for most of the passage.[1]

There was no indication of the origin of this information, and I had no reason to doubt its authenticity until

I examined *Cossack*'s surviving logs, which cover the period from June 1938 to December 1939.[2] These provided the following details which, when compared to the above, provide a substantially different view of the events around the *Cossack's* mission to Caldetas on 22 August.

On the morning of 16 August 1938 *Cossack*, in company with the battleship *Barham* (Flagship of Rear Admiral Ralph Leatham, 1st Battle Squadron and 2nd in command of the Mediterranean Fleet), sailed from Malta for Palma, Majorca. En route, *Cossack* practised oiling from *Barham* during which she collided with the battleship twice but suffered only superficial damage. The two ships arrived at Palma in the morning of 18th and at 11.30 'Mr Rapley interpreter joined from [HMS] Sussex'. The cruiser *Sussex* had arrived from Barcelona two days previously having spent most of the previous two weeks patrolling the Spanish Coast. Following the transfer of Mr Rapley, she sailed for Malta in company with HMS *Hood*. In the afternoon of the 18th and the morning of the 19th *Cossack* employed divers to check for damage resulting from her collisions with *Barham* but they obviously found nothing to interrupt her intended operations. During the afternoon of the 19th, she 'Embarked stores from *Barham* for Consul Barcelona' and, on the following day, a sick Russian seamen for transport to Marseille. At 18.00 on the 20th she sailed for Gandía,[3] 60 miles south of Valencia, arriving at 06.40 the following morning. She sailed at 17.00 the same day and anchored off Caldetas[4] at 08.18 on the 22nd. Also anchored off the port was the French destroyer *L'Indomptable*. At 14.30 *Cossack* embarked eight refugees for transfer to Marseille and at 15.00 the French destroyer *Le Malin* arrived. In the log, below the reference to *Le Malin*, appears the phrase 'Mr Rapley interpreter injured while entering whaler in the surf.' At 18.32 *Cossack* sailed for Marseille, and by 10.30 the following morning was secured alongside the Quai des Belges in that port. At 11.45 Mr Rapley was discharged to hospital, and 30 minutes later the eight refugees and the Russian seaman were disembarked. *Cossack* returned to Caldetas with stores and mail for the British Embassy on the 26th and then continued to Palma where she remained until 1st September when she sailed for Malta to dock for repairs to her collision damage. She had no further service in relation to the civil war.

The obvious and critical discrepancy between the quote from 'The Tribals' and the Log entries is the fact that the accident did not occur to the consul but to Mr Rapley. Moreover, there is no indication that it was intended to embark the consul or that *Cossack*'s original destination was Barcelona. In addition, the *Cossack*'s speed during her 'rush' to Marseille is given in the log as 12.5 knots and not 30 knots (confirmed by the time and distance involved).

The logs still seemed to provide inadequate coverage of

HMS *Cossack* leaving Portsmouth for the Mediterranean on 4 July 1938. She already has the red, white and blue stripes added to the gun shield of 'B' mounting that served to identify the British warships of the Mediterranean Fleet during the Spanish Civil War. (Author's collection)

HMS *Cossack*: Area of Operations August 1938

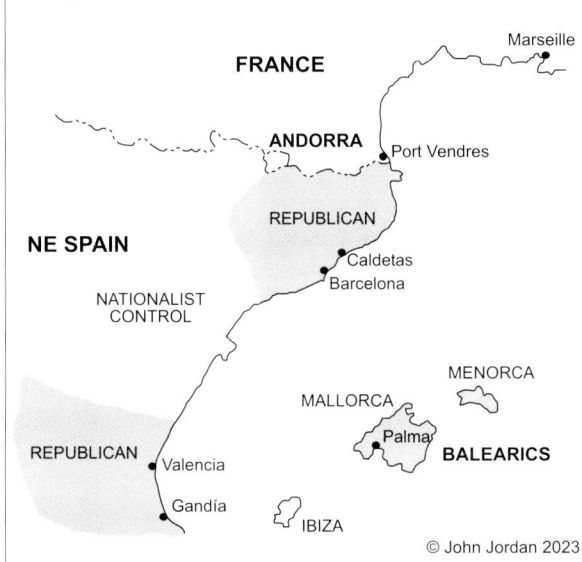

FRANCE

Marseille

ANDORRA

Port Vendres

REPUBLICAN

NE SPAIN

Caldetas
Barcelona

NATIONALIST
CONTROL

MENORCA

MALLORCA

Palma

BALEARICS

REPUBLICAN

Valencia

Gandía

IBIZA

© John Jordan 2023

events so, more recently, I caried out some further research – primarily on the internet and the British Newspaper archive. This revealed further discrepancies but no direct information regarding the accident to Mr Rapley. The coastal village of Caldetas was in fact the location of the British Embassy to the Spanish Republican Government in Barcelona. It was 25 miles northeast of Barcelona, something that had earlier raised a question in Parliament regarding its convenience for dealing with British interests in Barcelona. However, the distance does not seem to have been a serious problem given that there were several other foreign embassies and legations at the same location. Due to its primary purpose, the British embassy at Caldetas was generally referred to as the 'Barcelona Embassy'. Apart from this, a rather more cogent fact is that, apart from an air raid on 19 August, there was no 'fighting' in Barcelona in August 1938 despite this being given as the reason for *Cossack* going to Caldetas. Nationalist forces did not begin their assault on Catalonia until December 1938. Barcelona was taken by the Nationalists at the end of January 1939, at which time the British *chargé d'affaires* (Sir John H Leche) and his remaining staff were evacuated from the embassy at Caldetas in the cruiser *Devonshire*, and taken just across the border to Port Vendres in France. One could assume from this that the events at Caldetas in August 1938 have been confused with those of January 1939 except for the fact that *Cossack* was not involved in the final evacuation of embassy staff.

More importantly, I found that a document existed in the records of the Treasury held by the National Archives concerning an injury to AE Rapley.[5] My first attempt to see this failed, since the document had to be ordered in advance as it was not directly available at Kew. For various reasons, especially the disruption to research

caused by Covid, I did not follow this up until 2023. This document clarified the events of 22 August and added detail on both this, Mr Rapley and subsequent events.

Mr Rapley was initially employed during July–December 1936 to take charge of the embarkation of refugees at the quayside in Barcelona (possibly on behalf of the Foreign Office). Following this he was transferred to the Admiralty to serve on HM Ships as a civilian interpreter to assist the Navy in dealing with Spanish authorities and civilians. For this service he was paid 6 shillings (£0.3) a day plus a 7 shillings and 6 pence (£0.375) daily messing allowance. In the morning of 22 August1938 he went ashore from *Cossack*'s whaler in the company of Lieutenant-Commander B T Turner and one other officer of *Cossack* to consult with the Acting Consul-General, D J Rogers, regarding the evacuation of refugees. The following, written on 27th August by the *Cossack*'s Captain (Daniel de Pass) to Rear Admiral Leatham, describes events during Rapley's return to the ship.

> There was a considerable surf running and the method of landing and embarkation in use was to back the whaler in stern first to a small plank jetty (about 15 yards long and 18 inches broad) and to jump from the stern of the whaler to the jetty as the whaler rose on a wave.
>
> When re-embarking at about 1115 (Zone 1) Mr Rapley jumped onto the stern gratings of the whaler successfully but then slipped and was thrown over the jetty. In so doing he must have caught his foot between the whaler and the jetty.
>
> A party of officers from the French destroyer *L'Indomptable* were bathing nearby at the time; they ran into the water and carried Mr Rapley up the beach. Fortunately there was a doctor among them who did all that was possible in the circumstances and later in the same day, after Mr Rapley had been conveyed on board H.M.S. "COSSACK", visited him again. I have expressed my thanks by letter to the Captain of the French destroyer.

The injury seriously damaged the bones of his right ankle, together with the lower part of the two adjacent leg bones, his foot being distorted outward. At Marseille Rapley was transferred to the Clinique Montecelli and placed under the care of Doctor E Hawthorn. He was visited by Captain de Pass on 25 August, who relayed to Admiral Leatham that Rapley was concerned about his financial position given that his only income was from his role as volunteer interpreter. Leatham communicated this to the Admiralty, proposing that Rapley should continue to receive pay from 24 August (prior to this it was paid by *Cossack*) and the Admiralty should pay his hospital expenses. At the end of September 1938, the Treasury agreed to the Admiralty's proposals having concluded that the daily 6 shillings could be provided under existing regulations for 'sick pay'.[6]

While in hospital Rapley was visited by some of the Mediterranean Fleet's medical officers when their ships visited Marseille: *Barham* (30 Aug), *Arethusa* (21 Sep),

Hood (24 Oct and 5 Nov). Shortly after *Hood*'s last visit Rapley was discharged and on 12 November joined the destroyer *Hasty*. In March 1939 he was examined by Lt-Cdr J Crosfell, Squadron Medical Officer of the Mediterranean Destroyer Command in HMS *Galatea*. He concluded that Rapley had 60–70 per cent disability in his right leg and expected much of this to remain permanent. His movement was limited, and he had difficulty walking but could do so with a stick. On 9 April 1939 Rapley's employment by the Admiralty ended. It is not clear if this was due to the abolition of the post due to the end of Spain's civil war, or because of his disability or both. On 13 April 1939 Rear Admiral (D) John C Tovey, with his flag in *Galatea*, wrote to the C-in-C Mediterranean concerning Rapley's recent history and his examination by the Squadron Medical Officer. In relation to his disability, he stated that it would be:

> … impossible for Mr Rapley to resume his previous occupation in Barcelona, and his present disability, incurred on duty, will handicap him very severely in obtaining fresh employment.
>
> I feel very strongly that he should be awarded some compensation, and … that this should take the form of a gratuity rather than a pension.

By June 1939 Rapley was working for the British Chamber of Commerce in Barcelona. Steps were under way to process a claim for compensation but 'Unfortunately the case was lost through changes of staff and evacuation [presumably that from Caldetas] and traced through a reminder from the Commander in Chief, Mediterranean.' This from the last paragraph in a submission from Admiralty Secretary W Reid to the Treasury in March 1940 officially applying for Rapley's compensation. In April 1940 a doctor in Barcelona provided an up-to-date assessment of Rapley's condition in response to a request from the Admiralty. This confirmed that his right foot was deformed and that his walking was materially impaired. In September 1940 the Treasury agreed that Rapley should receive a gratuity of one year's pay under the Injury Warrant. Unfortunately, the existing warrant did not provide for payment in a case involving the abolition of the employee's work and it was necessary to make amendments to the warrant to make this possible. Meanwhile the Admiralty authorised an advance of £100 to Rapley in anticipation of the resolution of his claim to compensation. Modification of the

terms of the Injury Warrant eventually allowed the Admiralty to award Rapley £246.38 in May 1941 (*c*£9,600 today) less the £100 advance he had already received.

One final item in Brice's account is the reference to the person involved in the accident as an 'elderly gentlemen'. In fact, both Rapley and the Consul-General (JD Rogers) were 49 years of age which, even in 1938, would not deserve the description of elderly. However, Admiral Tovey described the re-embarkation to *Cossack*'s whaler when there was a considerable surf running as '… a hazardous feat for a man of middle age …'.

In some ways this story points to the danger of trusting a single source, although I am sure Martin Brice had little reason to distrust the information provided to him on the incident at Caldetas on 22 August 1938. The acceptance is understandable, especially given that the originator would gain little from the error and Brice would have found it difficult to verify. Apart from the fact that there was an accident that day and that *Cossack* took the person involved to Marseille, there is almost nothing in the quote above that is correct. It sounds like an old sailor's tale but even these usually have somewhat greater connection with reality. In general, this particular piece of history is of little importance, but I have a preference for correcting errors (including my own) when the opportunity arises.

Endnotes:

1 Martin H Brice, *The Tribals*, Ian Allen. 105.
2 HMS *Cossack* ship's logs ADM53/102009-102015 (1938) and 108114-108125 (1939), TNA (Kew).
3 British warships paid regular visits to the British-owned commercial port of Gandía during the Civil War. The author has not found the purpose of *Cossack*'s visit.
4 Caldetas is also known as Caldes d'Estrac (Valenciano).
5 'Rapley A E, Award under injury warrant …', T164/192/18, TNA. Note that Mr Rapley's actual initials appear to have been RA (Reginald Austin), a correction included in the file in 1939. However, both the correct and erroneous versions continue to appear throughout the document, probably due to those concerned referring to the file title or earlier correspondence which used the original erroneous initials.
6 The regulations allowed for a maximum of 10 weeks on full pay but this was increased to 12 weeks on the proviso that he did not receive the messing allowance. On 30 August Surgeon Commander RW Musson, 2nd Battle Squadron Medical Officer (HMS *Barham* at Marseille), following a visit to Mr Rapley, considered that he would not be fit for duties in HM Ships for at least 3 months.

AN OUTSIDER'S VIEW OF THE *MARINE NATIONALE*'S NAVAL CONSTRUCTION ORGANISATION IN 1939

Conrad Waters recounts the visit of a British Naval Constructor to France's *Service Technique des Constructions Navales* (STCN), the *Marine Nationale*'s naval construction branch, immediately before the start of the Second World War.

In the summer of 1939, Royal Navy Constructor Ernest Guy Kennett arranged to make an informal call on his French counterparts in Paris whilst travelling home from leave in the south of France.[1] Presenting himself at the French Admiralty on the Place de la Concorde on the morning of Wednesday 16 August, he was received by M André Lamouche, an *Ingénieur Général* of the second class and a senior member of STCN's management team.[2] The French naval engineer offered to

A 2012 view of the STCN's purpose-built offices at the Balard complex. The top storey was a postwar addition not present at the time of Constructor Kennett's visit. Note the large windows of the original top floor, a feature intended to assist the lighting of the design offices located at this level. (V Peheu under Creative Commons licence CC BY-SA 3.0)

arrange for Constructor Kennett to visit the STCN's design facilities, which were located in the Paris suburb of Grenelle. The Englishman regarded this as an opportunity that was too good to miss and the tour was arranged for 0930 the next day. On arrival, he was greeted by an unnamed senior member of STCN's management in charge of the Grenelle facility and various members of his staff. Constructor Kennett subsequently reported his impressions of his visit to Sir Stanley Goodall, the British Director of Naval Construction.

The STCN's design offices were housed in purpose-built accommodation at the corner of the Boulevard Victor and the Avenue de la Porte de Sèvres. They formed part of a wider complex – known as the Balard site – that had first been initiated with the completion of a purpose-built experimental tank for hydrodynamic trials under the auspices of the distinguished naval engineer Louis-Émile Bertin in 1906. The main office building was designed by the eminent architects Auguste and Gustave Perret in late 1928 as part of a programme to concentrate STCN's design and testing activities at the Balard site and had been completed between 1929 and 1932.[3] The two brothers were pioneers in the architectural use of reinforced concrete and the STCN offices were constructed from this material, with Kennett noting the advantages of its fireproof qualities. He described the offices as being

Ernest Guy Kennett, RCNC, was lead constructor for many Royal Navy light cruiser designs during the interwar period. (Courtesy of E G Kennett's family)

Gascogne (July 1939)

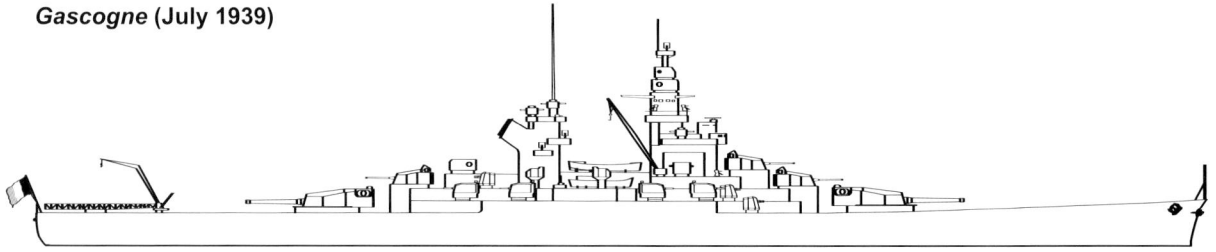

© John Jordan 2009

A general arrangement profile of the planned French battleship *Gascogne*. Constructor Kennett's visit to the STCN's design offices correctly reported the *Marine Nationale*'s intention to divide main armament forward and aft in future capital ship designs.

'... attractively modern but simple in style and most conveniently arranged'.

Adopting a 'U'-shaped plan and at that time occupying three storeys, the offices contained all of the STCN's staff except a handful of senior personnel located at the Admiralty. The organisation's records were stored on the ground floor. Here papers were kept in tiers of galvanised steel racks, with each being enclosed by a door. Again,

The STCN's *bassin de giration* was under construction at the time of Constructor Kennett's tour. Roofed-in post-war, it remained in use until the early years of the 21st century. This photograph shows a model of the aircraft carrier *Charles de Gaulle* undergoing turning trials in the facility. (Naval Group courtesy of Jean Moulin)

this reflected the importance of protection against fire. The floor above accommodated the STCN's directing staff. Each section head occupied a suite of three rooms: one for himself, one for a deputy and one for a secretary and current records. Moving upwards again, the top floor was occupied by the design rooms, with each section being arranged immediately above the corresponding section head's office suite. A lift for papers ran from each secretary's room to each design room above. Kennett noted that the design rooms were '... admirably arranged and lighted', albeit that their area would be far too small to house the equivalent design teams of the Royal Navy's Directorate of Naval Construction.

Constructor Kennett was given extensive access to STCN's facilities during his visit, noting various models of current and projected warship designs. He observed that various arrangements were being considered for the armament of new battleships, some with all turrets grouped forward as in *Dunkerque* and some with the main armament divided fore and aft. It was suggested to him that the latter arrangement was now more in favour, accurately reflecting its adoption in *Gascogne*, the planned but ultimately abortive fourth French 15in-gun battleship. He was also shown various aircraft carrier models, some with flush decks and some with island structures. One of the latter showed uptakes brought outside the hull above the waterline. Preferring discretion over valour, Kennett did not press for design details as he '... did not wish to embarrass [his] hosts with the possibility of having to decline to answer.'

The tour also encompassed the extensive workshops and testing laboratories that occupied much of the STCN's Balard site. The former included a large space for mock-ups, where a complete submarine bridge was in the course of being erected in wood. The latter housed, *inter alia*, a number of tensile, torsion and fatigue testing machines, some specially created to the STCN's requirements. As the main experimental tank built under Bertin's leadership was being cleaned, Kennett did not get to see it. However, he was shown an innovative circular tank then under construction that was intended for model turning trials. Some 65 metres in diameter, the *bassin de giration* was likewise fabricated from reinforced concrete and commonly attributed to the Perret brothers.[4] Kennett noted that consideration was being given to enclosing the then-open centre part of the basin, a project that was not ultimately undertaken until after the war.

All in all, Constructor Kennett appears to have been favourably impressed by the insight into the STCN organisation and facilities provided by his hosts. It seems likely he viewed the concentration of the STCN design and testing activities in one purpose-built complex as preferable to the more dispersed and, arguably, *ad hoc* arrangement of the Directorate of Naval Construction's own facilities. However, he appears to have been surprised by the comparatively small size of the organisation, assuming that much detailed design work must be left to the shipyards and overseers. That being said, he noted that the procedure of issuing building drawings from the STCN design department and submission of details by the construction yards was similar to the British model.[5]

Ominously, Kennett concluded his report by observing that, '... more than half of the staff were on leave and there was a complete absence of any visible concern with the political situation'. At the time, the outbreak of the Second World War was only two weeks away.

Sources: The National Archives
ADM 229/22: Director of Naval Construction: DNC's Reports, 397–400.

Sources: Published
Gaudard Valérie, 'Le "bassin des carènes" et le service technique des constructions navales à Balard: un exemple de cité scientifique à Paris', *In Situ: Revue des patrimoines 10/2009*, Ministère de la Culture (Paris, 2009). Published exclusively on line at: https://journals.openedition.org/insitu/3648.
Jordan John & Dumas Robert, *French Battleships 1922–1956*, Seaforth Publishing (Barnsley, 2009)

Endnotes:
1. Ernest Guy Kennett (1888–1961) began his career as an engineer cadet at Devonport Royal Navy Dockyard in 1904. His career as a Constructor was closely associated with British light cruiser design, commencing with the *Leander* class in 1928. This work also included the subsequent 'Town' class, as well as the preliminary designs for the 8,000-ton cruiser that formed the basis of the *Fiji* class. His final position was as Manager Constructive Department (MCD) – the overall head of construction activities – at Portsmouth before ill-health resulted in him being invalided out of the service in May 1946.
2. André Pierre Simon Lamouche (1888–1969) was a leading French naval engineer (architect) who entered France's Corps of Naval Engineers – *Le Corps du Génie Maritime* – in 1907. Kennett's report describes him as the '... French Directeur de Construction Navale ...' but this may have been a mistake, as other sources suggest he was an Assistant Director with the STCN's central management team at the time. Kennett's visit was facilitated by another leading, but by then retired French naval architect, Rolland Boris (1877–1957), but it is not known how the two were acquainted.
3. The Perret brothers' architectural practice was awarded contracts for much of STCN's development programme encompassing the Balard site that commenced in the late 1920s and which was not completed until after the end of the Second World War. The façade of the main office building – extended with an additional storey after the war – survives to this day. However, the other STCN facilities were cleared to make way for the Hexagone Balard – the new headquarters of the French Armed Forces – that was inaugurated in 2015.
4. The *bassin de giration* was designed under the Perret brothers' contracts with the STCN, but some sources attribute it to the government architect Camille Marion (1882–1964), who supervised the Perrets' work. Kennett's report estimated its diameter at 160ft but it was actually larger. It remained in use into the 21st Century, finally being demolished as part of the preparations for the Hexagone Balard in 2010.
5. Basic plans, which were much more detailed than preliminary sketches, were drawn up by the STCN in Paris. They were then sent to the naval arsenals or private shipyards, which drew up the detailed plans used for construction. The lead ship of a major class of warship was generally assigned to one of the arsenals. It, in turn, supplied a large number of detailed drawings to the private shipyards, which were therefore responsible mainly for the machinery arrangement. British procedures at the time were largely similar, although the leading private shipyards seemingly played a greater part in producing design drawings.

'ZOMBIES' IN WARSHIP HISTORY

Aidan Dodson looks at a few more of the 'zombie facts' that continue to stalk histories of the world's warships.

According to most 'standard sources', a wide range of surrendered German warships were scuttled after the Second World War 'loaded with poison gas'. These allegedly included the cruisers *Leipzig* and *Berlin*, the destroyers/torpedo boats *Z29*, *Z34*, *T21*, *T38*, *T156* and *T190*, the minesweepers *M16* and *M522*, and the MTBs *S7*, *S9* and *S12*. However, while there was certainly a programme to dispose of chemical munitions by sinking in ships, contemporary documents make it clear that the vessels involved were almost all cargo ships, whose holds could easily be packed with such munitions. The only warships mentioned in the official sources were incomplete units whose empty machinery spaces could easily substitute for cargo holds: the torpedo boats *T63* and *T65*, submarine chaser *UJ305* and *Sperrbrecher KSB3*.[1]

In all cases, these vessels, which were scuttled in a series of convoys organised by the UK and US occupation authorities, were sunk as gently as possible by opening seacocks, to avoid any danger of disturbing their cargoes. In contrast, the vessels mentioned by the standard sources were not only unsuited to the bulk loading of chemical ammunition (especially the MTBs!), but were dispatched by scuttling charges or even gunfire – diametrically opposite to the protocols applicable to the chemical weapons disposal convoys.

In fact, all were simply vessels incapable of repair lying in German ports which, under inter-Allied agreements, had to be destroyed by a strict deadline. Facilities simply did not exist in Germany to break them up in time, the solution adopted being to tow them out to sea and sink them in deep water. The fact they were towed out in convoys similar to those used for chemical weapon disposal simply led to a wholly-unwarranted assumption that *all* scuttlings of ex-German ships were part of the latter process. This is not simply a matter of detail in naval history, but has policy implications by greatly widening the number of wrecks regarded as potential environmental disasters beyond those which actually hold chemical munitions.[2]

The examples here and in *Warship* 2023 represent just a few instances of the many 'zombies' that stalk the literature of the history of warships. Sadly, although published refutations have in some cases existed for decades, undead 'facts' will doubtless continue to be recycled for years to come: it is hoped that these notes will help to make the real data better known. *Warship* would be pleased to hear from any readers who can suggest further 'zombies'. The annual can be contacted at the usual address.

Endnotes:
1. A Dodson and S Cant, *Spoils of War: The Fates of Enemy Fleets after the Two World Wars*, Seaforth Publishing (Barnsley, 2020), 196–97.
2. See the lists in P Lindström, *Vrak i Skagerrak: sammanfattning av kunskaperna kring miljöriskerna med läckande vrak i Skagerrak*, Forum Skagerrak II (Uddevalla, 2006), 24, which quote Gröner's volumes as their only source.

A's & A's

A MYSTERIOUS GERMAN BATTLECRUISER PROJECT
Kenneth Fraser makes an interesting observation regarding British knowledge of German battlecruiser projects during the First World War.

Some years ago, Norman Friedman published a collection of reprints from the 'Admiralty Confidential Books' that had been circulated to naval vessels during the First World War to summarise the Naval Intelligence Department's knowledge of the ships of the German fleet. An extract dated October 1918 speculated on the design of the 'battlecruisers' *Graf Spee* and *Prinz Eitel Friedrich*, believed to be under construction. Although their design was not certainly known, it was thought that they probably resembled a plan reproduced on pages 88–89 of the book. This depicted a two-funnelled ship with six 38cm (15in) guns in twin turrets, one forward and two superimposed aft.

In the comprehensive accounts of German battlecruisers provided by Gary Staff and Aidan Dodson, there can be found a proposed design D48 dating from June 1913, which was then being seriously considered for the ships that eventually became *Mackensen* and her sisters. It is remarkable that this plan resembles in almost every significant particular that provided in the 'Confidential Book' five years later. There must be a strong suspicion that the Naval Intelligence Department had somehow obtained a copy of design D48; but as it is believed that hardly any of their records survive, probably we shall never know how this could have been done.

Sources:
Norman Friedman (ed), *German Warships of World War I: the Royal Navy's Official Guide to the Capital Ships, Cruisers, Destroyers, Submarines and Small Craft, 1914–1918*, Greenhill Books (London, 1992).

Gary Staff, *German Battlecruisers of World War One*, Seaforth Publishing (Barnsley, 2014), 309.

Aidan Dodson, *The Kaiser's Battlefleet: German Capital Ships 1871–1918*, Seaforth Publishing (Barnsley, 2016), 228.

PATTERNS OF POLITICAL NOMENCLATURE IN THE RUSSIAN AND SOVIET NAVIES (*WARSHIP* 2023)
Regular contributor **Stephen McLaughlin** has supplied some additional information on the significance of the choice of the name *Varyag* (or *Variag*) for the Soviet and Russian Navies.

Variag holds a special place in Russian naval history. At the very start of the Russo-Japanese War she fought against overwhelming odds at the Battle of Chemulpo on 27 January/9 February 1904. Accompanied only by the

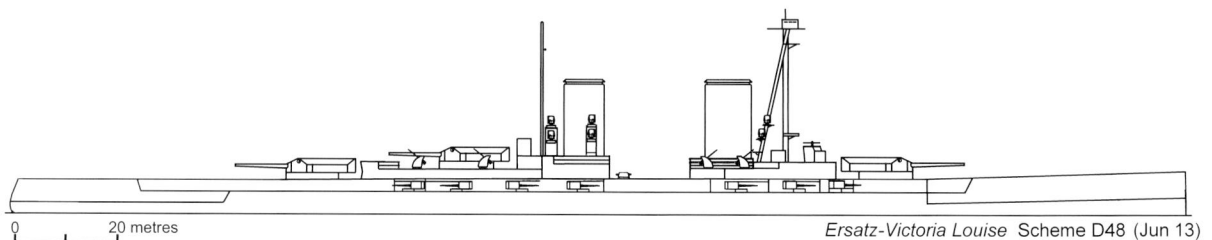

Ersatz-Victoria Louise Scheme D48 (Jun 13)

Sketch of the German D48 battlecruiser design of 1913, armed with six 38cm guns. (Aidan Dodson, *The Kaiser's Battlefleet*; drawing based on Grießmer)

The Russian protected cruiser *Variag* prior to the Russo-Japanese War. A photo of her 1960s counterpart, the missile cruiser *Variag*, is the first image in this year's Warship Gallery. (National History & Heritage Command, NH-94402)

gunboat *Koreets* she steamed boldly out to engage a much superior Japanese cruiser squadron. Although Japanese sources state that none of their ships was hit in the action, at the time the Russians believed they had damaged several of their opponents – and some in Russia still maintain this view. What is indisputable is that *Variag* was severely damaged in the battle, and both she and *Koreets* were forced to return to Chemulpo, where *Variag* was scuttled and *Koreets* blown up by her crew.

Variag's entire ship's company became national heroes, and her commander, Captain 1st Rank Vselvolod Fedorovich Rudnev, was awarded the order of St George and was appointed an *aide-de-camp* to the tsar. Rudnev died in 1913, but even after the October Revolution he remained a revered figure due to his relatively liberal views – he had reportedly been forcibly retired in 1905 because he failed to stop political meetings at Kronshtadt in the wake of the October Manifesto. Indeed, he was held in such high esteem that the Soviet regime erected a statue to him in 1956 in Tula, near the place of his birth. In 1946 a film was made about the cruiser's final battle, featuring *Avrora* with a false fourth funnel added, and in 1954 the surviving members of her crew were awarded a special medal 'For courage' (*Za otvagu*). Songs were written about *Variag*, and the Soviet postal service issued stamps depicting her.

Variag's name was given to a *Sverdlov* class cruiser (Project 68bis) which, however, never entered service, a victim of Khrushchev's defence cuts of the late 1950s; the name was then assigned to a 'Kynda' class (Project 58) missile cruiser. By the time she was stricken in 1991 the name had been transferred to the 'heavy aviation cruiser' (aircraft carrier) originally named *Riga*, which was building at Nikolaev, but the incomplete ship was sold to China, where she was eventually completed and commissioned as *Liaoning*. The name *Variag* is currently borne by a *Slava* class missile cruiser laid down in 1979 and originally named *Chervona Ukraina* ('Red Ukraine').

NAVAL BOOKS OF THE YEAR

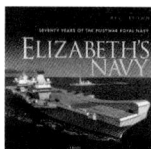

Paul Brown

Elizabeth's Navy: Seventy Years of the Postwar Royal Navy

Osprey Publishing, Oxford 2023; hardback, 344 pages, illustrated with numerous B&W and colour photographs; price £45.00.

ISBN 978-1472-8549-7

Paul Brown's 'coffee table'-sized book is a factually-based account of the senior service during the 70-year reign of the country's longest-serving monarch and is appropriately divided into seven, roughly decade-long chapters. Each comprises an introduction followed by pages of large monochrome or colour photographs of warships, with extended captions. These present the full range of RN warships and RFAs which comprised the fleet during the period in question, and are from a fairly wide range of sources; some might be familiar but certainly not all. The vast majority show the vessels underway; there are few close-ups and none from below decks.

Each chapter summary begins with a statistical analysis of the numerical and personnel strength of the Navy at the start of the decade, followed by a brief *résumé* and update on its infrastructure and standing commitments both at home and abroad. The rest of each summary deals with individual events both large and small. The 1980s chapter, for instance, is inevitably dominated by a succinct account of the Falklands War. However, in other cases the author selects a range of incidents to describe, which in turn reflect the significant political and social events of the era in question. Unfortunately, there is some repetition in content in the summaries and the captions which might have been avoided with more careful editing.

The author lends proper weight to the text with a considerable amount of data drawn from a mix of sources, both printed and electronic. Some originate from government statistics, others use information gleaned from the likes of *Navy News* or the website Navy Lookout. This lends a refreshing variety to the narrative. Interestingly, Brown often makes mention of collisions or the striking of underwater obstacles involving HM ships, both accidental or caused through negligence. Invariably these occurrences led to court martial, reprimand and in extreme cases dismissal from the service. Even the much-vaunted 'Perisher' submarine course was not exempt. However, the author does not appear to be criticising the RN for sloppy standards, rather it is a feature of his notably dispassionate, non-judgemental approach.

This work is timely. A spate of books published in the 1970s and '80s by the likes of Paul Kennedy, Eric Grove, Desmond Wettern, A Cecil Hampshire and others charted the decline of the Royal Navy at the drawdown of Empire marked by the withdrawal of all British armed forces from bases East of Suez. With the exception of John Roberts' neutral chronological record, *Safeguarding the Nation* (2006), in the intervening years there has been no similar, comprehensive assessment of the postwar Navy. This means that *Elizabeth's Navy* should have a valuable 'shelf-life' as a reference book.

The great theatrical tragedies normally either offer a modicum of hope for the future or else end dismally on a note of stoicism. Paul Brown's chronicle of the sad and possibly terminal decline of the once all-powerful Royal Navy belongs to the latter category, but nonetheless is a beautifully presented valedictory for the modern Elizabethan era.

Jon Wise

Mihály Krámli

Austro-Hungarian Battleships and Battleship Designs 1904–1914

Belvedere Meridionale, Szeged 2021; large format softback, 196 pages, copiously illustrated with plans and B&W photos; price $15.00 + postage [see below].

ISBN 978-615-6060-43-3

This unusual book, originally published in Hungarian and recently translated into English, is the result of two decades of research by the author, Mihály Krámli. The English-language version is a cut-down version of the original, focusing on the battleships designed for the Austro-Hungarian Navy from 1904: the three-ship *Radetzky* class, the four dreadnoughts of the *Tegetthoff* class, and the projected 'super-dreadnoughts' of the 'Improved *Tegetthoff*' class, which would have been armed with 35cm guns.

The design history and technical characteristics of each of these types are the subject of three separate chapters, which are followed by a general service history and a particularly detailed chapter on the Austro-Hungarian heavy guns, their mountings and their ammunition. The book is copiously illustrated with original plans and artwork, photographs (many from Hungarian sources and previously unpublished) and data tables.

The primary focus of the text is on the complex politics of the Dual Monarchy. Funding of the ambitious new naval programmes of the 1900s was hostage to trade-offs between the 'German' and the 'Hungarian' sides of the political establishment, with the orders for the ships and their armaments having to be shared out equitably between the respective shipyards and factories, and the chapters on design history give a comprehensive and detailed account of this often fraught process. The English translation is clumsy in places, and there are numerous grammatical errors and *non-sequitur*s; however, the meaning is generally clear.

The author has opted to reproduce the original plans of the ships rather than have them redrawn. The plans are not available elsewhere and show a high level of draughtsmanship; however, they have not always reproduced well: the large scale and faded nature of the originals means that linework is often faint and annotations illegible. The drawings of the guns and their turrets, on the other hand, are superb, and are reproduced at a good size with clear labelling. There is a particularly interesting (original) plan of the 35cm APC projectile on page 155.

This book is an important addition to the canon, and would be an excellent companion to the work with a similar title by Zvonimir Freivogl published by Despot Infinitus in 2018 and reviewed in *Warship* 2020. However, it is available only from the author. If you would like a paper copy, please contact Mihály Krámli at kramlimi@gmail.com. Alternatively, if you are happy with a low-resolution PDF, this can be freely downloaded from the publisher's website.

John Jordan

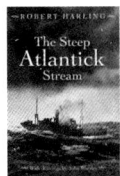

Robert Harling
The Steep Atlantick Stream: A Memoir of Convoys & Corvettes
Seaforth Publishing, Barnsley 2022, 218 pages, nine line drawings; price £14.99.
ISBN 978-13990-7288-5

Accounts of the Battle of the Atlantic are always in demand: general histories of the entire campaign, analyses of individual convoy battles, biographies of 'aces', technical histories dissecting every last detail of U-boat or escort construction and so on. Although the passing of time has brought some fresh information to light, this is a well-worn topic. It is therefore a delight to discover this gem, originally published in 1946 by someone with first-hand and very recent experience.

The Steep Atlantick Stream is a memoir of a period in the war, roughly 1941–43 (the author is inexact about dates), when the Battle of the Atlantic was at its height and the Allies struggled to cope with the U-boat wolf-packs. Harling served as first lieutenant aboard HMS *Tobias*, the fictional name for his 'Flower' class corvette. After a 'prologue' in London during the worst of the Blitz, Harling makes fast passage to Halifax, Nova Scotia, where *Tobias* is under construction, on board RMS *Queen Mary*. Assigned to a small convoy escort for the return crossing, the corvette's alert asdic rating detects a U-boat and holds the contact. To the utter chagrin of the crew, the convoy's Senior Officer arrives in his destroyer and takes over, ordering *Tobias* to resume her position on the convoy screen. They later learn that 'their' U-boat has been sunk.

This is the first and last time the ship has direct contact with the enemy while the author is on board. Thereafter, 'battles' always seem to take place on the far side of the convoy, Focke-Wulfs circle maddeningly just out of range and *Tobias* is sometimes left to pick up pitiful huddles of survivors from tankers and merchantmen as they endlessly criss-cross the North Atlantic. Visits are made to Iceland, Gibraltar and Freetown and each of these

wartime destinations is described in the writer's old-fashioned yet erudite style; even the book's title, with its archaic spelling of 'Atlantick', is a quotation from Milton's masque *Comus*. Robert Harling, both before and after the war, was a journalist, novelist and a renowned typographer. His fellow officers, with the exception of a RN Chief Engineer, are all RNVR and bring to a close-knit wardroom diverse life-experience and banter in equal measure.

Harling does not spare the reader in his descriptions of the grim monotony of life aboard these corvettes: the constant battle to stay upright or horizontal as the small ship rolls and lurches, of trying to suppress the ever-present threat of sea-sickness or simply the sheer misery of being forever wet or cold. *The Steep Atlantick Stream*, though, is at its heart a lament. Without giving away everything, the author must have felt compelled to complete his memoir so soon after the war end through a sense of personal loss. As a counter-balance to the many factually-based analyses of the Battle of the Atlantic, this work should be read for its many insights into the stark reality of war; it compares well with Montserrat's *The Cruel Sea* or Remarque's *All Quiet on the Western Front*. Congratulations to Seaforth Books for re-discovering and re-printing a classic.

Jon Wise

Michael Ellis, Gustaf von Hofsten and Derek Law (eds)
The Baltic Cauldron: Two Navies and the Fight for Freedom
Whittles Publishing, Dunbeath 2023; hardback, 290 pages, 211 illustrations (120 in colour), 12 maps, bibliography; price £30.00.
ISBN 978-1-84995-549-2

Sweden rarely appears in the annals of British naval history, as the two countries were hardly ever at war, but this collaborative work (first published in Swedish as part of the commemoration of the 500th anniversary of the foundation of the Swedish Navy) is devoted to the periods when there was a naval relationship between them. There are 21 chapters, of which ten concern the era after 1850. Sweden's prominence as a naval power had passed by then, but the earlier centuries had seen several wars with Denmark or Russia. England's main interest in the region had been to protect her Baltic trade, particularly important in naval stores such as timber, iron, flax and tar. It was this requirement that led to the first voyage of an English squadron to the Baltic, in Cromwell's time. This strategy would be repeated during several later wars; circumstances dictated that Britain was usually concerned to ward off threats to Swedish independence.

Perhaps Britain's most significant intervention came during the latter years of the Napoleonic Wars. The British squadron under Admiral Saumarez had the task of preventing Russia from invading Sweden, but preferably avoiding a full-scale war, while also encouraging Sweden to continue to break (covertly) the Emperor's

Continental Blockade, which Russia had compelled Sweden to support officially. In this the Admiral was entirely successful, and within a couple of years the Swedes had changed sides. Long afterwards King Karl Johan (the former French Marshal Bernadotte) declared to his son that Saumarez had saved Sweden.

A later chapter examines the naval ramifications of Swedish neutrality during the First World War, but several are devoted to Anglo-Swedish naval affairs during the Second. Some aspects of this will be known to British readers, such as the exploits of George Binney in ensuring the continuance of vital exports of Swedish ball bearings to Britain. Details are also given of the process by which the first news of the departure of the *Bismarck* was clandestinely passed to Britain by a Swedish intelligence officer. (One of those involved would, fifty years later, be a contributor to *Warship*.) Other significant episodes will be less well known in this country, notably the Royal Navy's brief confiscation in June 1940 of four small destroyers that Sweden had purchased from Italy, and Winston Churchill's plan, in the same month, for naval aircraft to lay mines off the port of Luleå to stop exports of iron ore to Germany – it was derailed only by the crisis on the French front.

Developments in the postwar era are covered in the concluding chapters, which include much information previously available only in Swedish. They reveal that Sweden, though a neutral country, had closer naval links with Britain than were known at the time. For instance, a distinguished officer of British coastal forces was seconded to help the Swedish Navy improve its tactics in that branch of warfare, and Sweden had assistance from Britain in the development of radar. Another sign of cooperation (which was publicly announced) was the sale of the last British midget submarine, *Stickleback*, to Sweden. By the 1990s some Swedish officers were attending Royal Navy training courses. These examples show how this wide-ranging work illuminates aspects of naval history that will often be unfamiliar to British readers.

Kenneth Fraser

Steve Kershaw
A Royal Navy Cold War Buccaneer Pilot: Flying the Famous Maritime Strike Aircraft

Air World, an imprint of Pen & Sword Books, Barnsley 2023; hardback, 201 pages, 36 B&W illustrations; price £20.00/$42.95
ISBN 978-1-39904-012-9

This 'autobiographical' book is unusual in that it was actually written by Steve Kershaw's son, Simon, who used his father's diaries and lovingly-retained letters to his family to create an account of his naval career in Steve's own words. He was killed in a tragic accident flying a Buccaneer in 1974 and, before going any further, the reviewer must declare an interest in that Steve and he served together and were friends. A number of other friends and contemporaries have contributed their own recollections of events to broaden Steve's story.

Setting aside the slight misgivings on the part of the reviewer engendered by this personal link, this is a totally absorbing book. It gives insight into the career of a naval pilot and captures the atmosphere of naval air squadrons as they operated fifty years ago in both the strike and commando helicopter roles from the aircraft carriers *Eagle*, *Hermes* and *Ark Royal*. Naval air squadrons have always been close-knit teams of professionals, and this book captures that atmosphere well. The reader is drawn into the everyday life of a carrier pilot with due emphasis laid on the things considered important at the time: the grading given to every deck landing (blue, green, amber or red); the evening film in the wardroom (Was it a new one and how good was it?); and eagerly-awaited mail from home, which was the only real contact with loved ones in the pre-digital era. The recollections inserted from other squadron members including Arthur Davies, his observer, are well-chosen and illustrate the respect in which Steve was held by his contemporaries.

Steve served in two Buccaneer squadrons, 800 and 809 NAS, and in a single commando helicopter squadron, 845 NAS, and besides illustrating day-to-day life in all three the narrative describes the contemporary roles that the units fulfilled, adding both human interest and operational detail to the historiography of an era in which the Cold War dominated defence thinking. Carriers were a vital national asset that were respected across the world; Steve's words bring back the feeling all shared that successive Governments failed to understand this.

The last chapter describes the accident in which Steve was killed but his observer, David Thompson, survived, when his Buccaneer impacted the sea surface during a low-level bombing run on a target in the Wash Bombing Range in 1974. It is based on evidence from the RN Board of Enquiry, Accident Investigation Unit and David Thompson's recollections. The result is both accurate and moving.

This is an excellent book that perfectly captures the enthusiasm and atmosphere of the Royal Navy's Fleet Air Arm all those years ago. I cannot recommend it highly enough to anyone with an interest in naval aviation.

David Hobbs

Angus Konstam (illustrations by Adam Tooby)
British Frigates and Escort Destroyers 1939–45

Osprey Publishing, Oxford 2023; paperback, 48 pages, 53 illustrations (12 in colour); price £12.99/$20.00.
ISBN 978-1-4728-5811-5

Nelson once said that when he died, 'want of frigates' would be found written on his heart, and Sir Dudley Pound might have thought the same about escort vessels. Before the Second World War, the Admiralty had foreseen the need for many more of them. Their response took two forms. One was a small destroyer (the 'Hunt' class), suitable for anti-aircraft or anti-submarine work rather than torpedo attacks. When the ships entered service it proved that they had been overloaded with weaponry and their stability was suspect; armament was reduced or (in later ships) beam increased, so there were several sub-groups.

In order to escort a transatlantic convoy it was necessary to reduce armament and speed to increase range. Corvettes had really been intended for coastal or short sea voyages, so were scarcely adequate for the Atlantic. The answer was the frigate, also built in mercantile shipyards, but larger, faster and better armed than the corvette. Later classes featured prefabricated construction and improved weapons.

The development of these various types is described in this book in a fashion typical of the Osprey 'New Vanguard' series (of which this is No 319), with appropriate lists of data, numerous photographs, and eight pages of detailed full-colour illustrations. The text is informative but occasionally in error, notably in claiming that the second group of 'River'-class frigates carried four 4in guns, which was not the case. There is a paragraph about the 'Captain'-class frigates (American destroyer escorts obtained by Lend-Lease) but nothing about the 'Colony' class, effectively US copies of the 'River' design with American armament. The numerous Australian and Canadian 'Rivers' have also had to be omitted due to lack of space. However, this useful little book presents a concise summary of the designs of two classes of ship that played a vital role in the Second World War.

Kenneth Fraser

Mark Stille (illustrations by Adam Tooby)
Essex Class Aircraft Carriers, 1945–91
Osprey Publishing, Oxford 2022; softback, 48 pages, illustrated with 42 photographs in B&W and colour plus 8 pages of full colour artwork; price £11.99.
ISBN 978-1-4728-4581-8

This book covers the development of the aircraft carriers of the *Essex* class after the successful conclusion of the war. It opens with a brief description of the design origins of the class, but then focuses on the various upgrade packages designed for and applied to the ships postwar. This enabled USS *Lexington* to continue in service through the late 1980s up to the 1991 of the title, albeit as a dedicated training carrier, her sisters being generally decommissioned in the late 1970s. Nevertheless this was an impressive record for the class as a whole given the accelerating design of aircraft over the same period.

This book is No 310 in Osprey's 'New Vanguard' series and follows their familiar layout, the text being interspersed with photographs and specially prepared colour images. Full-page deck and profile drawings illustrate the various modifications from 1951 onwards. The first of these is the only one not showing the addition of an angled flight deck, progressively added to most of the surviving ships and one of the features that allowed them to continue operating newer aircraft types. Other modifications were perhaps less radical, but included a 'storm bow' that closed the forward end of the hanger deck,

more powerful arresting gear and catapults, and electronic upgrades, much of it allied to the increasing complexity and size of the air component.

The book then covers the careers of every surviving unit of the class, making it a useful reference for the historian and a good source book for modellers.

WB Davies

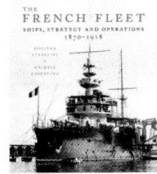

Ruggero Stanglini & Michele Cosentino
The French Fleet: Ships, Strategy and Operations 1870–1918
Seaforth Publishing, Barnsley 2022; large format hardback, 320 pages, copiously illustrated with B&W photos; price £45.00.
ISBN 978-1-5267-0131-2

The timing of this book is puzzling, as publication follows closely on from Stephen S Roberts' excellent *French Warships in the Age of Steam 1859–1914*, also from Seaforth. Photographic coverage is more extensive, but in terms of the technical data provided it is a step backwards, and there are errors that would probably not have been made had the authors had the opportunity to read Roberts' book.

The authors are not English native speakers. The general, introductory chapters (on French foreign policy and budget/shipbuilding programmes) read well, but the editing of the technical chapters leaves a lot to be desired, resulting in some odd phrasing and inappropriate terminology. Watertight traverse bulkheads are 'hinged on' the platform deck, the cruisers of the *Charner* class have 'horizontal-expansion' engines, and a number of ships have 'rammed bows'. The French *pont* is occasionally translated as 'bridge' rather than 'deck', and *machine alternative* as 'alternative' rather than 'reciprocating' engine; horsepower figures for the latter engines are given as 'shp', not 'ihp'.

The authors' grasp of the technology of the day is patchy and tenuous. This is particularly evident in the coverage of propulsion and protection. There were no examples of '5-cylinder' engines (p 174); moreover, high-pressure cylinders are not 'larger' than MP/IP and LP cylinders (p 60), and a 4-cylinder engine does not imply 'quadruple expansion' (p 220). There is no attempt to explain the relative advantages and disadvantages of small-tube vs large watertube boilers, despite this being key to the development of the French armoured cruisers. With regard to armour, the authors have assumed that the French adopted the Harvey face-hardening process from the outset, whereas the first ships with Harveyised steel entered service only from 1898; the battleship *Iéna* was the first to have a face-hardened waterline belt. The deeper belt of the later armoured ships was introduced by Bertin, and pre-dates Tsushima (p 62). There are similar confusions regarding the adoption of Barr & Stroud rangefinders: the first to be widely adopted was an 'export' model with a 2-metre base, not the later 9ft (2.74m) model (p 66).

There are also some bizarre errors: the battleships of the *Marceau* class had single, not twin, 34cm mountings (barbettes, not 'turrets') fore and aft (p 97), and it was the dreadnought battleship *Jean Bart*, not the armoured cruiser *Waldeck-Rousseau*, that was torpedoed by an Austro-Hungarian submarine in late 1914 (p 176).

Despite the above criticisms the book has its strengths. The photographs, many from Marius Bar, are superb and beautifully reproduced, and many of the images of the torpedo boats, destroyers and submarines are unfamiliar. The introductory chapters are generally well-researched and well-written, and the sections on the French development of wireless and the torpedo are excellent. There are no plans of the ships or their equipment; the authors have clearly consulted these, even though the information they provide has sometimes been misinterpreted. This means that the descriptions of individual ships and classes of ship has had to be rendered in prose; there are tables of characteristics, but these are selective in scope and the data is relatively basic.

It is good to see more books being published on the French Navy of the period, and this is a valuable addition to the canon. However, much of the analysis and the data presented here needs to be viewed with a critical eye.

John Jordan

Steve R Dunn

The Harwich Striking Force: The Royal Navy's Front Line in the North Sea 1914–1918

Seaforth Publishing, Barnsley 2022; 310 pages, 130 B&W images and 1 map; price £25:00/$42:95. ISBN 978-1-3990-1596-7

The author has written a number of books for Seaforth on First World War topics involving the Royal Navy, including *Southern Thunder* (the defence of merchant shipping trading between Scandinavia and the UK) and *Securing the Narrow Seas* (the Dover Patrol). This title is an extension of his interest in specific First World War naval commands and their achievements. The part played by Commodore Tyrwhitt, who commanded the Force throughout the conflict, is a central theme of the narrative, which focuses on the light cruisers and destroyers that served with the force; submarine, air and patrol forces based at Harwich are also covered but not in the same level of detail, as they were not under Tyrwhitt's direct command. The development of coastal motor boats as offensive weapons is described, as is Tyrwhitt's frustration that the naval staff did not share his own enthusiasm for their use against the enemy battle fleet in its harbours; the CMBs did not achieve their full potential until Baltic operations against the Bolsheviks in 1919/20.

The offensive and defensive use of mines are subjects in their own right, and Dunn makes the point that every aspect of naval warfare had changed in the years immediately prior to the outbreak of war and continued to do so as the conflict evolved. The inability to navigate

precisely in the stormy waters of the North Sea and to maintain accurate situational awareness of both friendly and hostile warship dispositions between 1914 and 1918 is almost as far removed from modern understanding as the skills needed to operate Nelson's sailing ships. Dunn's description of the small, open, canvas-screened bridge of a destroyer on which the captain, navigating officer and several ratings in wet oilskins or duffle-coats huddled between a chart table, flag locker and a gun at action stations helps the reader to appreciate the difficulty of operating contemporary warships. Tyrwhitt's decision to live on board until the war was over gives another very personal insight into the man and his beliefs. Key elements of evolving naval warfare drawn out by Dunn include the emergence of signals intelligence and its successes and failures when used by the Admiralty to cue operations with information derived from Room 40. Another element was the 'Beef Trip': escorting convoys of food and other materials from neutral countries after 1916. By November 1918 some 1,816 ships had been escorted with only seven lost.

There are some errors and imprecisions. In the early part of the book Tyrwhitt's ships are described as 'flag-ships'; however, commodores were not 'flag officers'; their rank was indicated by a broad pennant not a flag. More could also have been made of the development of lighters towed by Harwich Force destroyers from which Sopwith 2F.1 Camels could take off to provide air defence over operations in the Eastern North Sea, and of the routine operation of flying boats from similar lighters. Below a historic photograph of barge H.3 Dunn writes that the lighter was towed 'at the end of 600 fathoms of line' (equivalent to 3,600ft, or well over half a nautical mile!). Measurements taken from photographs comparing the length of the destroyer *Redoubt* with the position of its lighter show that the tow length was actually a much more practical 150 feet.

Overall, this readable book gives an interesting introduction to the Harwich Force and Commodore Tyrwhitt, and its recommended UK price is certainly attractive. It could have gone into greater detail about some aspects of operations from Harwich, and more detailed descriptions of the new technologies as they evolved would have been welcome, but the activities that it does cover will, hopefully, stimulate readers' interest to find out more.

David Hobbs

Maurizio Brescia & Augusto de Toro

Italian Heavy Cruisers: From *Trento* to *Bolzano*

Seaforth Publishing, Barnsley 2022; large format hardback, 232 pages, copiously illustrated with line drawings and photographs in B&W and colour; price £40.00. ISBN 978-1-3990-9885-4

Italian Heavy Cruisers (or '*Crusiers*', as it reads on the spine!), like its predecessor *Italian Battleships*, has its origins in two separate studies published between 2018 and 2020 in the series *STORIA Militare Briefing* and follows the same approach. This is problematic from a developmental point of view, as in the earlier technical chapters *Bolzano* is grouped with *Trento* and *Trieste*. While *Bolzano* was intended to operate tactically with the earlier two ships, and shared with them her exceptionally high speed and relatively light protection, in design terms she was a completely different ship, with the 203/53 Model Ansaldo 1929 in place of the earlier 203/50 Mod 1924 and a more modern propulsion system, and her external configuration and internal layout owed much to the contemporary *Zara* class. It is striking that Chapter 2, which has the general technical description of the *Trento*s and *Bolzano*, is illustrated by a comprehensive set of plans of the former, but has only a body plan and a plan of the propeller shaft arrangements for *Bolzano*. The merging of two separate studies also means that there is a degree of repetition, particularly with regard to the secondary armament.

In terms of its organisation the book begins with a chapter placing the Italian 10,000-ton cruisers within the broader context of the Washington Treaty and the resulting developments in Italian naval policy. Chapters 2 and 3 outline the technical qualities of the *Trento*s and *Zara*s respectively, while the next three chapters focus on weapons and fire control, aviation facilities, and camouflage. The book concludes with chapters on prewar and wartime careers, and some comparisons with their foreign 10,000-ton counterparts, with a particular focus on the British Royal Navy and the French Navy.

The book has strengths and weaknesses. The introductory chapter is muddled and imprecise. The following technical chapters are better, although the terminology, which is translated from the original Italian, is often confusing: 'transoms' for transverse bulkheads, 'ordinates' for frames, 'shrapnel' for splinters, the use of the dated 'Orlop deck', and a mysterious 'ice room' above the rudder. There is also the occasional conspicuous error: the table on page 85 translates *Omogeneo Duro* as 'KC', whereas the Italian terminology (and the modest thickness of the plates) suggests non-cemented nickel-chrome steel. The book is unconvincing in its attempted explanation for the overweight of the *Zara* class, the designed (standard) displacement of which largely exceeded Washington Treaty limits. Finally, the chapter on the ships' wartime careers has detailed accounts of all the major actions in which they were involved but no action diagrams or theatre maps.

On the plus side the plans and drawings are superb, particularly those of the superstructures by Roberto Maggi (Appendix D). These were handsome ships, and the photographic coverage is likewise of the first order. Full advantage has been taken of the photographs and plans in the layouts. The issue of dispersion of the main guns is comprehensively covered and the discussion is backed up by data tables. Interesting points are made concerning Italian Navy practices and there are some useful appendices. Chapter 8 ('Conclusions and Comparisons') has detailed comparisons of the two main types of Italian 10,000-ton cruiser with their British and French

contemporaries and makes some perceptive observations.

All in all, this is a very good book which could have been even better had it been written *ab initio* with a single volume in mind and if some of the 'rough edges' had been eliminated by more thorough editing.

John Jordan

Trent Hone
Mastering the Art of Command: Admiral Chester W Nimitz and Victory in the Pacific

Naval Institute Press, Annapolis 2022; hardback, 430 pages, 15 B&W illustrations, maps, diagrams; price $43.95/£41.95. ISBN 978-1-6824-7595-9

Chester W Nimitz was appointed to command of the US Navy's Pacific Fleet in the aftermath of Pearl Harbor, and retained that position until the Second World War ended. His was a new type of command, operating from a distant headquarters building with secure intelligence links, not a flagship engaged in combat. Trent Hone brings his expertise in interwar USN operational and tactical doctrine to bear on the issue of 'command'.

The key question is: How did an Admiral based ashore in Hawaii command the Pacific Theatre? The answer is complex. Nimitz was a perfectionist, a man-manager and a team builder. Taking command after running the Navy's Bureau of Personnel, he chose the best men and knew how to exploit their particular talents. His HQ remained small and agile, and above all he avoided the classic mistake of trying to manage risk out of operations by attempting to know everything. Seeking to out-think and out-manoeuvre a relatively static and increasingly predictable enemy, Nimitz acted quickly and took educated risks, notably those that led to Midway, one of many occasions when his ability to exploit intelligence proved critical. This intellectual agility reflected a mastery of maritime strategy: having studied Julian Corbett's work at the US Naval War College in the 1920s Nimitz understood how to integrate sea, land and air, rather than relying on mass.

Nimitz's reliance on the cerebral, sophisticated Raymond Spruance flowed from a shared understanding of strategy. Their campaigns were not linear or predictable: Nimitz worked from concepts, not detailed plans, making possible rapid responses to emerging opportunities. Accepting calculated risk enabled him to learn and act quickly, seeking the initiative, indoctrinating his operational commanders and knowing when to rotate or relieve tired individuals. He did not micro-manage from a distance, but did step in when he had superior insight, often based on his intelligence sources. He relied on the intelligence of key subordinates, a trust Spruance amply repaid, while 'Bill' Halsey's focus on battle and his reluctance to listen to key staff almost led to a disaster at Leyte Gulf. High tempo feedback loops processed every new experience and improved future operations.

Nimitz was not self-absorbed or concerned with his public image, in stark contrast to his fellow theatre commander General Douglas MacArthur, whose self-regard compromised American strategy. Nimitz justified the faith of President Roosevelt and Fleet Admiral 'Ernie' King, who directed the wider war from Washington. Above all Nimitz understood that war was a political act. When Washington wanted to hasten the advance across the Pacific he obliged, exploiting the dramatic increase in naval resources and logistics support that began to flow in 1943 to keep the Japanese on the defensive, ignoring key bastions like Truk. Anticipating victory in Europe in the autumn of 1944, Washington was anxious to conclude the Pacific conflict before public support ebbed; Nimitz accelerated his plans, securing the islands that enabled strategic bombing of Japan. Critically Hone emphasises that war and strategy are arts, not a science. Success cannot be guaranteed, because friction and human failure are ever present. Nimitz managed the shift of theatre command from the flagship at sea to the modern headquarters, and mastered the new methods this required.

Andrew Lambert

Vincent P O'Hara & Leonard R Heinz
Innovating Victory:
Naval Technology in Three Wars

US Naval Institute Press, Annapolis 2022; hardback, 301 pages, 26 B&W photographs and 11 figures; price $36.95/£38.95.
ISBN 978-1-6824-7732-8

Six key innovations in naval armaments are examined here: two weapons (mines and torpedoes); two tools (radio and radar), and two platforms (submarines and aircraft, especially carrier-borne). It is proposed that all have been the outcomes of a common innovatory process: a perceived need stimulates an innovation based on existing (sometimes new) technology. Successful deployment requires innovation in use, while experience of use, including of countermeasures and vulnerabilities, encourages further innovation. Also, new and unanticipated needs (and perhaps also new technology) may initiate new cycles of the process. The authors readily acknowledge that they are in new territory and that their backgrounds are non-technical. In general, they name key technological innovations without discussing how they were made; rather, their principal concerns are with the subsequent phases of the process.

Both mines and torpedoes were deployed before they were really effective, hence requiring further phases of innovation. Mines were used successfully in the Russo-Japanese War, and fear of mines (moored and dropped) influenced battlefleet tactics in the First World War. New needs were met by mine-laying submarines and by sowing large minefields in an attempt to bottle up U-boats. Influence mines were deployed in the Second World War, but a particular vulnerability and discouragement to use was that, once deployed, mines could be found and their secrets discovered.

From early on, torpedoes were deployed on large warships, while it took time to develop the platform for mass surface attacks, the destroyer. However, the influence of these torpedo threats on the composition and tactics of the Great War battlefleets receives little attention, the main focus being on the torpedo/submarine combination (see below). For the next war, the Japanese developed the very long range Type 93 torpedo as part of their strategy for decisive battle, though hitting at long range remained largely a matter of chance until homing torpedoes were developed.

The rapid development of radio was spurred by the need for communication with ships beyond visual range; surprisingly, the early involvement of Captain Henry Jackson RN is not mentioned here. Innovations in use as well as in technology were needed as the radio spectrum became ever more crowded. Inter-war, the development of naval radio benefited from the growth of commercial radio. In both world wars, the broadcast nature of radio made it vulnerable to direction-finding and dependent on encryption. Germany twice made the mistake of assuming that her communications were secure; in the second war, the centralised radio direction of the wolf packs contributed significantly to their eventual defeat.

The authors suggest that 'the concept of radar languished for three decades' and that 'naval professionals [were] unable to to embrace an invention thrust upon them by engineers and scientists' (page 115). However, while there were many early observations of radio reflections, the need of navies to measure ranges and bearings could not be met until key electronic components became available in the 1930s. By the end of the decade, navies had embraced the deployment of the first seaborne radar sets. Despite the example of the Chain Home system, the Allied navies had to relearn, mainly through 'bottom-up' innovations, that radar intelligence required an effective action information organisation.

Submarines firing torpedoes had some initial successes against warships in the early months of the First World War, but it soon emerged that the best protection was high speed and numerous escorts. Unrestricted submarine warfare against merchantmen was the great threat, only alleviated by the introduction of escorted convoys. The Royal Navy developed an effective depth charge in 1914–18 (it is unclear why this is called 'a lost opportunity' – page 190), but it was not until the next war that sonar provided the means to direct underwater weapons onto the target. This goes unmentioned, while sonar is dismissed as 'the solution to a problem ... based upon ... World War I experience' (page 176).

The chapter on aircraft is concerned principally with innovations in the use of aircraft aboard ships. The through-deck carrier HMS *Argus* could for the first time operate aircraft large enough to carry heavy weapons, initially torpedo-carrying Sopwith Cuckoos. Inter-war, further innovations in technology and use led to carriers equipped with strike and fighter aircraft. Innovations in aircraft (a vast subject) are discussed only selectively, notably the long ranges of Japanese carrier aircraft and the Royal Navy's need for multi-role aircraft due to the small capacities of its armoured-deck carriers. During the Second World War, carriers proved their ability to make decisive strikes against ships at sea and in harbour. By the end of the conflict, with all enemy battleships sunk or neutralised, the carrier ruled the waves, albeit from the centre of a protective task force.

Inevitably, given the impressively wide scope of this study, readers will find a few points on which they will disagree. However, its innovatory model applies to all six studies, though with minor qualifications in some cases. O'Hara and Heinz have provided a valuable conceptual framework that will surely stimulate further study of naval innovation, both broadly and specifically.

John Brooks

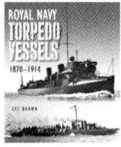

Les Brown
Royal Navy Torpedo Vessels 1870–1914

Seaforth Publishing, Barnsley 2023; 192 pages, numerous photographs & drawings; price £40.00.
ISBN 978-1-3990-2285-9

While the early destroyers of the Royal Navy were the subject of a seminal work by David Lyon back in 1996, the torpedo boats that preceded them and continued to be built in parallel with them down to 1909 have escaped their own book-length treatment. As the author sets out in his Introduction, the origins of this volume lie in his cataloguing of early annual reports from the torpedo school, HMS *Vernon*, with his research then expanded to embrace a wider set of published, archival and manuscript sources. The result is a very useful digest of what is known about these oft-forgotten vessels.

The volume begins with a chapter on the early history of non-locomotive torpedoes, especially spar torpedoes and the very first boats built to deploy them, constructed by such firms as Yarrow and Thornycroft for overseas customers. The next chapter covers the introduction of the locomotive (or Whitehead) torpedo and the first torpedo boats constructed for the Royal Navy. We then move to the *Vernon* and *Defiance* training and experimental establishments at Portsmouth and Devonport. While a broad overview of their responsibilities is given, the captions to the excellent photographs of both fail to date the images or identify most of the ships involved.

The chapter also covers the one-off torpedo vessels *Vesuvius* and *Polyphemus*, with a particularly extensive coverage of the latter, including an excellent set of images. Next come four successive chapters covering: first class TBs *1–79*; first class TBs *80–117*; the coastal destroyers (later TBs *1–36*); and the second class TBs. Each boat or group of boats is described, including modifications and fates. There is an account of key episodes in their careers, but it would have been useful to have more information on the boats' broader employment.

The final two chapters cover torpedo gunboats, and the torpedo boat parent ships *Hecla* and *Vulcan*. While the latter ship is described in detail, there is nothing to relate her design to the big protected cruisers of the *Blake* and *Edgar* classes. The volume closes with six appendices, covering the disposition of first class TBs, a report

on an experimental cruise in 1887, trials of second class TBs, quick-firing guns, the davits of *Vulcan*, and the turbine launch *Turbinia*.

The book quotes extensively from contemporary accounts, in particular from the journal *The Engineer*, and also official sources, all referenced as such. However, there are large sections of the main narrative that, from the archaic style, would seem also to be taken verbatim (or near-verbatim) from older accounts, but without any clear indication of the original source.

Illustrations are particularly comprehensive, and are derived from a range of published and unpublished sources. There are also a number of 'as fitted' drawings from the National Maritime Museum collection, including a number of sets in colour. These drawings have had their backgrounds cut away which, while making some aspects clearer, make others less satisfactory. In particular, it seems that the person making the changes was not wholly *au fait* with the nature of the drawings. For example, TB *98* seems to have a high superstructure running fore-and-aft (actually an awning with the negative space below not deleted), while the torpedo gunboats *Halcyon* and *Rattlesnake* seem to have their boats carried on wide sponsons. While an interesting experiment, it is to be hoped that Seaforth do not repeat this approach in other volumes reproducing NMM plans.

Aidan Dodson

Brian Lavery
Royal Yachts Under Sail

Seaforth Publishing, Barnsley 2022; large format hardback, 160 pages, many B&W and colour illustrations; price £50,00.
ISBN 978-1-3990-9291-3

Any book by Brian Lavery is bound to be well researched, well written and extremely well illustrated, and *Royal Yachts Under Sail* is no exception.

The book is organised into seven sections. The first, entitled 'Sailor Kings', starts by discussing the early instances of royalty going to sea, notably Henry VIII's trip to France for the Field of the Cloth of Gold. Splendid as they were, none of the ships used were 'royal yachts' and it is only when the young Charles Stuart, Prince of Wales, develops a fascination with the sea during his exile in Holland that we are introduced to the concept. Both he and his brother James enjoyed sailing, and occasionally racing their yachts.

The next section focuses on the yachts and their wider role in the wars of the 17th and 18th centuries; it covers the transport of royalty and government officials to and from the continent, service as tenders to larger ships, the yachts at war, and even their employment as tenders for the press.

The third section covers the Hanoverian period, when the yachts steadily became larger – more like small sloops than yachts. There is also a chapter on Deptford and Greenwich, the customary home of the royal yachts, while the fourth section, 'Marriage, Business and Leisure', covers the more limited uses of the yachts during the later Hanoverian period; it is followed by 'Uniting the Kingdom', which highlights George IV's use of the yachts for travel to Ireland and to Scotland.

The sixth section, 'Building and Sailing', is devoted to design, construction, sailing and manning, from the extravagantly-decorated yachts of the Stuarts to the more austere internal and external decoration of the Georgian age.

By the 1820s steam was making its appearance, and the last section of the book, 'Last Days Under Sail', covers the demise of the sailing yacht under Queen Victoria, who found steam a more comfortable mode of transport and commissioned the first steam yacht, *Victoria and Albert,* as well as smaller yachts. Her son Edward helped popularise a different type of royal yachting: his racing yacht *Britannia* had a total sail area of 10,000 square feet, and he was a regular attender at Cowes Week. Prince Philip continued this tradition, albeit on a much smaller scale, successfully racing small dinghies such as *Bluebottle* and *Coweslip* (taken out to Malta by the aircraft carrier HMS *Glory* when he and Princess Elizabeth were there).

This is an excellent book, covering an unusual topic and successfully combining details of the sailing yachts with descriptions of the major historical events, allowing the reader to put these vessels in context. Despite some very minor editing errors (most notably the aircraft carrier *Glory* becoming '*Glori*') the book is a joy to read.

Andy Field

Helmut Blocksdorf
Hitler's Secret Commandos: Operations of the *K-Verband*

Pen & Sword Military, Barnsley 2023; hardback, 188 pages, illustrated with 51 B&W photographs; price £19.99.
ISBN 978-1-84415-783-9

This book is a 2023 reprint of the 2008 original, and covers the activities of the *Kriegsmarine*'s *Kommando der Kleinkampfverbande* (*K-Verband* for short), a group using manned small craft for clandestine attacks on enemy vessels during the Second World War. It starts, however, with where the concept began in Europe: Italian small fast boats in the First World War, divers and 'charioteers' in the Second, and the Royal Navy's use of small submarines and improved chariots.

The book is conventionally set out with twelve chapters, an introduction and an epilogue, followed by an appendix listing badges and clasps. Chapter 1 covers the history, Chapter 2 the setting up of the *K-Verband*, while the remaining ten chapters cover individual types of craft and their operation, from midget submarines to manned torpedoes and exploding launches. This can be slightly confusing, as each chapter starts more or less from the beginning, leading to a degree of duplication. In addition the reader needs to be on top of the named individuals, as this is sometimes the only link between events and chapters.

For the historian this is a valuable book covering the little-known history of some very bold but ultimately unsuccessful missions. Indeed, although the author seeks to suggest that the personnel were mainly looking for action rather than being Nazi fanatics, one does wonder if boredom alone could make men persist with such near-fatal operations. Quite often the author has to report that such-and-such a craft and its crew's fate is unknown, or that out of nineteen craft launched on a mission sixteen failed to return. These were not strictly 'suicide missions', but many were extremely hazardous and survival was not a given. As with many such specialist groups, a postwar camaraderie seems to have developed, and some survivors went on to serve in the resurrected *Bundesmarine*.

WB Davies

Julia Jones
Uncommon Courage: The Yachtsmen Volunteers of World War II

Adlard Coles, Bloomsbury Publishing 2022; hardback, 320 pages; price £20.00.
ISBN 978-1-47298-710-5

What started as the chance discovery of the author's late father's cruise diary detailing his prewar adventures photographing German warships in the Baltic, has evolved into a comprehensive and compelling account of the heroism and sacrifice of the 'gentlemen sailors' of the Second World War.

At the heart of this book are the experiences of the 2,000 individuals who served in the little-known Royal Naval Volunteer Supplementary Reserve (RNVSR), hastily established in 1936 and subsequently subsumed into the Royal Naval Volunteer Reserve during the war. The aim was to utilise the talents of yachtsmen, with their navigational abilities and knowledge of the European coastline. These 'Gentlemen, who are interested in yachting, and were prepared to serve as officers in the Royal Navy in case of an Emergency' were initially without formal training, pay, rank or uniform. Yet this new organisation appealed to many patriotic but committed yachtsmen. Some volunteers were famous, others wealthy, but the majority were just 'ordinary' professionals who loved sailing. Few could have imagined what they would ultimately go on to experience.

Following the Munich Crisis, the Admiralty struggled to cope with the inrush of volunteers, and the RNVSR was not immediately mobilised. When it was, the boat handling skills and experience of the personnel was soon put to good use in the Patrol Service, manning the many small ships requisitioned for war service. As the war progressed these plucky volunteers stood against prejudice from the regular service and brought to situations an inventiveness that got results. In return they were given tasks vital to the war effort and demonstrated their worth under exacting circumstances.

In time RNVSR volunteers commanded destroyers and submarines. Some undertook the dangerous daily chore of minesweeping; others bomb disposal. A number fulfilled tasks such as landing spies and special forces in enemy territory, developing war-winning technology and assembling intelligence information on the enemy coast-lines that would ultimately allow the Allies to gain footholds in Europe. A few served in Ian Fleming's famous intelligence commandos, but the majority found their niche in the vast number of small boats and landing ships, engaging the enemy at high speed or conducting amphibious operations against defended shores.

By 1945 six years of war had taken its toll both of the yachts that had been laid up in 1939 and of the brave individuals who survived; many were burnt out, with symptoms of what would now be termed PTSD. The book's poignant final chapter describes how these men reintegrated with civilian life. While some returned to pre-war careers, others were irrevocably altered by the experience of war and opted for change. Some fell out of love with the sea and a few, unable to adjust after the unimaginable strain of wartime service, took their own lives.

The author is to be applauded for this masterful work, which gives a wonderful flavour of the RNVSR, with its lack of formality and its individuals capable of thinking and working 'outside the box'. The book is an engaging and entertaining read, combining first-hand accounts drawn from letters, journals and memoirs. All are empathetically woven together with a narrative that puts the experiences of these ordinary men who were placed in extraordinary circumstances into context.

Phil Russell

Brian Lane Herder, illustrated by Paul Wright
US Navy Armored Cruisers 1890–1933
Osprey Publishing, Oxford 2022; paperback, 48 pages,
34 photographs, 13 paintings & drawings; price £11.99.
ISBN 978-1-4728-5100-0

US Navy Protected Cruisers 1883–1918
Osprey Publishing, Oxford 2023; paperback, 48 pages,
34 photographs, 12 paintings & drawings; price £12.99.
ISBN 978-1-4728-5703-3

These two works together summarily cover the fifteen cruisers carrying side armour ('armored cruisers' according to the definitions of the time) commissioned by the US Navy between 1893 and 1908, and eleven of those with only protective decks ('protected cruisers') that entered service between 1886 and 1894. The latter volume does not include two ex-Brazilian vessels nor the eight vessels of the *Cincinatti* and *Denver* classes. (It is, however, noted in the Introduction that the *Cincinatti*s may be covered in a future volume). The books are in the standard 'New Vanguard' format, being the 311th and 320th of that series. The protected cruiser volume states that it is explicitly intended to supplement an earlier part in the series, *US Cruisers 1883–1904* (published 2008 as No 143), and to provide additional focus on vessels and events not covered in detail in that earlier work.

Both volumes begin with overviews of the origins of the 'modern' cruiser in the USA, which marked a very clear break with the old 'steam-and-sail' vessels of what came to be called the 'Old Navy', although some of the ships covered by the protected cruiser volume still carried canvas. The armoured cruiser story then looks at the Navy's first ideas for the type, including the intention that *Maine* be classified as an armoured cruiser, although she entered service as a battleship. The 'real' story is then kicked off with *New York*, which would not only be the first US ship to enter service as an 'armored cruiser', but the penultimate surviving example of the type as well. The justifications for the type are explored, as are the numerous doubts as to its true utility and consequential 'schizophrenic' reputation.

The technical basis for the ships is next addressed before each class is dealt with in turn, beginning with *New York*, and then moving through *Brooklyn* and the so-called 'big ten' of the *Pennsylvania* and *Tennessee* classes, before finishing with the 'semi-armored' *St Louis* class (so-called owing to their incomplete waterline armour belts). Operational history is then covered, beginning with the Spanish-American War – where the type's performance encouraged the building of the 'Big Ten' – and then service in the Atlantic down to 1917. Linked with this is a section describing the loss of *Memphis* (ex-*Tennessee*) to a freak wave at Santo Domingo in August 1916.

Next comes an account of Asiatic and Pacific service to 1917, with another story of accidental loss, of *Milwaukee*, stranded on the North California coast in January 1917. Focus then shifts to the use of 'Big Ten' vessels in the early development of US naval aviation. War service during 1917–18 is then described, including the loss of *San Diego* (ex-*California*) to a mine off New York in July 1918. The last part of the book considers the period between the end of the First World War and the disposal of the last US armoured cruiser, *Seattle* (ex-*Washington*), in 1946 after her final two decades as a receiving ship.

After its overall introduction, the protected cruiser volume looks at general matters of ship nomenclature, design and construction, protection, and weapons, before launching into a chronological account of the origins, building and careers of each of its subjects. Each entry includes the ships' specifications and a tabulation of key dates. The book then moves to overviews of the operations in which the vessels were involved, starting with the Squadron of Evolution of 1889–90, which took the first three protected cruisers to commission (*Atlanta*, *Boston* and *Chicago*), plus a gunboat, across the Atlantic and into the Mediterranean, both to show the flag and to undertake exercises to develop tactics under steam. Next comes the Spanish-American War of 1898, in which all but two of the ships covered took part. The period 1899 to 1923 is then covered quite briefly, including the 1918 Atlantic escort roles of the big *Colombia* and *Minneapolis*, *Olympia*'s Russian intervention deployment, and *San Francisco* and *Baltimore*'s work as minelayers in the North Sea. A final section takes a brief look at the preserved *Olympia*, the last surviving example of a protected cruiser.

Illustrations in both volumes are a mix of photographs, near-contemporary paintings/sketches, and specially-commissioned colour images, plus a few line drawings in the armoured cruisers volume (poorly reproduced from an uncredited contemporary reference work). The selection of photographs is generally good, covering many changes in ships' appearances, but all three of *Chicago* depict her after modernisation. The colour artwork covers a number of the ships, some contrasting their appearance in original white-and-buff and later grey liveries. These include an error, in that the white-and-buff (as built) and grey (1898) *New York* are shown as all-but-identical. However, this ship underwent a refit prior to 1898 which built up her after superstructure and stripped most of the platforms from her mainmast – as can be seen in a photograph published on page 34! The colour images include renderings of some of the more dramatic events in the ships' careers, such as the Battles of Santiago and Manila Bay, the loss of *Memphis*, and *Baltimore* minelaying in the North Sea.

Although there are areas one might have liked to see covered in more detail, with one or two important sources missing from the Bibliographies, these slim volumes provide a handy summary of the key facts about two important but often overlooked sets of ships of the US Navy.

Aidan Dodson

Leo Marriott
US Naval Aviation 1945–2003
Pen & Sword, Barnsley 2023; softback, 179 pages,
illustrated with 199 photographs; price £16.99.
ISBN 978-1-399-06257-2

This book, one of the 'Images of War' series, is the second from this author and features rare archive photographs of the US Navy's aviation. The previous title, subtitled *The Pioneering Years to the Second World War*, covered the period from 1898–1945, and the present volume brings the story up to date.

It is set out in ten chapters which are preceded by a useful glossary and list of abbreviations and a stand-alone introduction. The chapters cover specific periods or equipment types, for example; 'The Post war Legacy 1945–1950', and 'Helicopters and ASW', each starting with a couple of pages of introduction outlining the period/aspect covered, followed by several pages of photographs.

While manufacturer's and prototype photographs are always similar, it is possible that every photograph here is a first, thereby justifying the author's claim that they are rare. Each has a full caption offering concise but useful information. In places the author has managed to add something unusual, such as the suggestion that the Martin Mauler wasn't particularly easy to handle, or the story of Lt Denny Earl who managed to land on perfectly after a mission despite having both legs shattered by Vietnamese ground fire. Although primarily intended as a photograph album, the book is an interesting read and a worthy successor to the earlier volume.

WB Davies

Przemysław Budzbon, Jan Radziemski, Marek Twardowski
Warships of the Soviet Fleets 1939–1945: Volume I: Major Combatants
Seaforth Publishing, Barnsley 2022; large format hardback, 304 pages, copiously illustrated with data tables, B&W photos and line drawings; price £45.00.
ISBN 978-1-5267-0131-2

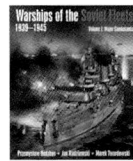

Building on the earlier work of Jürg Meister, Jürgen Rohwer, Siegfried Breyer, John Eriksson and Richard Pipes, the authors have benefited greatly from further work by Boris Lemachko and Vitalij Kostrichenko and the general opening up of the Soviet archives that took place with the advent of *Glasnost* to produce a definitive history of the Soviet Navy during the Second World War and a detailed reference for the various categories of surface ship and submarine designed and built from the mid-1920s.

The approach is essentially the same as that adopted by Stephen S Roberts in his recent *French Warships in the Age of Steam*. Introductory and explanatory chapters covering the historical context, the organisation of the book, the warships, building programmes, shipyards, naval bases and fleet organisation, and the mobilisation of civilian tonnage plus an evaluation of performance are followed by chapters for each of the categories of major combatant from ships of the line through monitors and gunboats down to motor torpedo boats. These chapters have an entry for each class, arranged in chronological order, that comprises a general description (including the design process), tables with building/deployment data and technical characteristics, plus a brief individual service history for each ship/submarine of the class.

The range and depth of research is impressive, and the quality and quantity of the data unmatched. The line drawings, by Tomasz Grotnik and Jerzy Lewandowski, are superb and cover the major variants of every class of warship. The photographs are inevitably a mixed bag, given the mediocre quality of Soviet optics during the period and the poor weather conditions prevailing in the northern Baltic and the Arctic regions during the winter months, but they are the best available; the collection reproduced here is comprehensive and each image benefits from an explanatory and informative caption that highlights the key points of interest and places it in context. The reproduction of both the photographs and the line drawings cannot be faulted. Where there are gaps in the data (particularly when it comes to the loss of a submarine), the authors are transparent regarding differences between the various accounts and are happy to point to the most likely explanation for the loss.

One constant theme that dominates the narrative is the impact of the complete break with the *ancien régime* that occurred after the October Revolution and its adverse consequences for the design and construction of the first generation of Soviet warships and submarines. The design experience painfully acquired by Tsarist Russia – which acquired a reputation for originality rather than competence – was almost completely lost, and the ambitious early Soviet 'guard ships', flotilla leaders and submarines all experienced serious technical problems when first completed. Obsolete vessels such as the destroyers of the *Novik* type, and battleships and scout cruisers whose design dated from the First World War, were not only retained into the 1940s but underwent piecemeal modernisations, with new weaponry simply 'bolted on'. There was also an increasing dependence on 'western' technology, initially from Fascist Italy and Nazi Germany, then from the Allied nations, which supplied motor torpedo boats and radars that were subsequently copied. However, some of these benefits would only be fully realised after the end of the conflict in 1945.

The authors of this series are native Poles. Nevertheless the quality of the English in the introductory chapters, including technical terminology, is impeccable. The only minor 'blip' is the slightly disconcerting omission of definite and indefinite articles in the 'technical' chapters, which was presumably influenced by a desire for concision and brevity in what is a comprehensive and ambitious work. Only occasionally does this cause confusion, and the book generally reads well despite its essentially encyclopaedic format. There can be little doubt that this will become the authoritative English-language account of the Soviet fleet to 1945.

John Jordan

Neil Datsun
The British Air Power Delusion 1906–1941

An Oxfordfolio publication, 2023, available through
www.theairpowerdelusion.com and Amazon; softback,
353 pages; price £18.95.
ISBN 978-1-9163-0998-2

Technological evolution has constantly changed the way the Royal Navy located and fought the nation's enemies. Most of this was absorbed within the service and, prior to 1918, the RNAS was so far ahead of other naval air arms that Captain GW Steele, USN, who had been tasked with evaluating its achievements, reported that 'any discussion of the subject must consider the Royal Navy's methods'. That changed in April 1918 when the RNAS and the Army's RFC were subsumed into a separate, independent Royal Air Force. For the following two decades the Air Staff convinced successive governments that, against all the experience gained in the First World War, bombers represented a new and decisive form of warfare, unrelated to the roles of navies and armies.

Neil Datsun observes that the Air Staff's insistence on bombing was intended to justify its independent existence, but made it the tool of an amateur strategist like Churchill. He may have been a great war leader but, from his time as First Lord of the Admiralty in 1914/15 onwards, he failed to comprehend the real value of aircraft and mistook them to be a weapon of strategic value that could be separated from other aspects of warfare. Their critically important tactical value in both anti-submarine warfare and army support was forgotten

by the Air Staff, and this delusion created an imbalanced air force that was incapable of opposing the tactical air offensives inflicted on British forces by Germany after 1939. Fleets, convoys and armies in the field had to operate without effective air cover because the RAF had the wrong types of aircraft flown by pilots with the wrong sort of training; Germany did not.

Quite rightly, Datsun explains that the decision taken in 1937 by the Minister for Defence Co-ordination, Sir Thomas Inskip, to return full administrative control of the Fleet Air Arm to the Admiralty was of critical importance in enabling the Royal Navy to fight the war at sea. The Royal Navy had the world's largest carrier building programme, but every attempt to provide these ships with sufficient aircraft had been opposed by the Air Staff. Datsun's logical appreciations reveal the depth of his research, and his comments about the Air Staff's failure to pay sufficient attention to the Royal Navy's needs are carefully chosen. Particularly damning is that on page 257 which quotes a conference held in July 1940 to discuss countermeasures to a possible invasion. C-in-C Nore, the senior RN officer present, asked for assurance that RAF fighters would defend warships intercepting German shipping in the Channel against enemy bombers. The Fighter Command representative replied that its role was the *defence of Britain* (Datsun's italics taken from the original document) and not defence of Britain's ships. Other examples are just as shocking.

This book adds a perceptive new dimension to the historiography of British war strategy and gives reasons for its weakness prior to 1941. Highly recommended.

David Hobbs

Les Brown
British Light Cruisers: *Leander, Amphion* and *Arethusa* Classes – ShipCraft 31

Seaforth Publishing, Barnsley 2023; softback,
64 pages, heavily illustrated in colour and B&W; price £16.99.
ISBN 978-1-3990-3056-4

Seaforth Publishing's popular 'ShipCraft' series will need little introduction to many readers. Targeted primarily at warship modellers but of value to the more general enthusiast, each aims to provide an overview of the design and history of a particular class or wider group of warships, as well as of the kits, accessories and visual references needed to recreate them in miniature. Largely produced to a standardised, 64-page paperback format, the books have benefited from a consistent and attractive design theme encompassing considerable use of both black and white and colour images, all at a relatively affordable price.

Les Brown's latest instalment in the series, encompassing the related British-built *Leander, Amphion* and *Arethusa* class light cruisers of the inter-war era, illustrates both the strengths and weaknesses of the ShipCraft format. Opening chapters that provide, first, a design history and then an overview of wartime actions and losses set the scene for the more detailed review of model-making products that follow. The latter comprises a wide-ranging examination of the various representations of the ships available to modellers in various scales, not least the outstanding 1/700 scale representations of many of these cruisers recently produced by China's Flyhawk. Options for detailing and otherwise further improving the kits are also covered. Inevitably, limited space means that analysis of product strengths and weaknesses is necessarily concise.

The subsequent 'Modelmakers' Showcase', which extends to a full quarter of the book's total page count, is arguably the highlight. Providing full-colour imagery of completed models in various scales, it demonstrates what can be achieved from the products previously described. Subsequent chapters covering camouflage and modifications – supplemented by various colour and line drawings – are intended to help ensure the accuracy of the modeller's endeavours.

There are two major weaknesses. The constrained page count means that hard choices have been required in terms of what and what not to include. This is particularly evident in the wartime overview. Here an understandable focus on better-known actions such as the Battle of the River Plate and the loss of HMAS *Sydney* means that many of these ships' most notable successes, for example against Italian forces in the Mediterranean, are entirely absent. More significantly, heavy reliance on secondary sources means that the book's accuracy is only as good as the information these contain. This is particularly problematic with respect to the camouflage chapter, where many descriptions are questionable and some very clearly erroneous. For a book claiming to be '… simply the best reference for any modelmaker …' this is a major drawback and one that needs attention if the series' reputation is to be maintained.

Conrad Waters

Vincent P O'Hara and Trent Hone (eds)

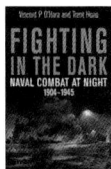

Fighting in the Dark: Naval Combat at Night, 1904–1944

US Naval Instutute Press, Annapolis 2023; hardback, 348 pages, illustrated with photographs, maps and diagrams in B&W; price $39.95/£25.00. ISBN 978-1-3990-3051-9

The title of this excellent collection of essays by distinguished naval historians can be read both literally and, especially for the earlier years, metaphorically. Initially the torpedo promised that small vessels could successfully attack large warships, especially at night. But, as Stephen McLaughlin demonstrates, in the Russo-Japanese War torpedo hits proved elusive, even against ships at anchor. The Japanese switched to firing torpedoes at short range just before Tsushima but, in their night attacks, made only four hits from 70 torpedoes. In the years before the Second World War they further embraced the idea of night attack to reduce their numerical inferiority prior to the anticipated 'decisive battle' with the US Navy.

Similarly, the Imperial German Navy expected its destroyer forces to deplete Britain's battlefleet. Leonard Heinz explains that these hopes were frustrated by the British abandonment of close blockade and because, after Jutland, the German flotillas failed to find the British fleet.

Before his untimely death, Rear Admiral James Goldrick, to whom this volume is dedicated, contributed a wide-ranging survey of British developments in night fighting from the aftermath of Jutland to 1939. Principal lessons from the battle were: firstly, that after a daytime battlefleet action the flotillas must be marshalled to attack soon after dark and, secondly, that fire must always be opened immediately. During the interwar period, there was continuing doctrinal development and experimentation in night fighting, with world-leading technical advances made in direction finding, plotting tables and tactics.

According to Vincent O'Hara and Enrico Cernuschi, the Italian *Regia Marina* (RM) was forced to fight in unanticipated ways, believing that battlefleet gunnery actions would be fought only by day, despite intelligence of frequent British night combat exercises in the Mediterranean. The RM's principal night actions are described in considerable detail. Initially, night searches were made in extended lines abreast from which concentration was difficult, and Italian torpedo vessels were often hit before they could reach the prescribed short firing range.

Jonathan Parshall relates how, after 1922, the IJN devised night fighting techniques that were second to none, and developed the Type 93 oxygen-fuelled torpedo. Thus, although the 'decisive battle' never happened, they were well prepared for the night actions that took place around the Solomon Islands. Parshall describes six major actions of 1942. The first two were outright Japanese victories, beginning when their superb optical instruments enabled them to sight the enemy first; however, thereafter the advantage was generally with American radar.

Trent Hone covers the Pacific War in 1943 and 1944. From November 1942, the Americans began to install Combat Information Centres based on radars with PPI displays, though it would take several months before commanders learned to use them properly. At Kolombangara the Japanese were able to execute 'a perfect torpedo attack' on American ships that were still in line. However, in the summer of 1943, Commander Arleigh Burke proposed that destroyers should operate independently in two divisions, one to make an initial stealthy torpedo attack, the second to follow up with gunfire. This plan was employed successfully at Vella Gulf and Cape St George. By Surigao Strait (October 1944) American supremacy was complete, thereby securing a complete victory.

Finally, Michael Whitby describes the 1944 Allied campaign to block German coastal traffic in the western Channel. Their ships were opposed mainly by German Type 39 torpedo boats; but, with limited numbers, the Germans had to resort to 'shoot and scoot' tactics, firing torpedoes as they withdrew. In 1943, only a few fleet destroyers were available to the C-in-C Plymouth. However, following the loss of the light cruiser *Charybdis* in October, the 10th Destroyer Flotilla was formed with five British and Canadian 'Tribal' class destroyers as a dedicated night fighting force, fitted more or less completely with Action Information Centres. The force initially achieved mixed results, but with the acquisition of experience through fighting together it successfully beat off desperate German attacks against the western flank of Operation 'Neptune' in June 1944, sinking one destroyer and driving another ashore.

John Brooks

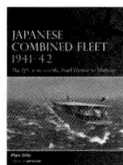

Mark Stille
Japanese Combined Fleet 1941–42
Osprey Publishing, Oxford 2023; softback,
80 pages, illustrated with B&W photographs and
colour artwork; price £15.99/$23.00.
ISBN 978-1-4728-5643-2

The first title in Osprey's new 'Fleet' series, covering some of history's most notable naval forces, focuses appropriately enough on the Imperial Japanese Navy during its brief but spectacular peak, from the attack on Pearl Harbor to the decisive defeat at Midway. That said, however, much of the book is actually concerned with the two preceding decades, during which the fleet was brought into being, while it also touches on the subsequent years, which saw its increasingly inevitable and complete demise.

Author Mark Stille has written a number of titles for Osprey on individual ship classes of the IJN, from capital ships to escort vessels, but here turns his attention to the fleet as a whole. He correctly highlights its undoubted strengths, in particular its ships, aircraft, training, optics and some of its weapons, while he also covers the numerous weaknesses, such as the submarine arm, logistics, organisation, fleet support and the gathering and use of intelligence. With these latter all being rather less high-profile areas, their relative neglect was not immediately apparent amidst the deluge of triumphs in the six-month period in question; the result was an empire – and a fleet – that ended up seriously over-extended, thus exacerbating these very problem areas and, in due course, contributing significantly to defeat.

The book is very well printed on substantial, high-quality paper, with good quality reproduction of the photographs (mostly well-known views of ships), colour artwork and maps. The colour artwork, by Jim Laurier, and featuring *Hiryu* and *Yubari* among other vessels, is fine but, to this reviewer, lacks character compared to, say, the box art of the classic 'Waterline' series of ship models.

The main problem the book has is simply that the IJN at its zenith is not just a fascinating subject, but also a huge and complex one which it is hard to cover thoroughly in eighty pages. This is where the 'Further Reading' section comes in; unfortunately its single page list runs to just eighteen entries (17 publications plus the long-standing and well-regarded 'Combined Fleet' web

site), of which nearly half are by Stille himself. Standard sources such as those by Watts, Jentchura *et al* and Evans and Peattie are present, but not Lacroix and Wells, nor *Warship*'s own Hans Lengerer, who must have published as much on the IJN as anybody.

Nevertheless, within the constraints of the format, this represents a decent and attractively produced one-stop starter reference on the IJN at the short-lived but dramatic peak of its power.

Stephen Dent

Alejandro A Vilches Alarcón
From Julietts to Yasens: Development and Operational History of Soviet Nuclear-Powered Cruise-Missile Submarines 1958–2022
Helion & Company, Warwick 2022; softback, 88 pages,
heavily illustrated in colour and B&W; price £19.95.
ISBN 978-1-915070-68-5

From Julietts to Yasens examines the series of specialised guided missile submarines developed by the Soviet Union from the 1950s onwards to combat the challenge posed by the numerous aircraft carrier-based task groups fielded by the United States and its principal naval allies. The book uses a softback, A4 format and is well illustrated with a wide range of photographs. These are supplemented by some large-scale colour profiles of the principal submarine designs covered.

The book's nine chapters adopt a broadly chronological approach. The opening sections focus on the strategic situation facing the Soviet Navy after the end of the Second World War and initial efforts to develop missile-carrying submarines through experimentation with adaptations of the Project 613 (NATO 'Whiskey') design. Each of the principal cruise-missile submarine classes is then accorded its own chapter, commencing with the diesel-electric (SSG) Project 651 ('Juliett') class and then encompassing the various nuclear-powered (SSGN) designs. Although not made entirely clear by the positioning of the chapters in the book, the first of the nuclear-powered boats were constructed before the Project 651s started to enter service. Despite the book's title, coverage of the post-Soviet Project 885/885M *Yasen* class is only brief, being considered an '… ideological successor to the SSGN of the USSR' rather than a true member of the design line.

Relatively compact in length, *From Julietts to Yasens* is nevertheless broad in its scope, encompassing technical, financial and operational considerations. The risks associated with bringing advanced technologies into service are particularly well illustrated by the coverage of the many accidents, some of them fatal, experienced by various boats of this series throughout their sometimes lengthy careers. The value of the book is enhanced by its lengthy endnotes and somewhat shorter bibliography. The latter is, however, largely limited to 'Western' sources and therefore excludes Russian-language books in this field published since the end of the Cold War.

All in all, *From Julietts to Yasens* provides a well-written and readable account of the asymmetric strategy adopted by the Soviet Union in fielding this bespoke series of submarine designs to combat NATO's naval strength. The understanding this provides is of particular current relevance given the renewed threat posed by Russia's powerful underwater forces following the return of East/West tensions.

Conrad Waters

Norman Friedman

British Coastal Forces: Two World Wars and After

Seaforth Publishing, Barnsley 2023; large format hardback, 400 pages, heavily illustrated with numerous B&W photographs and line drawings; price £50.00.
ISBN 978-1-3990-1858-6

The latest in his series of books assessing Royal Navy warship designs, Norman Friedman's *British Coastal Forces: Two World Wars and After* is ambitious in scope. Spanning most of the 20th century, its subject matter encompasses not only the motor torpedo boats (MTBs) and motor gunboats (MGBs) that formed the core of the Royal Navy's fast attack craft flotillas but also the various motor launches and specialised types that likewise played important combat and supporting roles. The breadth of this subject matter is reflected in a lengthy, 400-page count and an equally hefty price tag.

Structurally, the book is divided into 14 chapters with a single appendix, supported by detailed endnotes, an extensive bibliography and two data lists. The three opening chapters essentially set the scene for what follows: they examine the role of the fast attack craft in the Royal Navy, the main builders of these vessels, and the defining characteristics of the type. The remaining chapters are arranged in a loosely chronological order from the First World War onwards, with a particularly heavy emphasis on the craft of the Second World War period. Eight are devoted to this era, with vessels such as MGBs and motor launches being allocated their own chapter. The concluding chapter looks at postwar developments, while the appendix focuses on the Royal Air Force's air/sea rescue boats. These shared much technology with their Royal Navy coastal counterparts and were often built by the same yards.

Although various specific Royal Navy coastal craft were assessed in the two volumes on Allied Coastal Forces first written by John Lambert and Al Ross in the 1990s, this book is the first to encompass the entirety of the Royal Navy's fast attack craft and place them in a historical and organisational context. While, like other books in the series, it is primarily a warship design history, *British Coastal Forces* is also particularly valuable in extending its analysis beyond this remit. For example, it provides useful information on weaponry, electronics and other equipment, as well as a degree of operational assessment.

In addition to Dr Friedman's always insightful analysis and comprehensive referencing, *British Coastal Forces* also benefits from a very extensive selection of imagery. As well as numerous black and white photos, these include large numbers of line drawings of the most important designs described. Many of these are John Lambert's acclaimed illustrations, now in Seaforth Publishing's ownership. Regrettably, the layouts do not always do these drawings full justice, with the more complex being reproduced at too small a scale. Elements of the photographic design are also questionable: while the impact of many high quality images is diminished by their diminutive size, poorer quality photographs have sometimes been stretched beyond acceptable dimensions.

The book would also have benefited from tighter editing given its broad scope and sometimes complex subject matter. It is not entirely clear why the early Cold War period Seaward Defence Boats are dealt with as an extension to the chapter on Second World War motor launches rather than in the postwar section; some very short chapters might have been better as appendices; and greater use of tables within chapters would have been a valuable aid to clarity.

These minor criticisms aside, *British Coastal Forces* is a notable extension to Dr Friedman's work on Royal Navy warship types, and an essential purchase for anybody with an interest in this aspect of the service's history.

Conrad Waters

Steve R Dunn

The Petrol Navy: British, American and Other Naval Motor Boats at War, 1914–1920

Seaforth Publishing, Barnsley 2023; hardback, 320 pages, illustrated with B&W photographs, paintings and drawings, price £25.00/$49.95.
ISBN 978-1-3990-6285-5

It has to be said straight away that 'The Petrol Navy', the latest book from Steve Dunn for Seaforth covering some of the lesser-known aspects of the Royal Navy's operations during the First World War, is not the most eye-catching of titles; however, while the story told is not generally one of valour, action and drama, it does recount the activities of a largely unsung but vital element of the Royal Navy's force during and immediately after the conflict.

The tale begins with the volunteers of the Royal Naval Motor Boat Reserve and their craft. Dunn clearly feels a good deal of affinity with these wealthy, privileged individuals, for many of whom volunteering to go off to war in their own boats was merely something of an extension – albeit a significantly more dangerous one – of an expensive and novel hobby. Their undoubted enthusiasm and energy was not initially matched by much success, not least because their craft proved not to be up to the sort of tasks that the Admiralty had in mind. They were soon superseded by the numerous purpose-built Motor

Launches which, though small and basic, ended up having very busy wars indeed, participating in all manner of operations, many of which were probably far from the minds of those who conceived them. They were, ironically, generally far busier during the war than the mighty dreadnoughts that spent most of the time at anchor in Scapa Flow. As a writer Dunn is very much at home here, recounting the activities of these comparatively forgotten small ships and their crews, endlessly engaged in some activity while their much better known comrades-in-arms were often lying idle. Anti-submarine patrol was a particular speciality; tangible success was rare, but mere presence played an important part in the ongoing battle against U-boats. The highlight, if it can be called that, of the MLs' war were the raids on Ostend and Zeebrugge, where the boats and their crews acquitted themselves heroically amid the disastrous and futile bloodbath.

Along with the tale of the MLs Dunn also tells the story of the rather more glamorous and better-known CMBs, the American submarine chasers (known, perhaps unfairly, as 'splinter ships'), and of assorted similar vessels in the French, Italian, German and Austro-Hungarian Navies.

Dunn is a thoroughly readable author; however, the book would have benefited from more careful editing, as there are multiple instances of repetition. The illustrations encompass a wide range of photographs and paintings from both official sources and private collections, including the author's own. The paintings by Donald Maxwell of MLs in the eastern Mediterranean are a highlight. The positioning of plans of the various craft in an appendix is unfortunate, in that they are far removed from the descriptions of the boats in question. The conclusion is a strange mixture of clear-sighted analysis and a slight bitterness at the passing of an era now long gone. These minor criticisms apart, this is an important story, well told.

Stephen Dent

Kenneth L Privratsky
The Norwegian Merchant Fleet in the Second World War

Pen & Sword Books, Barnsley 2023; hardback, 200 pages, 52 B&W photographs, appendices, notes, bibliography, index; price £22.00.
ISBN 978-1-3990-4386-1

Despite the plethora of publications concerning convoys in the Second World War, little space is afforded to the story of the merchant vessels themselves beyond passing statistical references to numbers or tonnage lost. Kenneth Privratsky, by contrast, directs our attention to just one aspect: the vital role played by the Norwegian Merchant Fleet during six years of conflict. During the war itself, the significance of these ships was such that the Minister of War Transport remarked in 1942: 'Norwegian tankers are to the Battle of the Atlantic what the Spitfires were to the Battle of Britain in 1940'.

Norwegian shipping companies in the interwar years invested heavily in new-build dry cargo vessels and tankers that had greater capacity, and which were fast and comfortable and thus popular with their well-paid crews. Moreover, they specialised in efficient diesel-powered ships; by 1939 sixty per cent of the entire fleet was powered by diesels – a figure that far exceeded that of their nearest rivals, the UK and USA. At the outbreak of war, Norway possessed 272 tankers, which in turn comprised one fifth of the world's total tanker capacity.

Germany's unexpectedly rapid invasion of Norway in April 1940 caught everyone off guard, causing King Haakon and his entourage to flee to London, where they set up a government-in-exile. In the meantime, Norway's vast mercantile fleet was scattered worldwide; many of the ships had incomplete liner contracts to fulfil, others were trapped and isolated in foreign ports awaiting instructions regarding onward passage. The urgent requirement for some kind of central control led to the creation of 'Nortraship' (The Norwegian Shipping and Trade Mission), the story of which is the primary focus of Privratsky's book.

This was by no means an easy task. On the one hand there were myriad ships, existing contracts, cargoes, insurance and sailors' wages, not to mention individual shipowners' requirements to take into account; on the other, pressure from the British Government to gain access to this fleet as replacements for mounting losses during the most difficult months of the Battle of the Atlantic. It soon became necessary to open an additional Nortraship office in New York; however, this only added to the complications, as neutral American shipping companies were most reluctant to allow the precious Norwegian ships to endanger themselves by working war-affected routes.

A power struggle ensued within Nortraship itself as well as between the two offices, due mostly to the very strong personalities involved. Both the Norwegian and British Governments tried to impose their authority in a spirit of cooperation, and the situation eased considerably once the USA entered the war. However, almost inevitably, it was the ordinary sailor who lost out postwar as a disgracefully mismanaged 'Secret Fund' failed to deliver compensation promptly to the returning sailors or to war widows and those disabled in the fighting.

The Norwegian Merchant Fleet does not mention a single convoy action, nor does it follow the fate of any particular tanker or cargo ship. The course of the invasion of the country and the escape of the king are narrated in some detail in the early chapters, but only to provide context from a Norwegian point of view. Privratsky recounts the story of Nortraship operating under extreme wartime exigencies principally from a business perspective. This demands the close attention of readers who might not be conversant with such matters. That accepted, this book is an excellent, authoritative account of an overlooked subject which should add considerably to our understanding of the war at sea 1939–45.

Jon Wise

WARSHIP GALLERY

The Soviet Navy 1960–1990

From the late 1950s until 1990, the Soviet Navy attempted to build a blue-water fleet that could contest the waters around the USSR and counter Western carrier task groups and nuclear-powered submarines. This year's Gallery features images of the major types of surface ship completed during the period; the photos have been carefully selected from those issued by official government agencies to the Press, and come from the Editor's own personal collection. In order to make clear the technological developments made during the period in question and avoid the complexity of the Soviet numbering system, the NATO designations of weapons systems and radars are employed throughout.

Below: From the mid-1950s the Soviet Navy had experimented with enlargements of existing destroyer designs armed with bulky launchers fore and aft for anti-ship missiles ('Kildin' & 'Krupny' classes) to protect the seas surrounding the Soviet Union from attack by Allied carrier task forces. The four ships of the 'Kynda' class (*Proekt 58*), built between 1959 and 1962 at the Zhdanov Shipyard in the Baltic, were the first of a series of purpose-designed 'Rocket Cruisers' armed with the long-range SS-N-3B 'Shaddock' missile. There were two quadruple launchers, which could be trained and elevated, with reloads in the superstructures behind them. For self-defence they had a SA-N-1 'Goa' surface-to-air missile system forward and two twin 76.2mm DP mountings aft, together with RBU 6000 antisubmarine rocket launchers and 533mm torpedo tubes. This is the Pacific-based *Variag* in 1970. (US Navy)

Right: Typed 'Large Antisubmarine Ship' on completion, the 'Kashin' (*Proekt 61*) was a general-purpose design intended to conduct antisubmarine operations in the outer zone of the seas surrounding the Soviet Union. It was a 'double-ended' ship that employed the same generation of antiaircraft and A/S weaponry as the 'Kynda', but did not mount the bulky SS-N-3 SSM and its associated guidance radars, and had an altogether lower profile. A conspicuous innovation was the propulsion system, which comprised four large gas turbines exhausting through two pairs of large, canted funnels. Built at the Zhdanov Yard and the 61 Kommuna Yard, Nikolaev, between 1959 and 1973, no fewer than 20 units were completed for the Soviet Navy. On completion they were assigned to each of the four Soviet fleets, with deployments to the Mediterranean and the Indian Ocean. This is the Pacific-based *Strogii*, photographed by the Royal Australian Air Force. (RAAF)

Above: The final unit of the 'Kashin' class, *Sterzhannyi*, was completed in 1973 to a modified design (Proekt 61M). A raised helicopter platform was fitted above the stern, and four aft-facing launch canisters for SS-N-2C 'Styx' short-range anti-ship missiles were mounted abreast the second pair of funnels. Other modifications included CIWS with 'Bass Tilt' and improved electronics. Five of the earlier 'Kashins' were modified to the same standard between 1971 and 1981. The 'shoot and scoot' arrangement of the anti-ship missiles meant that ships of this type were often selected for the close shadowing of US Navy carriers during NATO exercises. *Stroinyi* is seen here in 1981. (MoD)

Right, upper: Laid down in 1962 and 1964 respectively at the Black Sea Shipyard, *Moskva* and *Leningrad* (*Proekt 1123*) were the Soviet Navy's first air-capable ships and introduced a new generation of weaponry and sensors. Designated 'Large Antisubmarine Ships', they could operate up to 18 Ka-25 'Hormone A' A/S helicopters from their broad flight deck aft; the flight deck and the hangar beneath were served by two narrow lifts. The SA-N-1 'Goa' surface-to-air missile system of earlier major surface units was superseded by the more advanced SA-N-3 'Goblet' system, with height finding and target indication provided by the distinctive 'Top Sail' radar, the massive antenna for which was located at the head of the tripod mast atop a tall superstructure. Forward of the launchers for 'Goblet' was a twin-arm SUW-N-1 launcher for the FRAS-1 antisubmarine missile, a ballistic rocket carrying a homing torpedo similar in conception to the US Navy's ASROC. For underwater detection there were hull ('Moose Jaw') and variable depth ('Mare Tail') sonars; the handling apparatus for the latter can be seen beneath the flight deck in this stern quarter view of *Leningrad*. (MoD)

Right, lower: The 'Kresta I' (*Proekt 1134*) was the 'Kynda' successor, and the four ships of the class were laid down at the Zhdanov shipyard in 1964–66. Athough initially accorded the same 'Rocket Cruiser' designation, this was a radically different design. It was, like the 'Kashin' class, double-ended, with SA-N-1 'Goa' SAM launchers and their associated 'Peel Group' guidance radars fore and aft. Only four elevating launchers for the SS-N-3 'Shaddock' SSM were mounted; these were located beneath the bridge wings and no reloads were provided. However, there was now a hangar and a small flight deck aft for a Ka-25 'Hormone B' helicopter that could provide an over-the-horizon guidance relay for the missiles. Powered, like the 'Kyndas', by high-pressure steam turbines, these ships appear to have been successful, as the design was adapted as the 'Kresta II' (*Proekt 1134A* – see below). This image of *Vitse-admiral Drozd* was taken in the Baltic prior to her half-life refit, probably around 1970. (Royal Danish Navy)

Left, upper: Built between 1966 and 1977 at the Zhdanov Yard, the ten ships of the 'Kresta II' type followed directly on from the 'Kresta I' and utilised the same basic design, replacing the first-generation SA-N-1 SAM with the more advanced SA-N-3 'Goblet' introduced in the two ships of the *Moskva* class, with missile tracking performed by the bulky and distinctive 'Head Lights' radar. There was initially some confusion regarding the nature of the quadruple missile launchers beneath the bridge wings, which were significantly shorter than those in the 'Kresta I' and were initially thought to house shorter-range anti-ship missiles. It soon became apparent that they contained a new antisubmarine missile comparable to the Australian Ikara and the French Malafon, the SS-N-14 'Silex', which featured a cruise missile body with a homing torpedo payload and a range of approximately 50km. The single helicopter was the Ka-25 'Hormone A' A/S variant embarked in the *Moskva*s, and the 'Kresta II' was likewise typed 'Large Antisubmarine Ship'. In addition to the now-standard 57mm/80 twin AA gun, the 'Kresta II' was the first Soviet major surface unit to embark a point-defence system to counter enemy anti-ship missiles. Four six-barrelled AK-630 gatling-type mountings, together with their 'Bass Tilt' fire control radar, were fitted abreast the bridge. (MoD, November 1975)

Left, lower: The 'Kara' (*Proekt 1134B*) replaced the 'Kashin' on the slipways at the 61 Kommuna Shipyard, Nikolayev, from 1969; seven units were completed by 1980. The ships were propelled, like their predecessors, by four large gas turbines designed and manufactured locally, the exhaust uptakes of which were however combined in a large square funnel. Similar in capabilities to the 'Kresta II', of which it can be regarded as the Black Sea counterpart, this was nevertheless a completely new design with a hull 15 metres longer. Advantage was taken of the longer hull to incorporate an intermediate air defence system, the SA-N-4 'Gecko', the missiles and their twin-arm 'pop-up' launcher being housed in two circular 'bins' located abreast the radar tower, and to increase the size of the SA-N-3 'Goblet' magazines to accommodate 36 rather than 22 missiles. The twin 57mm of the 'Kresta II' were replaced by twin 76mm mountings, and there was a variable depth sonar to complement the bow sonar, as in the *Moskva* class. This is *Kerch* in the Mediterranean in 1976. (MoD)

Above: The 'Krivak' (*Proekt 1135*) was a smaller, simplified 'Kara', built contemporaneously and intended for mass-production in smaller shipyards such as Kaliningrad and Kamysh-Burun. With the traditional 'Patrol Ship' designation, 21 units were completed between 1970 and 1982, and a further eleven of a modified variant ('Krivak II') with 100mm single mountings replacing the original twin 76mm between 1975 and 1982. The SS-N-14 A/S missile system was retained in the form of a 'flat four' launcher forward, as were the bow and variable depth sonars, but there was insufficient space for a helicopter. Two modular SA-N-4 SAM systems provided close-range air defence, the circular bins being disposed on either side of the forecastle behind the SS-N-14 launcher. The ships were propelled by two large gas turbines similar to those in the 'Kara'. They were deployed to each of the four main Soviet fleets. This is *Leningradskii Komsomolets* in 1982. (RDN)

Left: *Kirov* (*Proeky 1144*) was the first of four large nuclear-powered 'Rocket Cruisers' built at the Baltic Shipyard, which had previously built nuclear-powered icebreakers, between 1973 and 1998. The anti-ship counterpart of *Kiev*, she was intended, in combination with nuclear submarines armed with similar cruise missiles ('Oscar' class), to defend the SSBN bastions against the incursion of Western carrier task forces. Beneath the broad forecastle were twenty inclined launch tubes for the SS-N-19 'Shipwreck' missile. The other new major new system was the SA-N-6 'Grumble' SAM, an advanced Mach 6 missile capable of engaging aerial targets at ranges of 80–100km. The SA-N-6 employed vertical launch, and eight missiles were stowed in a rotating magazine ring beneath each of the twelve square hatches located forward of the SS-N-19 launchers, giving a total of 96 missiles. Guidance was provided by the distinctive 'Top Dome' radar. There was a plethora of missile and gun systems for fire support and close-in air defence, most of which had featured on the 'Kara' and the 'Krivak II'. Problems with the innovatory combined nuclear and steam (CONAS) propulsion plant led to the early retirement of three of the four ships. (MoD, October 1980)

Left: The appearance of *Kiev* (*Proekt 1143*), the Soviet Navy's first air-capable ship designed for fixed-wing operations, caused a major stir in Western naval circles in 1975. Classed by the Soviets as a 'Large Antisubmarine Cruiser', she was in effect a hybrid carrier/cruiser design intended to defend the Soviet ballistic submarine bastions in the Arctic and the Far East against incursions by NATO surface forces. The 'cruiser' weaponry, which was divided between the forecastle and the after end of the island, was essentially that of the '*Kara*', but with the SUW-N-1 A/S missile system of the *Moskva*s replacing the SS-N-14 'Silex' and eight elevating launchers for the successor to the long-range 'Shaddock' anti-ship missile, the SS-N-12 'Sandbox'. There was a full set of reloads for the latter beneath the forecastle, handled by a narrow centreline lift and athwartship rails. The flight deck was angled at 3.5 degrees to the ship's axis and had a hangar beneath served by two aircraft lifts. The air group comprised a single squadron of Yak-38 'Forger' VTOL aircraft, similar in conception to the British Harrier, and a larger squadron of 'Hormone A' A/S helicopters plus a couple of the 'B' missile relay variant. Four ships of the class, including the modified *Baku*, were built at the Black Sea Shipyard between 1970 and 1987. This is *Kiev* following a major refit in the spring of 1985. (MoD)

Above: *Slava* (*Proekt 1164*) was the 'little brother' of *Kirov*. Three units were built at the Kommuna 61 Shipyard in the Black Sea between 1976 and 1989, a fourth was never completed and a further four were cancelled. The hull was essentially an enlarged version of the '*Kara*', and as with all the earlier ships built at 61 Kommuna, propulsion was by gas turbines, the exhaust uptakes being combined in a single large square funnel. There were twelve SS-N-12 'Sandbox' missiles in fixed-angle launchers along the sides of the forward part of the ship, while abaft the funnel there was a cut-down SA-N-6 SAM installation, with eight circular bins each housing 8 missiles in a rotating magazine ring and a single 'Top Dome' radar. Apart from the new 130mm twin mounting forward, which also armed the later units of the *Kirov* class, the weapons systems were those of the '*Kara*', including the less-capable sonar outfit. This is the second ship, *Marshal Ustinov*, in 1987. *Slava* would later be renamed *Moskva* and lost during the war with Ukraine. (MoD)

Entering service within a few months of one another in 1980–81, the lead ships of these two complementary classes of large destroyer were of markedly different design. *Sovremennyi* (*Proekt 956*) was a general-purpose ship, armed with medium-range anti-ship missiles (SS-N-22 'Sunburn'), a medium-range area defence missile system comparable to the US Navy's Standard, the SA-N-7 'Gadfly', and the 130mm/70 DP gun mounting fitted in the larger 'Rocket Cruisers'. She also had a flight deck and hangar for a single helicopter. The *Udaloy* (*Proekt 1155*) was a specialist antisubmarine vessel fitted with two quad box launchers for the SS-N-14 missile and the powerful 'Horse Jaw' and 'Horse Tail' paired sonars first seen on the *Kirov* class, and was capable of operating two Ka-27 'Helix A' A/S helicopters, a development of the 'Hormone A' with more sophisticated avionics. For self-defence there were eight circular 'bins' (in two groups of four) for the SA-N-9 'Gauntlet' intermediate-range air defence system, which had superseded the SA-N-4, and the two DP guns forward were the standard 100mm model. Differences between the two types extended to the propulsion system. *Sovremennyi* was built at the Zhdanov Yard in the Baltic and had a conventional pressure-fired steam plant exhausting through a single square funnel whereas *Udaloy*, built

at Kaliningrad, had four large gas turbines and reverted to the paired uptakes of the earlier 'Kashin' class. No fewer than seventeen ships of the *Sovremennyi* class were built by Zhdanov for the Soviet Navy between 1976 and 1993; the photo is of the fifth unit of the class, *Bezuprechnyi*, shortly after completion. The lead yard for the *Udaloy* class was Kaliningrad, with additional units built at Zhdanov; twelve units were completed between 1977 and 1991; depicted here is the second ship, *Vitse-admiral Kulakov*, in 1985. (MoD)